Doing the Impossible

George E. Mueller and the Management of NASA's Human Spaceflight
Program

Arthur L. Slotkin

Doing the Impossible

George E. Mueller and the Management of NASA's Human Spaceflight Program

 Springer

Published in association with
Praxis Publishing
Chichester, UK

Arthur L. Slotkin
Atlanta
Georgia
USA

SPRINGER–PRAXIS BOOKS IN SPACE EXPLORATION

ISBN 978-1-4614-3700-0   ISBN 978-1-4614-3701-7 (eBook)
DOI 10.1007/978-1-4614-3701-7
Springer New York Heidelberg Dordrecht London

Library of Congress Control Number: 2012937724

Cover design: Jim Wilkie
Project copy editor: David M. Harland
Typesetting: BookEns, Royston, Herts., UK

Printed on acid-free paper

Springer is part of Springer Science+Business Media (www.springer.com)

Contents

Illustrations

Tables

Foreword

It is amazing to me to see how much we have learned about the system within which we live during my lifetime. From developing the technology to fly across the oceans to the technology to land Men on the Moon! From developing the technology to send messages across a city to the technology to send television to every home on Earth. To developing the technology to feed five times the number of people – 1.5 billion in 1918 to 7 billion today. To the discovery of the structure of the cells of our body and the ability to define the interfaces between these cells that make our life possible.

It astonishes me that having done all that, we have not built on our capability to expand mankind's reach throughout our own Solar System. We clearly have the need to establish new frontiers to challenge and reward the next generations of our race. Frontiers that will provide the stimulus needed to spur the further understanding of the systems within which we exist. Frontiers that will provide the basis for the survival of mankind into the unknown future.

But, frontiers are established and conquered as a result of the dreams of people and the availability of transport.

If we are to open the frontiers of space, we too will need a dream, a dream shared by many, and a means of transportation capable of being used by masses of people.

Our nation needs such a dream! A vision that will guide its future! A dream worth striving for! A vision that can inspire our people to regain their pioneering spirit!

The first step towards enunciating that dream was taken in 1989 by President George H. W. Bush when he said on the occasion of the 20th anniversary of the Apollo lunar landing: "For the new century, back to the Moon, back to the future, and this time back to stay. And then, a journey into tomorrow, a journey to another planet, a manned mission to Mars."

However, tomorrow has yet to come!

George E. Mueller
2012

Acknowledgments

I first met Dr. George E. Mueller in early 1970. A graduate in aerospace engineering, I knew his name and reputation, though I never imagined meeting him. He had just come to New York City as vice president of the General Dynamics Corporation, and because I was the newly appointed head of the AIAA's student programs, Executive Secretary James J. Harford asked me to help to organize the International Astronautical Federation's first student conference. Mueller chaired the promotion of membership committee of the IAF, and the student conference was their idea. We held the conference in Brussels that fall as a part of the International Astronautical Congress, and Mueller served as the honorary chair. During the following year I saw him from time-to-time at AIAA meetings, and when he moved to the System Development Corporation in Santa Monica in May 1971 he asked me to visit the next time I went to California. Eager to follow-up on this invitation, I arranged a trip to Los Angeles, which led to a job offer – and in February 1972, I became his special assistant, a position that I held for several years. Promoted into a management position at SDC several years later, I continued supporting him, in addition to my other duties. And after he retired in 1983, I continued working with him on a voluntary basis until the mid-1990s when my career took us in different directions. We stayed in touch, and when I retired from industry to pursue my passion for the history of technology at Georgia Tech in 2002, I asked him if I could write about his work at NASA for the dissertation that I planned to write. I spent the summer of 2004 in Washington, DC at the NASA headquarters historical archives, researching his work. And since Mueller sent most of his personal papers to the Library of Congress, after completing my classwork at Georgia Tech in 2006 I received a NASA grant to continue this research thanks to then-NASA administrator Michael D. Griffin. Subsequently, I spent three years researching Mueller's and Sam Phillips' careers at the Library of Congress, did additional research at NASA headquarters, and conducted a series of interviews (2009 to 2011) with Mueller at his home in Kirkland, Washington. Most gracious with his time, he answered all of the many questions that I posed. Both he and his wife Darla made my visits to Kirkland productive and enjoyable.

There are many other people to thank. John Krige, the Kranzberg Professor of the History of Technology at Georgia Tech started me out on the adventure which led to writing this book. The staff at the NASA History Office assisted in a number of ways. Jane Odom, Colin Fries and others in the headquarters archives provided valuable assistance and answered many questions numerous times. The archivists at the Library of Congress manuscript reading room in the Madison building – Jennifer Brathovde, Joseph Jackson, Lia Kerwin, Patrick Kerwin, Bruce Kirby, and Lewis Wyman – and their leader, Jeffrey Flannery, were most helpful and fulfilled all of my many requests for information. When Dr. Mueller asked to see how his papers were stored, Leonard Bruno, manuscript historian for science and technology prepared an impressive display of rare and historic documents that not only included his papers, but some of our nation's most important political and technological leaders, from George Washington's first surveying notebook to Alexander Graham Bell's original sketch of the telephone and Orville Wright's signed photo of the first flight, which he developed himself. Spending several hours looking at these rare documents is one of the highlights of many visits.

I asked a number of people to read parts of the manuscript as I prepared it. Roger Launius, formerly NASA's historian and currently curator of space history at the Smithsonian National Air and Space Museum, provided some very useful advice. Steve Garber of the NASA History Office convinced me to reduce the size of the manuscript, which helped to improve the focus of the book. Jim Skaggs who worked with Mueller at NASA, General Dynamics and SDC, read the manuscript and provided important missing details. Bob Freeman, who I worked with at SDC, read some early chapters and made good suggestions as well. I have to also thank my brother Steve Slotkin who is the only person who has read all of the chapters, several times, my wife Marcella who insisted on the inclusion of pictures, and my daughter Chandra Slotkin Townsend, a graduate in visual communications, who advised me about the layout of the book. (And of course, Marcella deserves additional credit for her patience over a number of years while I researched and wrote this book.)

The production of the book was aided by Linda Andes of Buckhead Transcription Service in Atlanta, Georgia who interpreted and typed the many hours of interviews conducted with Mueller. I must also thank Maury Solomon of Springer in New York, Clive Horwood of Praxis in England, and my editor David M. Harland in Scotland, himself an author with multiple books published in the space field.

In closing, I want to acknowledge some of the people who helped make my work on this book enjoyable, the members of my immediate family – Marcella, my two daughters and five grandchildren. During frequent trips to Washington, Chandra's daughter Devin-Ann asked how many pages, or how many words were in the book; or how many pages did I write today. She, her mother and her brother Alex, along with my other daughter Jennifer Slotkin-Linn and her three boys, Dylan, Luke and Jack, are the people for whom I wrote this book. I hope someday when the children are old enough to read it, they will look up to the sky and realize we can explore the planets. Perhaps their generation will be the ones to do it...

And finally, I dedicate this book to Dr. George E. Mueller, my leader, mentor and

A-1   George E. Mueller and Arthur L. Slotkin at the Museum of Flight, Seattle, Washington, July 16, 2008.

friend, the person who in six short years during the 1960s accomplished the greatest technological achievement of the twentieth century: successfully sending American astronauts to the Moon and returning them safely to Earth.

Introduction

"Always the epitome of politeness, but you know down deep he's just as hard as steel."

John H. Disher on Mueller, April 15, 1971

On July 16, 1969, inside NASA's launch control center at Kennedy Space Center, Florida, four men smiled and congratulated each other after witnessing the successful launch of Apollo 11. In the middle is George E. Mueller (pronounced "Miller"), age fifty-one, NASA associate administrator for manned space flight, along with a few of his closest associates at the agency – Wernher von Braun, director of the Marshall Space Flight Center, Samuel C. Phillips, director of the Apollo Program Office, and Charles W. Mathews, Mueller's principal deputy. Who is Mueller and why did NASA choose him to manage Apollo? What did he contribute? And did he make a difference? In all likelihood the United States would have *eventually* landed an astronaut on the Moon with or without Mueller, but he made it possible to do so within the decade of the 1960s, thereby completing the goal established by John F. Kennedy in May 1961.

Mueller came to NASA in September 1963 to lead the human spaceflight program because D. Brainerd Holmes, his predecessor, found himself in difficulty with NASA administrator James E. Webb after appealing directly to the president for additional funding. An eminently qualified technical manager, Holmes ignored agency politics and underestimated the administrator's ego. Webb did not appreciate the attention the media gave to his subordinate, who *Time* magazine called "the czar of Apollo" when publishing his picture on their cover. All this contributed to Holmes' early departure, and the controversy found its way into the press, thus tarnishing NASA's image and raising questions about its ability to manage the human spaceflight program. The air force and their supporters fed those concerns, suggesting the military might be better suited to lead the civilian space program. After all, NASA had limited experience of managing such a large and complex program, whereas the air force had successfully developed, tested and deployed a series of ballistic missiles and satellites since the early 1950s, and had also provided the launch vehicles used by Mercury and Gemini. Whether to leave the Apollo Program with the agency came

I-1 Apollo 11 liftoff, Charles W. Mathews, Wernher von Braun, George E. Mueller, and Samuel C. Phillips, July 16, 1969. (NASA photo)

into question. So Webb realized that his new director of human spaceflight would have to be acceptable to the air force and their supporters.

In 1963 there was doubt about the nation's ability to achieve Kennedy's lunar landing objective. And because the journey to the Moon and back served as a surrogate for war with the USSR, the president considered achieving this goal to be one of the nation's top objectives, second only to national defense. People asked how the US could build the necessary equipment in the six plus years remaining in the decade, and how could they do it before the Soviet Union. Because of previous Soviet space spectaculars, many thought them to be well ahead of the US in space technology. It was not a question of bravery or courage, it came down to whether US scientist and engineers could develop the technology and build the equipment needed to match the feats of their Cold War adversaries, and beat them to the Moon.

This is when Mueller entered the picture. An expert in satellite communications with experience in managing research and development for the air force ballistic missile program, he had a good reputation and the right résumé. He lacked human spaceflight experience, but few in the US outside of the Mercury Program had such experience. However, he had the support of the air force, and from a political point of view Webb needed someone like him – although when he brought Mueller into the agency the administrator got more than he realized. Not only did he hire a capable, well regarded technical manager, Mueller had the ability to analyze and understand systems, and knew what it took to build complicated space vehicles. In particular, he could visualize a total system – hardware, software, people and processes – and everything necessary to accomplish the task at hand. A system engineer, Mueller knew how to apply it to management, using "system management." After arriving at the agency's Washington headquarters, he recognized the importance of building and maintaining relationships with Congress and the White House. External relationships became a key part of his new environment, and an important part of the system that he would manage.

Many excellent books about human spaceflight, written from different points of view already exist. Charles Murray and Catherine Bly Cox tell the story of Apollo from the perspective of the people who worked in the program, including some who are not well known to the general public. Walter A. McDougall wrote a Pulitzer Prize winning political history of the program, and there are histories of technology, such as Roger E. Bilstein's excellent book about the development of the Saturn launch vehicles. And lunar science is addressed by authors like W. David Compton. Charles D. Benson and William Barnaby Flaherty wrote a history of one of the centers during the Apollo Program, and there are excellent histories about the other centers as well. Accounts of astronauts, both autobiographies and biographies are plentiful, and one of the more interesting texts was written by James R. Hansen who had special access to Neil A. Armstrong. Even the women who did not fly in space are addressed by Margaret A. Weitekamp in her narrative about the first women who attempted to become astronauts. Christopher C. Kraft, Jr. wrote a memoir describing the story of human spaceflight from his personal perspective in flight operations, one of several by the flight controllers. Some of the leaders of NASA, Webb's predecessor T. Keith Glennan and his deputy Robert C. Seamans, Jr., wrote autobiographies, and Seamans also wrote a monograph about Apollo. Webb did not write an autobiography, but W. Henry Lambright published a well-researched biography of him. Former NASA historians Roger D. Launius and Howard E. McCurdy wrote interesting books and essays about different aspects of the space program that were published jointly and individually. Simon Ramo wrote about system engineering and matrix management. System management is addressed directly by Stephen B. Johnson in two of his books, one about how the air force developed system management for the ballistic missile program and the other partially about its use in the Apollo Program. Concurrent with the development of Apollo, Mueller managed Gemini and developed a follow-on program to Apollo. Barton C. Hacker and James M. Grimwood's history of Gemini provides interesting insights into the details of that program, and there are others addressing different parts of the Apollo Applications Program; including books about the Skylab space station by David Hitt, Owen Garriott and Joseph P. Kerwin, and by W. David Compton and Charles D. Benson.[1]

Despite the many books about the space programs of the 1960s, a void in the literature remains. Exactly how did the methods developed in the air force ballistic missile program get modified and used in the Apollo Program? How did Mueller, with the help of Phillips and others, manage the programs? How did they overlay the

---

[1]   See Bibliography.

system program office methodology developed by the air force on an agency that was organized functionally as standalone research centers? How did the centers, coming from federal agencies with individual cultures, adapt to the new structured approach imposed from Washington?

In this book, I set myself two main objectives. First, to answer the question: what specific contributions did Mueller make to the management of human spaceflight? Many books mention him and give him credit for important decisions, but the key role he played by introducing the air force's system program office methodology is at times lost in the details. Thus, I set out to define in one place the part Mueller played in shaping human spaceflight in the 1960s. This book mainly addresses the six years 1963 to 1969, when Mueller led the human spaceflight program. His education as an electrical engineer and physicist, a researcher on airborne radar during World War II, a teacher and researcher in electronics and system engineering as a college professor in the post-war years, and as an R&D manager during the heydays of the air force ballistic missile program of the 1950s and early 1960s prepared him for the NASA job. Based on his experiences prior to joining the agency, he became a practitioner of system management, which he used to organize and manage the human spaceflight program.

My second objective was to interweave the story of Mueller's work on Gemini, Apollo, and the Apollo follow-on programs into a single book, just as he lived them. On any single day he might address Gemini issues, and deal with Apollo and Apollo Applications, while coordinating with Congress, the White House and the Pentagon. Mueller rightfully gives Mathews credit for managing the Gemini Program, but he played an important role himself. And from the day he arrived at NASA, he planned Apollo follow-on programs. He took ideas about space stations and space shuttles and, like many other innovations, borrowed from earlier work to become a leading advocate for a new space transportation system. He saw this new system – a railroad in space, he called it – as the pathway to planetary exploration. He was emphatic that just building the shuttle and space station was not the goal.

I started this book in 2004, first reviewing documents at the NASA History Office headquarters archives and then spending almost three years reading Mueller's and Phillips' papers, mainly at the manuscript reading room of the Library of Congress. I used interviews conducted by others since the 1960s, which provide insights into the thinking of key participants. Of particular value were the interviews conducted by the Smithsonian National Air and Space Museum in the 1980s as part of the Glennan-Webb-Seamans Oral History Project. Another set of interesting interviews are those recorded by *Time-Life* journalist Robert L. Sherrod for a book that he never wrote. Many of these took place contemporaneously, and some were not for attribution. I used other interviews contained in NASA's archives, and between 2009 and 2011 conducted a series of interviews with Mueller to fill in gaps in the record. Mueller gave me access to some personal papers and speeches that he had not given to the Library of Congress and which the NASA archives did not have. I have read all of Mueller's public speeches, ninety-one during his six years at NASA, and about a hundred others since. Using these speeches and congressional testimony, one can follow his thoughts. He emphasized different subjects at different times, but at

NASA he always talked about the post-Apollo period and toward the end of his time at the agency it was almost all that he spoke about. A key thread to this story is Mueller's application of system management to Apollo. Although this book is biographical, it is not a biography; his personal life is mentioned when relevant but not dwelled upon, and I leave it to others to further explore him as a man.

Beginning at American Telephone & Telegraph's Bell Telephone Laboratories, where he helped to design one of the first airborne radar systems, Mueller recognized the need to visualize the total system in order to design it properly. He said "you have to have all of the facets covered. So it was just the natural way of doing things." But system engineering had not yet become a discipline – it remained an approach to understanding all aspects of a system. After arriving at Ohio State University in 1946 he taught his first system engineering course, a difficult course to teach and more difficult to learn. Some students got it but, as he later said, "not many people look at all the facets of a problem naturally." His approach to teaching system engineering involved creating problems and illustrating how to solve them by using this new approach. These were not simple problems. Students could not look up the solution somewhere, they had to go out and invent the solution. Taking what he knew about system engineering with him to Ramo-Wooldridge, Incorporated in 1956, he applied it to his work on the Able spacecraft of the early Pioneer missions and in the air force ballistic missile program. He said there was nothing magical about applying system engineering to management problems, and engineers at the company were already applying it to the development of the ballistic missiles. He honed his management skills in the air force ballistic missile program by applying system management. The transition from system engineer to system manager came easy, he reflected, because "engineers are really managing things in any event. So that's a semantic problem, not a real problem," and the words are interchangeable.[2]

Mueller's application of system management became a key factor in the success of Apollo, but exactly what is it? Simon Ramo, a co-founder of Ramo-Wooldridge, defined the term *system engineer* as "a peculiar form of generalist [with] the faculty of understanding enough of each of the pieces and [is] good at communications." In other words, a system engineer is an "integrating negotiator" who understands how to make trade-offs, and can visualize what has to be added or subtracted to "make the thing compatible [and] harmonious." Mueller called *system engineering* "a discipline which involves all of engineering and ... has to be applied to a particular system." And thus system management is "a structure for visualizing all the factors involved as an integrated whole, much as system engineering visualizes all of the physical aspects of a problem."[3]

Mueller made his only extended remarks about the use of system management with Apollo in a speech that he gave to a joint meeting of the American Institute of Aeronautics and Astronautics and the Canadian Aeronautics and Space Institute in

[2]   Mueller Interview, Slotkin, 2/23/11.
[3]   Ramo Interview, NASM, 6/27/88; Mueller Interview, Slotkin, 9/9/09; Joint AIAA/CASI Meeting, Montreal, Canada, 7/8/68.

Montreal, Canada on July 8, 1968. As he explained, NASA needed an organizational and management system to define the work to be done and monitor its achievement. System management, he said, is really system engineering applied to management, and permits the system manager to "recognize the nature and interaction of complex procedures in advance of their becoming problems." He explained that the criteria necessary to manage large scale R&D programs required the objective to be within the state of the art without having to reach for performance, and be achievable in a reasonable timescale. Activities required to be clearly defined for each organizational element, with engineering and design aiming for high reliability, and manufacturing providing adequate inspection and quality control. But he cautioned that *reliability has to be designed into an object*; it "cannot be built into a badly engineered design," it can only be achieved by having a sound design to begin with, followed by careful manufacturing and testing. Since change is part of R&D, the design must tolerate changes. On the other hand, he warned, "we must guard against change leading to overdesign."[4]

The successful management of Apollo required extensive project planning, a program oriented organization, and the division of responsibilities into manageable segments. The projects were divided into phases, with adequate review and analysis at the end of each phase. Mueller implemented phased project planning consisting of preliminary studies, followed by more detailed studies and preliminary design aimed at selecting the best approach from the options available. The next phases involved the detailed definition, system design, scheduling, and costing. And the final phases included design, development and fabrication, followed by test and flight operations. During the Apollo Program, the agency had to conduct further reviews and analyses following the Apollo 204 fire in which three astronauts lost their lives, resulting in additional changes in hardware and procedures.[5]

Mueller superimposed a program office organization on the existing institutional structure, and both remained throughout the Apollo Program. Overlaying program management on the NASA field centers created a matrix to facilitate achievement of both program and institutional objectives. The permanent institutional structure remained after completion of the program, but the program structure lasted only as long as the program. Nevertheless, the agency maintained continuity by assigning experienced program managers to new programs. And by using program managers from the ballistic missile program, NASA was able to gain program management skills from the military. In addition, specialized operational knowledge gained on Mercury was carried over to Gemini, and then the lessons from that program were applied to Apollo. Contractor organizations responded to project offices at the centers, which in turn followed the direction of the Apollo Program Office in Washington. To perform the actual work necessary to achieve program objectives, NASA divided tasks into manageable work packages with their individual cost and schedule requirements. As a replacement for the traditional step by step approach to

---

4   Ibid.
5   Ibid.

flight testing, Mueller introduced "all-up" testing – a practice successfully applied by the air force to the Minuteman Program – in which on the first flight test every stage of the launch vehicle was live and a complete spacecraft was carried. And he defined the success criteria as two successful flights in an all-up mode before rating the space vehicle consisting of the launch vehicle and the spacecraft ready for human flight. The all-up testing concept became a special consideration which significantly reduced the high cost of flight testing in return for extensive ground testing. Mueller said, "The 'all-up' concept provides for the earliest possible readiness of the system and makes it possible to capitalize on success." As another special consideration, Apollo adopted an "open-ended" mission concept to allow the achievement of as many objectives on each flight as possible. This approach treated each mission as a research experiment, and planned alternative procedures which could be changed during flight. At predetermined points, NASA reviewed key systems, checked on expendables, and looked at overall status before committing to the next milestone. This allowed the agency to gain the maximum amount of knowledge on each mission without affecting flight safety. Other special considerations included redundancy for increased reliability, and reuse of proven technology to increase safety, reliability and quality. However, planning, problem resolution, and schedule discipline were necessary for success in managing large R&D programs. Mueller introduced planning and scheduling with daily milestones, and insisted, "Today's work must be done today." The only way to keep programs on track involved establishing daily schedules at the contractor level because, he said, "unless they knew what had to be done each day they weren't able to stay on top of it. So you just had to have a fine enough grained schedule so that you became aware of the slip before it got out of hand."[6]

What is the best background for a program manager? As Mueller put it, some engineers make good managers, while others do not. He believed that few scientists were effective managers because they had "a different mindset." Scientists are more interested in the search for knowledge as opposed to applying knowledge, and for the most part they do not have an interest in applications. However, Mueller began as a scientist, earning a doctorate in physics and spending much of his early career "doing science, real science." He always remained interested in the application of science, "building things. Not a pure scientist." So, in a sense, he became a system engineer who brought a unique set of skills to NASA. He said, "I was able to understand the scientists and to some extent they were able to understand me. And that was also true of the engineers … But the challenge was to be able to really meld all of that into a useful pattern."[7]

Through the efforts of men like Mueller, NASA rapidly produced dependable space vehicles and the techniques which landed astronauts on the Moon and returned them safely to Earth, at a predictable cost and within a scheduled period.

---

[6]   Ibid; Mueller, Aviation/Space Writers Association, Washington, DC, 5/1/68; Mueller Interviews: Slotkin, 9/9/09 (second session) and 6/8/10.
[7]   Mueller Interview, Slotkin, 2/24/10.

Mueller, an engineer, manager, and scientist, acted like a military officer when he needed to, demanding rapid progress. He possessed the scientist's interest in novelty and new knowledge, but sacrificed it when necessary to meet the mission objective. He had the engineer's interest in dependability, but took risks when necessary – "reasoned" risk as he liked to call them. He had the manager's interest in cost and schedule, though he sacrificed the former to achieve the latter. Mueller claims a "varied industry background," but "not an industrial background in the normal sense." He worked as a researcher and educator, not always as a manager; though during his six years at Ramo-Wooldridge/Space Technology Laboratories he managed R&D for the air force ballistic missile program. However, as he said, management jobs at Space Technology Laboratories "are different because you're dealing with a set of people who are highly motivated and very, very competent. So you really only needed to point out what needed to be done and not [involve yourself in] making them do it. So it was more collaborative than directing them." His philosophy of management began with the people immediately around him. He made sure they could handle what they needed to do and then made sure that the next layer was tied into the first layer, and so on. For a program as large as Apollo, establishing communications up and down the line became a critical success factor.[8]

Recognized by those who study human spaceflight in the 1960s for his leadership, Mueller is not well known today. As Murray and Cox wrote, he is "one of the most elusive figures in the Apollo story. For the most part, Apollo was led by men of great ability, often colorful and occasionally eccentric," but because of Mueller's reserve and lack of outward emotion, they said, these men "bled when they were pricked. With Mueller, it is hard to be sure." However, he was "almost inhumanly rational, unmoved by any argument that was not scrupulously logical and grounded in data," said NASA's John H. Disher. Furthermore, Mueller was "always the epitome of politeness, but you know down deep he's just as hard as steel." Nonetheless, Disher called him "one of the few authentic geniuses I have known. And he was a good manager. He knew what was going on." Phillips said that he "could never understand why the histories of Apollo paid so little attention to Mueller, and asked why "he hadn't gotten the credit that he really deserves for the success of Apollo." Mathews called him "a very brilliant person, though extremely controversial." New to the human spaceflight program, he "came up in the space world rapidly, without a large amount of direct experience in operations." Mathews called Mueller "a super egotist," but also "a humble man – with the courage of his convictions." A contradiction in terms? Perhaps. Although Mueller did not promote himself he fought for his interests, which some interpreted as egotism. Proud, but not arrogant, his mild manner and modest appearance created the image of humility. According to many, Mueller favored the Marshall Space Flight Center in Huntsville over the Manned Spacecraft Center in Houston, which led to a reputation of his being divisive, and some accused him of playing the two centers against each other. George

---

[8]   Ibid., 6/15/09 and 9/9/09.

M. Low also called him divisive. Joseph F. Shea, himself a brilliant but controversial man, described him as "one of the most strong-willed guys I'll ever meet ... He's a tenacious son of a bitch, he really is."[9]

Not a colorful figure, Mueller did not seek personal publicity, and did not follow his performance at NASA with other high government or academic appointments, preferring instead to return to private industry. He continued to promote the goal of reusable space systems, but after the success of Apollo the US lost interest in space spectaculars. He led the System Development Corporation for about fourteen years before retiring in 1983 at age sixty-five. However, he did not end his career there. After being semi-retired for a number of years, something that he called "boring," he consulted, served on government boards and committees, and became president of the AIAA and then the International Academy of Astronautics, thereby gaining a platform to speak out about spaceflight. Then at the age of seventy-eight, in 1995 he returned to full time employment at Kistler Aerospace Corporation, the developer of the K-1 reusable launch vehicle, in a final attempt to attain his dream of developing a reusable space transportation system. Following several years of ill health, he stepped down in 2004 at the age of eighty-six, and Kistler, after years of stop and go effort, and crippled by financial problems, failed to complete its K-1.

Mueller remains in retirement in Kirkland, Washington with his wife Darla, and continues to receive recognition for his contributions to human spaceflight, with the latest being the Smithsonian National Air and Space Museum's lifetime achievement trophy in April 2011 (see the Epilogue). He achieved great things during his time at NASA, and went on to a long and successful business career, but his love for the challenges of human spaceflight never left him. At age ninety-two, having recovered from earlier health problems, he said that with a billion dollars he could complete the goal he started out with at Kistler and develop a fully reusable launch vehicle. All he needs is a billion dollars to do it. The dream goes on.[10]

Arthur L. Slotkin
Atlanta, Georgia

---

[9]   Murray and Cox, *Apollo*, 158-160; Disher Interview, Sherrod, 4/15/71; Mathews Interview, Sherrod, 2/17/70; Low Interview, Sherrod, 9/7/72; Shea Interview, Sherrod, 3/10/73.
[10]   Mueller Interview, Slotkin, 2/22/11.

1

Reorganizing

"I will report for duty on September 1."
Mueller to James E. Webb, July 18, 1963

After firing D. Brainerd Holmes as the head of human spaceflight, NASA administrator James E. Webb spoke to a few people who were of the view that the air force should take over the program, telling them that since it was so important to the nation he planned to hire someone that they could trust and respect, in the hope that this would enable them to support the civilian space agency in accomplishing the lunar landing goal. After seeking the advice of friends, the administrator ended up asking TRW's chief executive officer, J. David Wright, for help. Mueller has speculated that Webb talked to many people before calling Wright and asking him to volunteer someone acceptable to the air force to take over the Apollo Program. Wright asked Simon Ramo, TRW's executive vice president and co-founder of Ramo-Wooldridge, Incorporated and Space Technology Laboratories for recommendations, who suggested Ruben Mettler the Labs' president. But Mettler had no interest. Next, he suggested two other company executives, Edward C. Doll, head of ballistic missile programs and Mueller, in charge of R&D. Mueller thought that Bernard A. Schriever, commander of the Air Force Systems Command, also had a hand in the selection, and the general knew Doll better than he knew Mueller, although he had confidence in both men based on Ramo's recommendation. And as Mueller remembered nearly fifty years later, "They asked me to go back and talk to Jim Webb and see whether we were interested." Adding, "I was volunteered."[1]

After agreeing to interview, Mueller spent time talking with people at NASA. He knew many from his work with the agency since 1958, including Joseph F. Shea who also worked with him at Space Technology Laboratories. Shea had become one of Holmes' deputies, and from these conversations he learned more about the structure,

---

[1] Mueller Interviews: Sherrod, 3/20/73, JSC, 8/27/98, NASM, 2/15/88, Lambright, 9/21/90 and Slotkin, 2/24/10.

plans and problems in the Office of Manned Space Flight. Shea said Holmes did not have a good working relationship with Webb, never tried to work with Robert C. Seamans, Jr., NASA's associate administrator, and did not allow them to participate in the management or understanding of OMSF. Without relationships at the top, he did not know the extent of his own authority and tried to make the center directors allies by avoiding conflict with them, and this forced him to compromise to keep their support. However, they never accepted his leadership. Shea told Mueller that Wernher von Braun, director of the Marshall Space Flight Center in Huntsville, Alabama used a democratic management system with dozens of people reporting to him and his technical deputy Eberhard F. M. Rees. Robert R. Gilruth's principal deputies at the Manned Spacecraft Center in Houston, Texas, Walter C. Williams and James C. Elms, were at loggerheads fighting for control. Shea said that the human spaceflight program ran "open-loop." Specifications were incomplete. Milestones existed, but were not believed. The schedule was not integrated, costs were out of control, and center management thought R&D programs could not be priced or scheduled. The agency ran programs on a level of effort basis, working until completed. To make matters worse, all contractor relationships had troubles. General Electric's Apollo integration contract created major problems because the centers did not understand their role, and even where this was understood it was "resisted and resented." He had ideas about reorganizing, telling Mueller it needed better program control, or it would not meet Kennedy's goal; and he said that the most severe problem at NASA concerned "communications, real communications, between the operating elements."[2]

Mueller also had discussions with Harry J. Goett, director of the Goddard Space Flight Center, whom he knew fairly well from his work on the Able spacecraft. He considered Goett "an unheralded manager," who established Goddard and initially led NASA's Space Task Group until outmaneuvered by Gilruth. Goett thought highly of George M. Low, another of Holmes' deputies, but did not consider Gilruth a team player. And without working as a team, he said the agency could not pull everything together. He expressed pessimism about Apollo, citing the many diverse agendas, lack of direction, and the conflict between the different organizations. And he told Mueller that the agency had "overruns that were just not stopping;" the warlords, as he called the center directors, ran their centers as independent fiefdoms and "hardly talked to one another."[3]

Mueller had a candid conversation with air force General Osmond J. Ritland, Schriever's deputy commander for human spaceflight. Ritland knew the agency well, and made a number of cogent recommendations. A bit more than a year from retirement, the general said that Gemini ought to be decentralized by placing its management in Houston. He argued that NASA headquarters confused the problem

---

[2]   Mueller Interviews: NASM, 2/15/88, Sherrod, 4/21/71 and JSC, 1/20/99; Lilly Interview, Lambright, 7/18/90; "Notes on Discussion with Joe Shea," Meeting Notes & Transcriptions, 1968-1969, GEM-79-15.

[3]   Mueller Interviews: Slotkin, 2/22/11, NASM, 2/15/88, Sherrod, 4/21/71 and JSC, 1/20/99.

without adding to its solution, and the air force had a good working relationship with Houston on Gemini. On the other hand, the Apollo Program was so massive that control should be at headquarters, a need he said the centers recognized. And doing this would focus the roles of the headquarters staff. He told Schriever the air force should scale back support to NASA, concentrating on OMSF. He suggested Mueller hire a deputy, and cautioned him not to trust either the air force or the Department of Defense because they had their own interests. Ritland had strong feelings that Apollo was in trouble, and without significant improvements it was likely that responsibility for human spaceflight would be transferred to the defense department. While NASA had some of the skills that Space Technology Laboratories possessed, it lacked the capabilities the air force brought to the table. Ideally the agency should subcontract some development to the air force, but it was not politically feasible to do that, so Mueller had to find a way to bring air force talent into the agency. And his final piece of advice was to create a buffer between the "political environment ... and actually carrying forth the program."[4]

Before Webb had to decide between the two candidates, Doll bowed out. Mueller then flew to Washington, where the administrator explained his view of the problems the agency faced and what he hoped to accomplish. He took Mueller home for dinner to discuss what should be done and asked whether he would undertake the job if the agency were reorganized. Mueller had already decided he would take the job if Webb agreed to make the necessary changes. However, their first one-on-one meeting did not give him the assurances he wanted. "Jim is a very charming person and a very voluble person, and he generally ... covers a very wide range of subjects, and only occasionally touches on the thing that you are there to talk about," Mueller recalled. And as a consequence, they did not discuss all the key issues. "I think he was really trying to decide whether he had the confidence in me to do what he needed ... I guess he decided that he did." Mueller's concerns centered on controlling the program, but the administrator spoke about salary, perquisites and benefits, and because taking the job would be a financial strain Mueller had to consider whether he could afford it. He had mixed feelings from this meeting because, he said, "it wasn't clear that Jim was willing to make a clear decision to allow me to run the program ... But he was sufficiently under the gun so that he was willing to accept almost anything in order to get somebody back there to run the program." Mueller needed authority over the center directors, and Webb finally agreed. Ultimately Mueller accepted the job because it presented a challenging opportunity, and he judged it "very important for the future of the country in terms of our international relations, because at that time it was fairly obvious that if we didn't lead, why the Russians would."[5]

---

[4] "Notes on Conversation with Gen. Ritland," Meeting Notes & Transcriptions, 1968-69, GEM-79-15.

[5] Seamans, *Aiming at Targets*, 109; Doll to Mueller, 12/13/63, GEM-41-1; "Off Target," *Newsweek*, 9/23/63, Articles, GEM-83-3; Mueller Interviews: NASM, 2/15/88 and Slotkin, 6/15/09. Mueller frequently used the term "the Russians" when referring to the USSR.

Webb offered Mueller the position of director of manned space flight programs in a letter dated July 18, 1963. Reporting to Seamans, he would have line authority over the center directors, as well as over the program management group at headquarters. Responding a few days after his forty-fifth birthday, he accepted the job with "some trepidation," describing it as a challenge; later calling it "something that needed to be done. [But n]ot the least of the challenges was that it cut my salary in half." He wrote, "It is with great pleasure and anticipation that I look forward to working for you on the NASA Manned Space Program ... I will report for duty on September 1."[6]

After NASA announced Mueller's appointment, hundreds of letters flooded into his office in Los Angeles congratulating him on the appointment, including two from the air force chief of staff General Curtis E. LeMay and Secretary of the Air Force Eugene M. Zuckert. The Associated Press called Mueller a "dedicated fighter ... When outspoken Holmes quit as chief of the United States moon program, a lot of people predicted his successor would have to be a quiet plugger who never argued with space boss James E. Webb." However, they wrote, "Mueller ... is quiet, but there the resemblance to the forecast image ends." He is "a brilliant scientist ... and when convinced he's right, friends say, he'll do more than argue – he'll fight until he drops." The news article spoke of his work ethic, and they were right on target as people soon discovered.[7]

Before relocating to Washington, Mueller sent letters to the center directors. While each contained similar messages, he wrote them with different tones. To von Braun with whom he had a working relationship, he said, "Just a note to tell you how much I am looking forward to being able to work with you on the Manned Space Program." Adding, the "task we have to accomplish is the most challenging our nation has ever undertaken and to carry it out will require the utmost strength and continuing cooperation of all the members of this team." He appealed for assistance, saying he could not succeed without his "support, help, guidance and counsel," and counted on him "for all these and more." He did not know Gilruth very well, though in deference to his long government service and leadership of the Space Task Group, he wrote, "I do believe that the program to date has been the most successful large scale program ever carried out in this nation," and its success going forward depends on "the continuation of the wonderful team that has done so well." Kurt H. Debus, who was director of the new Launch Operations Center at Cape Canaveral, Florida, had previously worked for von Braun, and he received a similar, but less deferential letter. Mueller also wrote his daughter Karen Ann Mueller, a Stanford undergraduate studying in Europe that summer, telling her, "I am starting a new job in September. It is a job which will take a great deal of time

---

[6]   Webb to Mueller, 7/18/63, GEM-29-8; "The Apollo Spacecraft-A Chronology," vol. II, Foreword; Mueller Interview, NASM, 2/15/88; Mueller to Webb, 7/26/63, GEM-29-8.

[7]   "NASA Names New Head for Manned Space Flight; Succeeds Homes," Press Release 63-162, NASA Archives, 7/23/63; LeMay to Mueller, 7/29/63, GEM-28-4; Zuckert to Mueller, 7/30/63, GEM-29-6.

and energy, and one with a heavy load of responsibility as well as challenge."
Candidly he added, "At the moment I don't know enough about it to do more than
recognize that there are problems as well as rewards in its execution."[8]

In early August, Mueller shared some thoughts with Eugene R. Spangler, co-
author of his 1963 book *Communications Satellites*, asking for help in gathering
information about alternative management and organizational structures, and
telling him to hold the information in the strictest of confidence. In notes to
Spangler, he outlined the basic concept of an Apollo Program Office, "similar to the
combination of the Air Force and STL Minuteman program office," which would
have "sole responsibility and authority" over the Apollo Program. He described a
headquarters group with branches at the centers for each major system and
subsystem. Headquarters would have responsibility for overall program control,
system engineering, test planning, and reliability and quality assurance. The centers
would have similar functions, plus engineering and manufacturing. Describing a
matrix organization similar to the one used in the air force ballistic missile program,
he proposed separating the project offices from direct control of the centers and
having them report operationally to the Apollo program director, and adminis-
tratively to the center director. Continuing, he outlined a steering committee of
center directors and himself as the senior policy group.[9]

Mueller said he needed the people doing the work to report to him, "or else you
just weren't going to be able to get there from here." He took note of Holmes'
organizational problems, and proposed a new organization to avoid them. The
Apollo Program Office structure reflected the Minuteman Program Office, but
instead of assigning contracting and program control to the air force, with Space
Technology Laboratories providing system engineering and technical direction,
NASA fulfilled both roles with the help of support contractors. Space Technology
Laboratories' technical directors had broad expertise, using resources in their
laboratories to solve problems. So that is how he organized Apollo. He wanted
strong program offices, while maintaining the technical competence in the basic
center structure. He planned to use the "good parts" of the ballistic missile program
organization, and move it over "modified for the circumstances" to change from an
institutional to program organization, because they had been "all mixed up before."
The center directors needed to be concerned with both institutional and program
objectives, making it necessary to establish a strong program management
organization in parallel with the institutional chain of command to insure that
communications flowed up and down both chains. Thus, before leaving Los Angeles,
Mueller made decisions about major features of the OMSF reorganization. He knew
the centers were not working together as a team, but in order to achieve program

---

8   Mueller to von Braun, 7/24/63, GEM-27-2; Mueller to Gilruth, 7/24/63, GEM-27-10;
    McCurdy, *Inside NASA*. 19; Mueller to Debus, 7/24/63, GEM-27-5; George to Karen
    Mueller, 7/31/63, GEM-28-1.
9   Mueller to Spangler, 8/2/63, "Management Concept," GEM-36-7.

objectives that had to change. He needed to improve communications, from the shop floor "all the way up to the people that were managing." Improving communications was a prerequisite to enabling "all of us to understand what was going on throughout the program."[10]

Mueller invited the center directors to meet with him so that he could learn their views, and they went to Los Angeles in August. Meeting one-on-one, each expressed the view, "Headquarters was really interfering with progress, and they needed more money and less direction," he recalled. They did not trust people in Washington or at the other centers, and considered "jockeying for position" between the centers and headquarters to be a serious problem. Each thought the other center directors "great people," although they did not understand the overall program and NASA needed a better way to coordinate programs. They knew about their own problems, but did not know how to solve the total problem, which frustrated them. Gilruth found himself "up to his eyebrows" trying to create MSC, while von Braun wanted "to figure out how much control he could continue to exercise," as Mueller recalled. And Debus, the most straightforward of the three, told him, "Look, we've got a real problem here. These center directors aren't talking to each other nor to me, and I'm supposed to fly these things when they get here. You've got to do something so at least, we're going to build something that's flyable."[11]

Gilruth and von Braun felt they knew how to run things better than anybody in Washington; after all, headquarters was out of the mainstream. Both suggested that headquarters be moved to their center, and informed Mueller of difficulties working with the other center directors. They had gotten direction from Holmes, reported to Seamans, and the money came from elsewhere. Confusion reigned, and that hurt the program. Mueller wanted everyone focused on the same objectives and to avoid just looking out for their own center. He assessed the difficulty he could expect, while realizing they viewed him with skepticism because they did not know what he had in mind, and he called these first set of meetings "sparring matches, more or less." As he sized up the center directors, he recognized the need to do something drastic to get their attention. Placing them under him would only be a first step, and so while he continued his fact finding he began discussing with them the program office structure that would be separated from the administration of the centers.[12]

As Mueller fleshed out his understanding of NASA, and discussed ideas about reorganization, the people he confided in told him he "couldn't do it . . . it wouldn't work because the center directors wouldn't agree to it, and they would sabotage it." However, he did not think any other way would work, so he continued planning to centralize control with strong program offices. Yet he worried this change might be too traumatic. He understood change would be difficult because, as he later wrote, "No one who has been operating autonomously likes to have authority taken away, and in this case we were dealing with three strong-minded, highly capable men with

---

[10]   Mueller Interviews: NASM, 2/15/88, Sherrod, 4/21/71 and Slotkin, 6/15/09.
[11]   Mueller Interviews: NASM, 2/15/88 and Putnam, 6/27/67.
[12]   Mueller Interviews: Sherrod, 3/20/73 and NASM, 2/15/88.

international reputations. I certainly couldn't predict how Wernher von Braun, Kurt Debus, and Robert Gilruth would react to my reorganization plans, but it would have been naïve not to expect strong – and loud opposition." However he added, "Those first meetings ... proved to be very productive. By the end of the third or fourth meeting with each of them and their staffs, we weren't discussing if the program should be reorganized, but how."[13]

II

Mueller spent his first few days in Washington hidden away at his new home with the center directors going through their problems, followed by briefings at NASA headquarters on August 28, 29 and 30, although he did not become a government employee until September 1, then deputy administrator Hugh L. Dryden swore him in on September 3. He then scheduled visits to the three centers with Seamans, telling the center directors he wanted frank discussions of the problems they faced.[14]

Keeping a daily diary during his first few months at NASA, Mueller jotted down the issues he faced. He kept this diary to record what was going on because he knew he would forget otherwise. Although as his workday expanded he discontinued the daily practice, he continued to make notes and keep action items throughout his six years at the agency because he considered it "important to have some continuity ... to convince people I was really interested." In one of his first diary entries, he outlined what he called the "criteria that must be met by the manned space organization," writing that it should use all of the resources of the government and immediately identify schedule slips and cost overruns. He considered capabilities of the centers, external inputs, politics and public relations to be issues of immediate concern, and wrote that the new organization "should provide proper basis for the addition [of] programs and the orderly development of these programs from conception through completion."[15]

Mueller thought about what he had to do and how to get things done, and tried to evaluate the necessary interfaces. He dived into the issues, making notes about what he had to do. Although he formulated many ideas before arriving in Washington, he did not immediately implement them; instead he listened as people expressed their thoughts and told him what they considered the shortcomings which kept them from doing their jobs well. He found things "pretty much a mess," and used this as an opportunity to make changes. Had the crisis not existed, it would have been more difficult to impose controls over the centers. However, he said the center directors

---

[13] Ibid.; Mueller to Teague, 7/20/78, Committee on Science and Technology, GEM-161-2.
[14] Mueller Interview, Putnam, 6/27/67; Bothmer to Center Directors and Mueller to Center Directors, 8/30/63, GEM-79-3.
[15] Mueller Interviews: Slotkin, 9/9/09 and 2/22/11, and NASM, 2/15/88; Criteria That Must Be Met by the Manned Space Organization, 9/4/63, GEM-84-11.

1-1   Hugh L. Dryden swearing in George E. Mueller, September 3, 1963. (NASA photo)

were "concerned enough so that they were willing to give up some of their autonomy since it looked like they would lose the program and … respect of the country unless they did something different." Summing up the situation, he later said, "There wasn't a difficulty you could imagine that wasn't in the forefront of things at that time … we had problems with everybody – the Bureau of the Budget, PSAC [the President's Science Advisory Committee], you name it." Then paraphrasing

Benjamin Franklin prior to signing the Declaration of Independence, he told the center directors "if we didn't work [hang] together, we were sure going to be hung apart."[16]

As he developed his plans, Mueller met with other Washington insiders. On September 4 he learned the importance of establishing the proper relationship with Olin E. "Tiger" Teague representative from Texas, who was chair of the House Subcommittee on Manned Space Flight and a "big supporter" of Holmes, he wrote. Mueller knew that Holmes had considerable support in Congress, and anticipating major difficulties he met with Teague, who did not "pull any punches." The chairman said without any preamble, "I don't like what happened to Brainerd Holmes, but I believe in supporting the job, not the man. I don't have any personal opinion about you, but as long as you do the job, I'll support the office." Adding, "If you double-cross me once, it's your fault. If you double-cross me twice, it's my fault – and I never have that problem … Those few words certainly cleared the air, and from that moment forward, Tiger Teague never wavered in his support for me or my program." Mueller called Teague, "The most interesting character I have ever run across … He's a typical Texan and a typical product of Texas A&M … He wasn't brilliant in any way. He wasn't a charismatic leader as such, but he really was good at knowing what to do on each interface with each other congressman … So if you could get him convinced about something he could get it done in Congress … And he was probably in the House almost as effective as Lyndon Johnson in the Senate. And we were fortunate in having him on our side."[17]

Mueller met with his new staff to talk through alternative approaches for the program offices. They argued that dual reporting had been tried by Holmes, although without success due to lack of support from the center directors. However, Mueller explained to them that the program needed strong program offices in addition to the management council, because those councils tended to become "debating societies" rather than effective management tools. Outlining ideas in his diary, he wrote that he wanted part of the organization devoted to external affairs as a focal point for the scientific policies of the human spaceflight program, and he considered appointing a chief scientist for OMSF and creating a scientific advisory board. He needed strong program directors, and decided to abolish functional organizations at headquarters and organize by program, giving the program offices responsibility for managing spacecraft, launchers and flight operations for their programs. He summarized the political, public relations and management constraints, and wrote that the changes he made "must not erode or drastically change the role of Debus, Gilruth or von Braun. They must not be such as to cause any of the Directors to have a legitimate reason for complaint, either to Congress or the press." Moreover, the organizational changes "must create the impression of

---

[16] Mueller Interviews: Slotkin, 6/15/09 and 2/22/11, Sherrod, 3/20/73 and Putnam, 6/27/67.
[17] Mueller Interviews: Sherrod, 11/19/69 and 3/20/73 and Slotkin, 2/25/10; Meeting Notes, 9/4-5/63, GEM-79-12; Mueller to Teague, 7/20/78, Committee on Science and Technology, GEM-161-2.

utilizing the experience of Polaris and the ballistic missile program ... [and] must reduce DOD/NASA tensions." The changes had to be consistent with the lunar landing goal, increase communication with Congress, gain the support of the scientific community, and deal carefully with the press. He had to make the organization simple, straightforward, and easily understood. In addition, the new organization needed to be decentralized as much as possible.[18]

Mueller had no problem decentralizing those areas where it made sense. However, making the program run properly required a central authority to assure things worked together. He returned to the theme of organization in a diary entry on September 8, beginning: "Without much improved management ... we will not achieve the lunar goal prior to 1972-1975 ... at a cost of $35 billion or more," a premise he knew would not be acceptable. He had to improve program management and satisfy NASA's critics. His notes say "the duality of the reporting and responsibility lines, and the geographic dispersion present serious problems." The Washington staff felt that his plan would not work without Seamans' and Webb's full support and the complete cooperation of the center directors, and they had divided opinions about whether he could achieve that consensus.[19]

MSC derived from the National Advisory Committee for Aeronautics research centers, and it retained much of the cultural orientation of the old committee. Under the NACA structure, the centers acted semi-autonomously and did not receive orders from headquarters. And Gilruth ran MSC much as in the NACA days, referring to NASA headquarters as the "Washington office," whose primary job involved dealing with Congress and budgeting the funding for the centers. The advisory committee headquarters did not have operational responsibility, and its research centers carried out small R&D projects that were managed locally. It had no cross-center programs that required inter-center coordination, so the problem of program versus institution never arose. And when Mueller made the program office "king," it ran counter to Gilruth's experience. NACA also took pride in not being directly affiliated with the military; although much of their work benefited the armed forces. The advisory committee remained doggedly independent of the War Department and later DOD. However, von Braun and his team began their work at Peenemünde as a division of the German army. After the war they affiliated with the US Army, and did not join NASA until 1960. While von Braun's engineers approached development and testing in the same way as those from NACA, other differences stood in the way of close cooperation between the two centers.[20]

Mueller called his relationship with Gilruth cordial, but said, "Gilruth was always reserved and quite different than Wernher. And his folks were very much the same way." He considered them both quite capable, but thought NACA veterans were not "very amenable to external thought," although "the MSFC folks were at least as

---

[18]   "Organizational Meeting," 9/4/63, GEM-84-11.

[19]   Ibid.; Mueller Interview, 9/9/09.

[20]   Hodge Interview, Lambright, 10/5/91; McCurdy, *Inside NASA,* 18; "Langley way," Murray & Cox, *Apollo,* 25-28.

capable but more pliable with respect to adopting new thoughts." He considered one of Gilruth's strongest characteristic also to be one of his weaknesses. As a father figure, people at MSC "had a tremendous affection for him and he had a great affection for them probably to the detriment of the program as a whole because it tended to make him more insular and separate from the mainstream of activities." Mueller characterized von Braun as an "outstanding engineer, [and an] outstanding charismatic leader," calling him "the most capable" center director. He considered Gilruth to be a good engineer but not a good manager, and found him more difficult to work with than von Braun. As he described it many years later, "Wernher and I thought much more alike than Bob [and I] did. Our thought process was similar." As he remembered, "Gilruth did not take direction very readily, it took a fair amount of leaning to get him to move," and he "would go around you if he could." Nonetheless, Gilruth's staff greatly admired him, particularly the original members of the Space Task Group, so Mueller found it infeasible to remove him. Instead, he supported the center director by assigning strong managers like Shea and Low to MSC.[21]

Shortly after his first meeting with Teague, Mueller had to appear before the House space committee to defend the Mercury Program. MSC's Williams prepared a report "describing in some detail the horrors of the Mercury program and published it," and a copy found its way to Congress, Mueller said. Williams "convinced himself and everybody else that the contractors were completely incompetent," and thought that recording the lessons learned during Mercury would help Gemini and Apollo. However, he had no sense for politics, and did not understand the impact such a report would have in Washington. Consequently, shortly after arriving at NASA, Mueller found himself defending McDonnell Aircraft, builder of the Mercury and Gemini spacecraft, as competent, conscious of safety, and fully understanding the importance of good workmanship. With Williams at his side, Mueller testified "this was just an internal document that we'd written to make sure we didn't repeat any errors in the future, and of course, we didn't write about all of the good things that happened and all of the things that were right. All we wrote about were the things that were wrong." The subcommittee showed no sympathy, he said, "but after appropriate discussion of the shortcomings of management of NASA ... why they went on to other things." The lesson learned from this experience was, "if it's on paper, forget it," it exists, and someone is probably already off publishing it.[22]

Mueller wanted to add a chief scientist to OMSF, and over lunch with Seamans they discussed the need. The general manager suggested that Homer E. Newell, about to be appointed associate administrator for space science, would effectively fill the role for NASA. Mueller did not like the suggestion, but he was unable to sell the idea to the general manager. Seamans also told Mueller of Webb's "great desire" to

---

[21]   Mueller Interview, Slotkin, 6/9-15/09; Meeting Notes, Nick Golovin, 9/11/63, GEM-79-12.

[22]   Mueller Interviews: Slotkin, 9/9/09 (second session) and 2/22/11, and NASM, 6/22/88.

bring air force officers into positions of responsibility in the Apollo Program, and Mueller suggested appointing them to program control, one of the key roles that they were performing very effectively in the ballistic missile program. Then after reviewing the reorganization of OMSF, Seamans agreed with it.[23]

By September 21, when Mueller again discussed the OMSF reorganization with the center directors, he simplified it but maintained the basic ideas he brought to Washington. In his diary he noted that he had "reached essential agreement ... on the general concept of program office structure." Then again on September 28, he discussed the changes at a meeting which "went smoothly" and "without major controversy." However he added, it "is not clear in retrospect as to whether or not the center directors really understood the implication of the changes that were being introduced." Concluding, "[It is] essential to get out a reasoned document describing the general outlines of the concepts ... and being sure that these are general enough and yet specific enough so that people will understand the inter-relationships."[24]

III

Mueller had concerns about the cost and schedule of the Apollo Program. Based on experience in the ballistic missile program, he extrapolated the time to deploy the Saturn S-IVB stage to at least March 1965, while the S-II stage would not be available until June 1966 (Table 1-1). Then the agency would have to add about four years to those dates before the stages became fully operational, which would put a 1969 lunar landing out of reach. In pondering how to speed up development, the question was whether to reduce the amount of work or to seek alternate sources for some of North American Aviation, Incorporated's Apollo contracts. NASA had awarded the contract for the Apollo spacecraft to the company's space division in 1961. The same division was building the S-II stage, and the Rocketdyne division was developing most of Saturn's engines. By 1963, Low and others had concluded that the agency had overloaded the NAA space division by awarding them two major Apollo contracts. However, Webb thought the problems came about because they had weak management. He said the company had the capability to manage two major contracts, but without the right people in place they could not be effective. Mueller agreed that the space division lacked strong management, making it necessary for him to spend a great amount of time insuring their "managers were managing," he said. As he explained, the company's "upper management ... were not very detail oriented, and were used to building airplanes" using "a craftsman approach," as he called it. He argued that building complex spacecraft required a

---

[23] For the Record, 9/11/63, Diaries – Notes and Reminders, GEM-70-8; Mueller Interview, Slotkin, 9/9/09.

[24] Organizational Meeting, 9/4/63, and Organizational Concept, O&M, GEM-81-11; Constraints on Organizational Changes, undated, O&M, GEM-85-6; Diary for 9/10, 14, 21, 27/63, GEM-70-9.

more rigorous approach. And while he agreed that the space division was overloaded, "At least it simplified the interfaces somewhat."[25]

Table 1-1: Saturn launch vehicles and stages[26]

SA	NASA designated Saturn missions based on the launch vehicle and flight number. SA-501 indicates the first Saturn V launch vehicle. The first Saturn IB launch vehicle is designated SA-201, while the Saturn I missions were designated SA, followed by a single digit.
S-I (Chrysler*)	Saturn I first stage.
S-IC (Boeing*)	Saturn V first stage.
S-II (North American*)	Saturn V second stage.
S-IV (Douglas*)	Saturn I second stage.
S-IVB (Douglas*)	Saturn V third stage and Saturn IB second stage.
Saturn I	Two-stage vehicle, with eight H-1 engines in first stage and six RL10 engines in second stage. H-1 engines built by Rocketdyne and RL 10 built by Pratt & Whitney division of United Aircraft.
Saturn IB	Two-stage vehicle, with eight H-1 engines in first stage and single J-2 engine in second stage. Engines built by Rocketdyne.
Saturn V	Three stage vehicle, with five F-1 engines in first stage, five J-2 engines in second stage, and single J-2 engine in third stage. Engines built by Rocketdyne.
Command Module (North American*)	Contained the crew quarters, plus controls for the various in-flight maneuvers including extensive guidance and navigation system and a stabilization and control system. Incorporates an ablative heat shield for re-entry. (The Apollo-Saturn was designated AS-2nn** or AS-5nn.)
Service Module (North American*)	Contained a stop-and-restart engine, and the primary electrical power supply for the spacecraft.
Lunar Excursion Module/ Lunar Module (Grumman*)	Flight unit that detached from the orbiting CSM and descend to the Moon's surface. Had complete guidance, propulsion, computer, control, communications and environmental control systems. A two stage vehicle, the lower stage contained the decent engine and landing gear.

Notes: *Prime Contractor
**nn equals the number of the mission, with the first mission designated 01, the second 02, etc.

---

[25] Action Items, 9/5-9/9/63, GEM-47-9; Low Interview, Sherrod, 8/12/70; Webb Interview, Sherrod, 4/28/71; Webb and Paine Interview, Sherrod, 10/7/70; Mueller Interviews: Slotkin, 9/9-10/09 and NASM, 6/22/88; The official name of the NAA space division was the Information and Space Systems Division.

[26] "Saturn Terminology," NASA, http://history.nasa.gov/MHR-5/glossary.htm, July 1, 2009; Technology Club of Syracuse, Syracuse, NY, 2/1/64; University of Sydney Summer School, Sydney, Australia, Lecture III, 1/11/67.

J. Leland "Lee" Atwood, NAA's CEO had a hands-off style of management which conflicted with Mueller's hands-on approach. Over dinner with Ramo shortly after joining NASA, Mueller learned that Atwood believed "the government is not capable of managing large programs ... [and] industry has to do the job." Atwood told Ramo that putting a few people like Mueller into government "will not have a lasting effect on the course of major programs. And the proper course for a contractor to follow is to agree with the customer, but do what the contractor thinks needs to be done, and not worry about getting very good people on the job because it doesn't make any difference in the long run anyhow ... [And] don't worry about spending too much money." Atwood would tell the customer whatever schedule he wanted and then keep working until the product was delivered.[27]

In late September, Mueller met with Jerome B. Wiesner at the White House to review his reorganization of OMSF. The president's science advisor did not have much interest in the reorganization, and he did not support establishing a science advisory committee for OMSF, calling it a bad idea because a chief scientist should report to the administrator. He told Mueller the scientific community had a negative view of NASA. He also adamantly opposed the lunar orbit rendezvous approach to landing on the Moon, calling it "the wrong decision." Introduced to Kennedy in a prearranged rendezvous in a White House corridor, the president welcomed Mueller to Washington, and said "we expect you to do some marvelous things," as he recalled almost fifty years later. And he called the young president quite a charismatic figure. Wiesner took Mueller to lunch in the White House mess, and as he noted in his diary, although the president's advisor "listened attentively," the lunch was "anticlimactic at best."[28]

Mueller met with a former colleague from Bell Labs, MIT Provost Charles H. Townes, who promised to do whatever he could to help. He supported Apollo, and recommended Mueller do "something overt to improve our communications with the scientific community." A Nobel Laureate in physics, Townes suggested that a space science advisory committee "might be worthwhile," but cautioned him to establish "the right kind" of committee. At the time, many scientist opposed Apollo, and while a majority in Congress supported human spaceflight, opposition from scientists had won enough support to cut the fiscal year 1964 NASA appropriations.[29]

After introductory meetings at the three centers, Mueller returned to Huntsville without Seamans and his entourage. As he recalled, he went to a meeting "where Wernher gave me one of his impassioned speeches about how we can't change the basic organization of Marshall." After politely listening, Mueller told him, "Marshall was going to have to change; the laboratories were going to have to

---

[27] Atwood Interview, NASM, 8/25/89; "Notes from discussion with Si Ramo at dinner on Sunday (11/11 [1963]), Notes & Transcriptions, GEM-79-15.

[28] Undated Diary for Oct.? 1963, GEM-70-9; Mueller Interview, Slotkin, 9/9/09.

[29] Notes, 11/20/63?, Notes & Transcriptions, GEM-79-16; Compton, *Where No Man Has Gone Before*, 9-10; Mueller Interview, Sherrod, 5/20/73.

become support to the program offices or else we weren't going to get there from here." And he also recalled "in the early days when they were trying to do their usual snow job on the people from Headquarters," he got into an argument at MSFC about stratification of hydrogen in flight. They discussed strategies to test it, and wanted to build a vehicle to check it out. He said, "It was the funniest thing because you get this big sales pitch about how you just have to do this and you can start anywhere in the presentation and say well, 'now explain this to me' – and they get some expert on that layer deeper and then he quite promptly proved that he didn't understand the problem so they would rush out and get some more experts ... It turned out that it was just sheer sales pitch without any foundation ... They only did this about three times and they took to reviewing everything." As a result, they avoided unnecessary differences between the Saturn I and Saturn V. He had similar meetings at MSC, but with Shea moving there he felt they would reorganize in a sound way. To make the program office structure work, he created new organizations at each center, put the right working relationships between the centers in place, and held them "long enough so that communications could grow," he said.[30]

Shea moved to Houston in October as manager of the Apollo Spacecraft Project Office, and Low became Mueller's primary deputy. Nevertheless, Low told journalist Sherrod that he did not hit it off very well with Mueller. "I had nothing personal or otherwise in those days that made me want to leave except that Mueller did not need or use a deputy," he said. Adding that although Mueller needed staff to do things, he did not want an alter ego or to have someone to work on their own. While some people need deputies, Mueller did not, and, as Low discovered, "he didn't need me." Mueller remained unaware of Low's dissatisfaction, but admitted circulating a lot of people through the post of deputy, having eleven deputies of one sort or another while at NASA (Table 1-2). He did not want a double, he wanted people to help him to "manage the systems," as he termed it. He needed somebody to handle day-to-day interfaces. However, he acknowledges that sometimes this was a source of some frustration for his deputies.[31]

By the beginning of October, after a month in office, Mueller had crystallized the shape of his organization and sent information packages to the center directors. He included new organization charts for the centers, and concept papers with detailed plans. He provided functional statements, or overall road maps for the program offices and the organizations supporting them. On October 4, he sent an information bulletin to OMSF's Washington employees announcing the organizational study, saying the initial results showed "some decentralization of authority may result from the overall NASA study," and that his "thoughts and concepts have changed since

---

[30] Mueller Interview, Putnam, 6/27/67; History, 3/13/69, Mueller, Notes & Transcriptions, GEM-79-15.

[31] Low Interview, Sherrod, 7/5/72; Mueller Interview, Slotkin, 9/9/09 (second session).

Table 1-2: OMSF directors 1963-1969

Deputy Associate Administrator (DAA)
George M. Low, 11/63-5/64
James C. Elms, 9/65-9/66
Edgar M. Cortright, 10/67-4/68
Charles W. Mathews, 5/68

DAA, Management
William B. Rieke, 11/64-6/65
Frank A. Bogart, 9/65

DAA, Programs (dropped 5/67)
David M. Jones, 11/64

DAA, Technical
Joseph F. Shea, 4-7/67
Harold T. Luskin, 3-4/68
Charles J. Donlan, 5/68

DAA, Flight Operations (dropped 4/64)
Walter C. Williams, 11/63-4/64

Field Center Development
Robert F. Freitag

Program Control
William E. Lilly
Frank A. Bogart,* 3/67
Maynard E. White, 6/67
Jerald R. Kubat, 1/68
Frank A. Bogart,* 6/69
Charles E. Koenig, 11/69

Management Operations
Clyde Bothmer
Frank A. Bogart, 2/65
Paul E. Cotton, 9/65
Maynard E. White, 1/68
Frank A. Bogart,* fall 1969

Space Medicine
George M. Knauf*
W. Randolph Lovelace, II, 4/64
Jack Bollerud,* 2/66
James W. Humphreys, Jr., 6/67

Gemini (downgraded 1967, eliminated 1968)
George M. Low*
George E. Mueller,* 65-68

Apollo
George E. Mueller*
Samuel C. Phillips, 10/64
Rocco A. Petrone, 8/69

Advanced Manned Missions
Edward Z. Gray
George S. Trimble, 4/67
Douglas R. Lord,* 10/67
Edgar M. Cortright,* early 1968
Charles J. Donlan,* 5/68

Mission Operations (added 1/65)
Everett E. Christiansen, 1/65
John D. Stevenson, 2/67

Apollo/Saturn Applications (added 4/65; changed to **Apollo/Skylab** in 1969)
Harold G. Russell, 4/65
David M. Jones,* mid-65
Charles W. Mathews, 12/66
Harold T. Luskin, 5/68
William C. Schneider

Manned Space Flight Safety (added 6/67)
Jerome F. Lederer

Note: *Acting. Date started. (No date for first person means incumbent had job when Mueller arrived. No date for last person indicates incumbent in place when he departed.)

he came on board." But he promised to continue to receive input from others prior to making any changes.[32]

The briefings which Mueller received gave him a clear picture of where the programs stood and what the agency hoped to accomplish. They showed the need to restructure, for otherwise "we wouldn't stand a chance at all of coming out on schedule and particularly within the cost," he said. He asked Low to suggest the names of two experienced managers who could provide a "candid assessment of the real status of the program, schedule wise." Low recommended John H. Disher, and Adelbert O. Tischler, assistant directors for spacecraft development and propulsion respectively, "two old hands" in OMSF, to "figure out how we could get to the Moon in time" based on the plans in place at the time. After about two weeks they told him there was "no way you're going to be able to do that" in the time or at the cost estimated. Their projections totaled upwards of $40 billion, which was twice the original program estimate. As Disher remembered, Mueller "wanted a realistic and conservative estimate." At that point, a conservative assessment did not show a high degree of confidence in landing on the Moon before the end of the decade because there were just too many problems to overcome. After listening, Mueller took them to see Seamans. Disher recalled, "Seamans looked at the results, which probably were, as I say, overly conservative, and he was pretty quiet there in the meeting." The

---

[32]   Organizational Package, Last Week of 9/63, GEM-84-11; OMSF Information Bulletin, 10/4/63, GEM-84-12.

general manager asked to meet privately with Mueller, and when Mueller returned he told them "Dr. Seamans would like to have you destroy that material," and they did. Seamans was surprised by the results, and told Mueller to get rid of the report and "find out how to do it," meaning achieve the program within time and budget. There was a lot of confusion about the schedule at the time, and Mueller considered it important for the general manager to understand the real schedule, rather than some optimistic assessment. Reasoning that if *he* did not know the true status, how could others with less knowledge understand? And he felt it "essential that everybody understand the problem."[33]

Mueller said he destroyed his copy of the report, but one resurfaced in 1976 in the files of the agency's historian, Eugene M. Emme (1959-1979). In about two dozen briefing charts, the two old hands laid out the facts, showing Apollo ready for piloted flight on a Saturn IB in November 1966. However, the assessment "modified by judgment" had a one year slip. Scheduled to last forty months, Apollo spacecraft development would take more than forty-nine months with only a five percent probability of success. A sixty-three month schedule lifted the probability to fifty percent, and it required more than seventy months to achieve a ninety percent probability. To Disher, the report was "just one indicator of how shrewd George was. He got on record the real status of the program as a couple of old hands would look at it when he was taking over. So he had a point of departure to work from." Over dinner with the administrator that evening, Mueller broached the idea of cutting back on the Saturn I by terminating flights at the tenth vehicle and diverting resources to the Saturn IB and Saturn V. Mueller later said "the Saturn I didn't do anything for anybody. So that was an easy decision to make." The only reason for it was "because that was the way Marshall did things – Saturn I and Saturn IB and left to their own devices" they would build "the Saturn II . . . as well as the Saturn III and IV and V."[34]

Von Braun briefed Mueller on October 17 about the impact of canceling Saturn I. Preliminary analysis showed a net reduction of $289 million over a five years period beginning in FY 1964, saving $150 million in the first two years. However, von Braun said cancellation would have no impact on the schedule, although it would increase confidence in meeting milestones. The same day that he met with von Braun, Mueller testified before the Senate space committee and said cuts in NASA's budget impacted the development of the Saturn, Apollo and Gemini. He told them the agency was looking at ways to save money, and one alternative involved the early phase-out of the Saturn I.[35]

[33] Mueller Interview, Slotkin, 9/9/09; Seamans, *Project Apollo*, 49-51.

[34] Presentation to Mueller, 9/28/63, Del Tischler and John Disher, NASA Archives, RN 13286; Disher Interview, Ertel, 1/27/67; Diary for 9/28/63, GEM-70-9; Mueller Interview, Slotkin, 6/15/09.

[35] "Impact of Elimination of Saturn I Flights," MSFC Presentation to Mueller, 10/17/63, GEM-91-8; "Replacement of Scheduled Manned Flights on Saturn I," 10/18/63, GEM-91-8; Notes for Mueller. . .(Senate) Subcommittee on Independent Offices Committee on Appropriations, 10/18/63, Mueller Speeches, NASA Archives.

After the Disher-Tischler study and von Braun's briefing, Mueller had a way to improve the schedule. Upon getting Seamans' support, he submitted a plan to Webb on October 26 and it was approved two days later, albeit with expressions of concern that major contractor layoffs could become a political issue. NASA's congressional relations staff informed members from the affected districts and Mueller personally called members of the House and Senate space committees so that they would not be surprised. Then after checking with DOD, NASA announced the cancellation of the Saturn I. Leaving some savings as a cushion, the agency said, "This realignment of the program saves $50 million in FY 1964, a step which helps to stay within [our]" congressional authorization.[36]

Mueller published his organizational plan for OMSF, which was to take effect on the same day as an overall NASA reorganization on November 1, 1963. He divided institutional and programmatic roles, and established three program offices for Apollo, Gemini and advanced missions. The program and project offices included five functional groups: program control, system engineering, testing, flight operations and reliability and quality assurance, the latter called R&QA. Explaining the dual reporting relationship, the announcement said, "Just as the Associate Administrator for Manned Space Flight in Washington has delegated the responsibilities for all aspects of the Apollo Program to the Apollo Program Director, so has the Center Director delegated responsibilities for all aspects of, for instance, the Saturn V vehicle to the Saturn V Project Manager." He established a Program Management Council composed of the three center directors and himself, with the three program directors reporting to it. Although he gave the council a new name and charter, it continued to be called the "management council."[37]

Mueller did not make up his mind about implementing all-up testing until he saw the results of the Disher-Tischler study. Then at a management council meeting on October 29 he announced he wanted to minimize testing non-operational equipment, and to maximize all-up flight tests. He directed contractors to deliver completed vehicles to the Cape, and that henceforth schedules must specify equipment delivery and launch dates. During Mercury, when contractors shipped hardware to the Launch Operations Center, the "Germans" took everything apart, retested the components and put them back together again. Requiring final assembly at the contractor's plant and not reassembling everything following receipt would improve launch schedules. He issued a directive stating that recent "schedule and budget reviews have resulted in a deletion of the Saturn I manned flight program and realignment of schedules and flight mission assignments on the Saturn IB and Saturn V programs." He went on, "It is my desire that 'all-up' spacecraft and launch vehicle flights be made as early as possible in the program." Consequently, the first Saturn

---

36 Mueller Interview, JSC, 1/20/99; Recommended list of members to be notified..., 11/12/63, GEM-91-11; Replacement of Scheduled Manned Flights on Saturn I, 10/18/63, Mueller to Webb, 10/26/63, and Brown to Mueller, 10/29/63, GEM-91-8-9.
37 Reorganization of the OMSF, 10/23/63, GEM-84-12.

IB and V flight tests would utilize live stages and carry complete spacecraft. He also announced that the agency required two successful all-up test flights before they could be flown with astronauts aboard. Mueller had to convince the center directors to accept all-up testing, and even after they concurred it took months for them to be intellectually convinced. However, had the schedule crisis not existed they never would have accepted it. As von Braun later wrote, with the traditional step-by-step approach "at least ten unmanned flights with the huge new rocket [Saturn V] would be required before anyone would muster the courage to launch a crew with it . . . The first manned Apollo flights would be limited to low Earth orbits. Gradually we would inch our way closer to the Moon, and flight no. 17, perhaps, would bring the first lunar landing."[38]

Mueller also introduced parallel or concurrent development of some *subsystems*, usually where the agency had critical design problems. But building two different major systems was no longer possible at the end of 1963. However, NASA developed duplicate subsystems, such as two lunar excursion module guidance systems, either of which was capable of carrying out the critical lunar orbit rendezvous. Similarly, the agency developed two lunar excursion module ascent engines, and nine months before flying to the Moon the primary manufacture of the engine, Bell Aerosystems, had problems due to engine instability; however, Rocketdyne's alternative fuel injector solved the problem. Mueller carried the parallel development of the engines all the way through the program and developed "a strategic plan of logical decision points," as he called them, which he later used when the program hit road blocks.[39]

When he arrived in Washington, Mueller found some people in the wrong jobs. While impressed with many of the key people at headquarters, the centers had sent "a cadre of people that were not gifted" to fill lower level positions, as he recalled, and arranged to ship them back. This caused quite a trauma, creating controversy which led to the formation of an engineers union at headquarters. Nonetheless he called it "a constructive thing to do," because it allowed him to bring in well qualified new people. NASA also hired some people directly from industry, although there were not many willing to take a large pay cut for the privilege. Nonetheless, he noted, "You couldn't get all those good people working collectively and correctly together unless you had an overriding goal because you've got to have people concentrating on what needs to be done if you're going to get it done." Furthermore, they needed more than a goal; they needed commitments "to making it happen."[40]

---

[38] Mueller Interview, Slotkin, 9/9/09; "Washington Redraws Management Lines," Benson, *Moonport*, Ch. 7-8; Mueller to Director MSC, et al., Revised Manned Space Flight Schedule, 10/31/63, "All-Up" Decision File, NASA Archives; Seamans Interview, Van Ripper, 12/6/66; Cortright, *Apollo Expeditions to the Moon*, section 8-6.

[39] Mueller Interviews: NASM, 2/15/88; Sherrod, 4/21/72, JSC, 1/20/99 and Slotkin, 9/9/09; Freitag Interview, Lambright, 6/27/91; Phillips to Mueller, LM Assent Engine Selection, 9/21/68, SCP-81-5-7; Orloff and Harland, *Apollo*, 55.

[40] Mueller Interviews: NASM, 6/22/88 and Slotkin, 2/22/11; US Civil Service Commission to the AFTE, 8/5/64, GEM-42-11.

IV

Mueller sought a balance between the inputs from the centers' engineering laboratories and the project offices. Under his approach to program management, the centers' project managers reported to both the center director and the Apollo program director, something the center directors found "difficult to understand for some period of time," he said. They asked him to explain their role if the program director had control from Washington, and it took time before they accepted it, because they did not like "anyone having direct access to 'their resources,'" he said.[41]

NASA had a few system engineers, although Mueller located more engineers with the "right thinking process" at headquarters and the centers. He thought people were born as systems engineers, and found it difficult to train them. It is "something that you feel," he said. Adding, "It requires you to really understand all of the forces that are brought to bear on a particular system and you've got to take account of 'whatever' or else the system won't work the way it's supposed to." A system engineer had to visualize the entire system, and fortunately each center had some people with the right skills, allowing Low and von Braun to identify and assign them to their system engineering groups.[42]

Initially taking the Apollo program director's role for himself, Mueller knew that unless he assigned someone to lead each of the five functions in the program and project offices, he would never find out what went on in the program. He established five separate lines of communications, assigning people at each level; and this came to be known as the five box organization or "GEM boxes" (a play on his initials). While similar to the air force's approach to system program offices or SPOs, he organized the five functions somewhat differently; although he ensured that both included system engineering, configuration management, program control, testing and R&QA in each program and project office. However, the SPOs did not include flight operations, and the GEM boxes had procurement and production at the center level, not in the program office. In Mueller's model, system engineering included configuration management; whereas in the SPOs, engineering encompassed system engineering, system integration and "technical excellence of engineering" (Table 3-1).[43]

Mueller said he put program control in place because there "wasn't any method for determining the costs or schedules or configuration," and he had to teach the centers how to use it. It took time to get people into key positions throughout the program, and to understand and work in that framework. He called program control "an active element" that forecasted where problems would occur by looking at program and budget trends – what actually happened in engineering, analyzing changes in the program. He considered system engineering the other side of the coin.

---

[41] Mueller Interviews: Sherrod, 4/21/71 and NASM, 5/2/88.
[42] Mueller Interviews: Putnam, 6/6/67, NASM, 2/15/88 and Slotkin, 6/15/09.
[43] Mueller Interview, NASM, 5/2/88; Space Systems Management: The Methods of NASA and USAF, James P. Cann, 6/67, SCP-79-7.

System engineers influenced what actually happened, assured the adequacy of designs, and examined contingencies and trade-offs. Program control and system engineering were two key elements of program management, they were different skills working together to understand the program and its implications.[44]

When Mueller arrived in Washington, Bellcomm, Incorporated supported NASA headquarters in "a somewhat ambiguous role, partly because people didn't really understand system engineering as such and partly because they were an outsider," he recalled. The company floated around because no one defined their role. However, he needed their technical skills, and required people with sufficient technical strength to determine what was going on and what needed to be done. He gave Bellcomm line responsibility for system engineering at headquarters. This created some difficulty, because the centers did not understand how a contractor could work on the overall design. As it turned out, Bellcomm did not do system engineering per se; rather, the company caused the centers to develop their own system engineering talent, and assured that the system engineering got done with the proper set of trade-offs. The company provided people who helped everyone to understand what needed to be done, and so in a sense performed a training role. Bellcomm established an elaborate structure in support of program direction, although they did not direct contractors. Instead, they observed, and gained knowledge of what went on in the program. They influenced and caused others to understand the problems at different levels. Their engineers were not specialists who solved specific technical problems. That came from the engineering laboratories, which had always been the centers' strength. But Bellcomm engineers understood the whole system and looked at how to make it work as a system. They brought the pieces together and reviewed software and interfaces to insure they worked properly. But, he said, "they did it without any fanfare. They just did it to help." Bellcomm became the glue that held it together, although "they had to do it in such a way that they weren't actually telling you what to do." They did not direct, they convinced people to do what they needed to do, because the centers would not do it any other way. However, their work remained underappreciated and according to Mueller the company did not get the credit they deserved.[45]

The R&QA function made sure the engineers did what they needed to do to assure the reliability of equipment. They established standards "from the acceptance of components to the final approval of designs – and then through ... review and assurance of the testing." Mueller established an independent test group because he wanted the design engineers to know they had to pay attention to testability, so that equipment would perform as required. So the test group acted as the counterpoint to system engineering, making sure engineers could test components once built. And as he pointed out, "so often you waited until it was built to figure out how to test it and

---

44   Mueller Interview, NASM, 6/22/88.
45   Mueller Interviews: NASM, 5/2-6/22/88 and Slotkin, 6/15/09; Boeing TIE Contract, Presentation to Congress, 6/27/68, Tab 5, Manpower and Cost, Bellcomm, SCP-81-2.

it was important to figure out how to test it at the same time you were building it." The test manager had personal responsibility to assure proper tests were defined, observed, and understood before declaring a test complete. These two groups – R&QA and testing – validated the design concepts, assuring their reasonableness.[46]

Mueller included flight operations as one of the five GEM box because, he said, "people were forgetting about they're going to have to fly this thing and ... the people doing the design and development [have to] know what the operations people need and how they have to work." They needed to know at the beginning of the design process, not at the end when people usually think about it. But the Apollo Program Office never fully implemented flight operations. Mueller attributed this to the time required to establish the function and the nature of the operations people involved. Nonetheless, he argued, "they played a very key role ... in influencing the design." He involved the astronauts in design decisions, but did not allow them to control those decisions. He considered most of the astronauts good engineers, though not system engineers. They did not have the depth of understanding of changes, nor did they understand the impact on the overall system, and many of their proposals were matters of convenience, or cosmetic in nature, not fundamental improvements.[47]

The program director and the five functional managers in the Apollo Program Office regularly reported to the management council. Because of the time involved in bringing everyone together, Mueller moved the management council meetings around, and conducted some via telephone. He personally spent about half his time traveling around visiting the centers and contractors "just seeing what was going on," he said. And he focused on "identifying weak spots ... making sure sufficient attention was being paid to them." The center and program directors and their people also went out to see what the problems were, and that got the attention of the contractors and made sure that problems got worked on. The program and project offices became "sensors" which kept him "abreast of where the problems or likely problems were going to be," he added.[48]

Mueller considered it important to establish the GEM boxes at three levels – program, center and contractor because, he said, "the thing that kills programs is not knowing that this small piece over here failed a test yesterday, not last month. And ... daily communications down those five parallel lines is probably the most significant contribution to getting the program done that I know of." He created the five box management structure "to focus, early on in the program, on the fact that you were going to test things, and you ought to design so you can test them. And you are going to have to have reliability, so you have to design for reliability." One of his challenges involved getting the program offices at the centers and contractors to "realize that all of those disciplines were essential and that they had to be addressed early on." But he

---

[46] Mueller Interviews: NASM, 5/2-6/22/88 and Slotkin, 2/24/10.
[47] Mueller Interviews: Slotkin 9/9/09 (second session) and 2/25/10, and NASM, 6/22/88.
[48] Mueller Interview, NASM, 6/22/88.

established the GEM boxes wherever it made sense to have independent lines of communication. "The important thing was to have threads of communications that went all the way up, not authority but communications; so that you had a counterpart who could follow what was going on," he said. Such an organization was not feasible for small projects, although in a large program like Apollo duplicate lines of communications were needed in order to eliminate gaps. Establishing the five GEM boxes, he explained, involved a "change in culture. You wouldn't need them if you had the ideal grouping, why you would have it all there anyhow, but that's not the way organizations tend to work. You need a little forcing function to make the things come together properly." It was equally important to have the best people occupy those boxes, "You've got to have somebody in it who knows enough to be able to discern what needs to be done and how to do it." Mueller had the support of NASA management to implement this new organization. Seamans understood matrix management, so Mueller did not have difficulty in explaining the organizational plan to him, and he recalled, "Mr. Webb was so concerned about the external world that he didn't spend a great deal of time on the internal world, except when there was something in the newspapers that caused a flap."[49]

However, project management already existed at NASA before Mueller arrived and both Gilruth and von Braun had their own management systems. Because the people at each center knew each other, shared common backgrounds and had worked together for a long time, they did not need the formality of Mueller's program office model, but von Braun's approach did not work with Gilruth's and vice versa. When a program "gets to be very large you have to have a more formal system" than the collegial one practiced at the centers, Mueller said. MSFC built launchers in-house for some time, and von Braun had a good project structure, but Huntsville did not manage contractors, nor had they developed a multistage rocket built by different contractors working on three or four stages. In the case of the Saturn V, they had five prime contractors working remotely, while in the past they built Jupiter's stages locally, so managing contractors building a multistage rocket at multiple locations had to be learned. MSC was a whole different story. In 1963, some of their work remained at Goddard and Langley, while most of MSC's team was spread out in facilities in and around Houston. Mueller called MSC at that time "an amorphous thing and they didn't have much of any structure, just some really highly dedicated people who were out there working their tails off." Most of the MSC managers had no large project experience with widely dispersed elements like Apollo. But it took more than structure and organization; they had to understand the total system being built in "depth and detail." Thus it was "equally essential and necessary to really understand at every level in the organization the technology you're dealing with and make the right engineering decisions. You can't really do that with a structure. You've got to have people," he explained.[50]

---

[49] Mueller Interviews: Slotkin, 6/15/09, 2/25/10 and 2/22/11, JSC, 1/20/99 and NASM, 5/2/88; "Apollo Program Oral History," 7/21/89, NASA Archives, RN 18924.

[50] Mueller Interviews: Slotkin, 9/9/09 and NASM, 5/2/88.

V

After Shea went to Houston, Low became Mueller's primary deputy and acting Gemini program director. Mueller initially took the Apollo program director's role for himself, and after two months he had many of the people in place who managed human spaceflight at NASA for the next six years (Table 1-2). Although some people came and went, the basic organization did not change. The management council became the apex of Mueller's communication system. Under Holmes, the council grew to fourteen members, although Mueller said that Holmes "really didn't have a management council per se ... What he did have was a once-a-month meeting of managers. And there's a difference, more in attitude." Mueller cut the membership to himself and the three center directors, and made it a real part of the decision making process. The center directors could not stand aside, and "If they didn't like the way it was going they had to stand up and say so." Despite earlier misgivings, he found it essential to build a strong management council where the center directors reviewed the programs, and "everybody knew what everybody was doing." The council gave the center directors key roles, not only providing resources for the programs – they were responsible for the quality and quantity of work. Prior to each meeting they would conduct pre-reviews to facilitate communications and identify problems. The pre-reviews became a forcing function that kept everyone informed about problems, and the council became an effective tool – a substitute for center autonomy. Having the program directors report to the council forced the center directors "to step back a pace and allow the program offices to work," he said. Operating like a board of directors, the program directors frequently presented, while the council members solved problems, made decisions, and divided the responsibilities between the centers.[51]

A study of Apollo program management commissioned by Webb and conducted by Eugene E. Drucker of Syracuse University made the point, "With the management council mechanism all of the hierarchical levels of the Apollo organization were collapsed into two, the council and the program organization ... [T]he three centers were brought together such that problems specific to centers and those that ranged across centers could be discussed and solutions arrived at." Mueller controlled both the institutional resources and the program offices; and according to this study, "by using the management council, [he] was in a position to evaluate the 'fit' between program and institutional activities." This control assured he became aware of problems arising from a lack of coordination between the two elements. Drucker credited the council for the success of Apollo, because it "provided the necessary authority to make the matrix concept work." Controlling

---

[51] Seamans, *Aiming at Targets,* 110; "NASA Realigns OMSF," NASA Press Release, 10/28/63, O&M, GEM-84-12; "The Apollo Spacecraft-A Chronology," vol. II, "Foreword; Minutes of OMSF Management Council Meeting, 9/24/63, NASA Archives; Mueller Interviews: NASM, 11/8/88 and 6/22/88, Sherrod, 4/21/71 and Putnam, 6/27/67; Seamans Interview, JSC 2, 11/20/98.

both sides of the matrix, Mueller could "force cooperation between the dimensions of the matrix at any level," the study said. But he rarely used force. The "Program people were given final authority in terms of schedule, cost, and performance criteria ... [T]he system really worked because ... program managers finally established very close working relationships with the head of each center," according to the study.[52]

A "first rate engineer," wrote Seamans, Mueller worked "seven days a week and expected others to do so." He did not appear concerned if his "decisions or manner of making them ruffled the feathers of his subordinates." He flew back and forth to Los Angeles for meetings with contractors, often taking the "red-eye" and sleeping on airplanes. "So many charts were presented at his meetings," the general manager said, they "were eventually referred to as 'pasteurized' – one chart after another going *past your eyes*." Staff complained that Mueller pushed too hard, and because the center directors reported to him, some expressed concern that they were too low in the organization. Seamans called Mueller "a double whirlwind ... The days of the week meant nothing to him. There were meetings on Saturdays and Sundays. George was indefatigable. I mean he traveled everywhere." However, Mueller kept Webb informed, was very careful about what he said about him, and did not seek personal publicity. He recalled, "I tended to not to want to get publicity," believing "I could be much more effective" that way. "And I think when all was said and done that worked in terms of getting it done."[53]

When NASA announced the agency-wide reorganization on November 1, they named three new associate administrators, including Mueller who became associate administrator for manned space flight. The centers reported to one of the three, who in turn reported to Seamans. On October 31, Walter L. Lingle, Jr., Seamans' chief of staff said that the reorganization would achieve a balance between delegation of responsibilities and maintaining control in order to assure that policies and objectives were adhered to. He called the new associate administrators the agency's three operating officers, and said Seamans, while maintaining authority to perform his role, will "delegate as much of his own general management responsibility as possible to each of these three operating Associate Administrators." Each center reported to one person who had "final authority for establishing and controlling all communications channels and procedures."[54]

Mueller also announced the all-up testing decision and laid out a new schedule for the Apollo Program on November 1. He said all-up meant "simultaneous, all at once," and would combine "launch vehicle and spacecraft development flights ... with all elements active and as close to lunar configuration as possible, beginning with the very first flight." It involved "learning from failure ... [and] capitalized on

---

[52]   Eugene E. Drucker et al., "Project Management in the Apollo Program: An Interdisciplinary Study," Syracuse University, 1972, 22-23.

[53]   Seamans Interview, JSC 2, 11/20/98; Mueller Interview, Slotkin, 6/15/09.

[54]   "NASA Announces Reorganization," NASA Release, 10/9/63, and Remarks by Walter L. Lingle, Jr., Moffett Field, CA, 10/31/63, NASA Archives. The other two associate administrators were Newell, OSSA and Raymond L Bisplinghoff, OART.

success," reasoning that the second stage of the launch vehicle was as likely to fail as the first. He ended incremental testing and condensed the test schedule, noting, "You can plan for disaster or you can plan for success. You might as well plan for success because you will have the disasters anyway." He had implemented a number of key decisions during his first few months at NASA, establishing a new schedule and setting up new interfaces with the external world. He considered all-up testing an important decision because, without it, the agency would not be able to achieve the new schedule. Establishing the schedule, new working relationships and the program management structure made the difference. Implementing all-up testing became a key part of that, but not necessarily the most important part.[55]

---

[55] Mueller Interviews: Van Ripper, 12/7/66 and Slotkin, 9/9/09 and 2/22/11; "The Apollo Spacecraft-A Chronology, vol. II, Foreword; Neufeld, *Schriever;* Murray & Cox, A*pollo*, 158.

2

Program management

"The Moon is our Paris."
Mueller comparing Charles A. Lindbergh's flight with Apollo, April 9, 1964

After nine weeks on the job, Mueller returned to a familiar platform to make his first public speech as a NASA official. Speaking at an American Institute of Aeronautics and Astronautics meeting in Washington on November 6, 1963, hundreds turned out to hear him call it a good time for a detailed status report. About 200,000 people worked in US human spaceflight programs, and within eighteen months another 100,000 would join them. Fully seventy percent of the money NASA spent went for salaries and benefits, while the remainder bought consumables and paid for training and developing new technologies. Consumables, less than eight percent of the total, represented the only part thrown away. About ten percent of the total went to develop technology, and another twelve percent provided permanent infrastructure. In all, ninety-two percent of the investment in human spaceflight provided the people, infrastructure, and technology to give the US the capability to explore and exploit space. Mueller said that NASA's recent reorganization established geographically dispersed program offices, with parts located at headquarters and the centers, with each responsible for their portion of the effort. Thus, he claimed, "the total job will be accomplished with the maximum delegation of authority and responsibility and minimum duplication." The GEM boxes, he added, would enhance "the free flow of information" between program offices in Washington and the project offices at the centers, coordinating work and assuring it was performed expeditiously. And these program offices would have full responsibility for performance, budget, schedule, and cost control, as well as design, test, and R&QA.[1]

With the major Apollo contracts in place, and after reexamining the cost, Mueller claimed "it is still possible to complete the program within the $20 billion estimate and within the time allotted." However to do that Congress had to appropriate the

---

[1] "Manned Space Flight – Where Do We Stand?" Mueller, AIAA, Washington, DC, 11/6/63.

full amount authorized for the fiscal year, and continue funding at "an adequate level" in future years. In taking into account cutbacks which had already been imposed, NASA had made adjustments based on these funding limitations. He discussed the recently announced changes, and said instead of flying astronauts on the Saturn I the agency would use this booster for scientific missions and proceed to the first piloted flight of an Apollo spacecraft on the Saturn IB. And Gemini would provide the operational experience the astronauts would have gained on the canceled Saturn I flights. Savings from the canceled flights would free up resources for other phases of the flight test program, and flying the first Saturn IB and the Saturn V flights in an all-up condition would result in additional cost savings.[2]

As another means of meeting schedules, Mueller said that NASA would institute "alternate efforts in the development of some systems and subsystems – those about which there may be reason for concern that they will not be ready to meet flight dates." He spoke of parallel or concurrent development of critical subsystems, considering this to be the best way to achieve the program objectives in a cost-effective manner because competition reduces time and cost; as he reminded his listeners, "time is money." Furthermore, having competing contractors working on separate approaches to the same problem creates the opportunity to select the best solution, an approach the air force dramatically demonstrated by developing five ballistic missiles in parallel, which produced five usable launch vehicles. However, he noted, the Apollo Program did not have that luxury. Although developing alternatives for critical subsystems came at a relatively low cost in a program spending more than $10 million per day, it did not take many days of delay to exceed the cost of developing these subsystems in parallel. Nevertheless, he insisted that Apollo was not a crash program, citing the beginnings of the Saturn in 1958, and the F-1 and J-2 engines in 1959 and 1960 respectively. During his six years at the agency he would continue to speak out against calling Apollo a crash program because, he pointed out, NASA developed the Apollo spacecraft "on a timescale four years longer than [the] Mercury spacecraft, two years longer than the B-58 bomber, and one year longer than the X-15." And the time between decision and the expected lunar landing spanned eight years, which was longer than any major R&D program in the US history, including the Manhattan Project.[3]

On a visit to California shortly thereafter, Mueller gave his first address aimed at the general public. "Within the next few years, if we carry on these endeavors with as much success as in the past, the United States will assume first place in space," he said. Using competition with the Soviet Union as justification for the Apollo Program, he pointed out that the US had not yet achieved preeminence in space, nor would it be achieved easily. It would take hard work to match their launch capability but, he said, "regardless of Soviet activities, we must establish and adhere to firm national goals. The goal of a manned Moon landing in this decade is within our

---

[2]   Ibid.

[3]   Ibid.; Mueller Interviews: NASM, 6/22/88 and Slotkin, 9/9/09; *The Lockheed Digest*, 6/9/ 64, GEM-42-7; Disher to Thompson, 2/17/64, SCP-46-17.

grasp. The team is organized. The installations are under construction. The astronauts are in training. The spacecraft are being tested. The rockets are being fired. The price is within our means."[4]

In these first two of ninety-one public speeches made while at NASA, Mueller defended the space program using various rationales, including economic benefits, knowledge gained, and competition with the Soviet Union. He later spoke of national security, spinoff economic benefits, human destiny, and discovering new scientific knowledge to justify the lunar landing and post-Apollo programs. He considered making speeches to be an important part of his job. "If you don't make speeches you don't get appropriations," he once said, because "people remember speeches." He had spoken at technical meetings before joining NASA, and afterwards expanded his audiences to include Congress, academia, business and industry, as well as the general public. About forty percent of his speeches while with the agency were made to technical groups, and twenty percent each to academia, business and industry. He frequently spoke to business groups when canvassing support for funding; and to scientists and technologists when he initially sought their support for Apollo, and again when promoting post-Apollo programs.[5]

When speaking at an internal Saturn launch vehicle review attended by Webb, Dryden and Seamans on November 9, Mueller reminded the triad that cancellation of the Saturn I flights avoided "dead-end" testing. By beginning human Apollo flights on the Saturn IB, he said, "we have probably advanced the time at which we would be able to fly manned vehicles in the same configuration as we would eventually want to fly to the Moon itself." And by employing all-up testing, NASA eliminated some flights and provided additional time for vehicle integration and checkout, thus reducing unplanned delays. The new schedule permitted the first Saturn IB to carry a live Apollo spacecraft, although the lunar excursion module would not be ready for that flight. Nonetheless, the first Saturn IB flight allowed earlier spacecraft testing "in essentially the configuration with which they will land on the Moon." Completing these tests earlier would save money, and testing the first Saturn V in an all-up mode would put the program ahead of where it would otherwise have been by step-by-step testing. The new launch schedule called for a rapid launch rate, completing the training of astronauts and checking out equipment earlier than would otherwise have been possible, thus allowing NASA "to capitalize on success" and provide "enough equipment and enough people to handle the problems when they occur."[6]

Under Mueller's program management plan the centers controlled the contractors with headquarters oversight. As he explained, no contractor deliberately did things "inimical to the success of the program," but without adequate management and oversight, contractors would emphasize what they thought best, which might not be

---

[4]   "Dedication of Satellite 6, LAX," Mueller, 11/10/63.
[5]   Dick, *Critical Issues*, 44; Mueller Interview, Sherrod, 9/20/68; Bibliography.
[6]   Mueller, Saturn Launch Vehicle Program Review, 11/9/63, NASA Archives, RN 17369.

best for the program as a whole. To keep contractors informed, he regularly met with their senior executives, providing his view of what the others were doing, and what the overall problems were. At the highest level, he established a NASA-industry Apollo executives group, a technique borrowed from the Minuteman Program. He asked the chief executive officers of each of the major Apollo contractors to serve as members of this new advisory board, establishing a communications link between them and the program. At the level below the CEO, the NASA program directors and project managers met with their industry counterparts before the CEO meetings, keeping everyone informed and allowing them to brief their bosses. His strategy involved giving everyone the "opportunity to recognize that what they were doing was going to get visibility by their boss's boss's boss." And to make Apollo a success, he recognized that it would take more than establishing an organization, it required support from the top to get contractors totally involved. By establishing the Apollo executives group and a similar one for Gemini, he got the CEOs immersed in the program and understanding their firm's responsibilities, which increased interest and support.[7]

Robert F. Freitag, a senior staffer at headquarters remembered establishing the Apollo executives group as "one of the first things Mueller did when he took over at OMSF." He kept the contractor structure in place, and created the executives group as a board of directors, asking each contractor to send their "top man" to a quarterly meeting. The center directors participated and occasionally Seamans or Teague would attend. Freitag called these meetings "a fascinating event to watch. These guys behaved as true statesmen. They spoke openly," resulting in "integration at the top." The CEOs learned they held more than contracts, they had become part of a national program. They "toured around to see what other contractors were doing. It was show time. They became wrapped up in the activity. And the side effects, or spin-offs, were considerable ... They were working to the same sheet of music ... [and] singing the same song," he said. Because of these meetings, Disher recalled, Mueller would not hesitate to call William M. Allen, CEO of Boeing, to ask him "what happened to that valve in the engine?" The first time this happened, Allen had no idea. He knew few details of Boeing's contract, much less what Mueller had called about. Although Disher added, "he quickly learned about it." The Apollo executives group succeeded in involving the CEOs because Mueller said he "really wanted them to get interested enough to go solve the problem." He not only met with the CEOs at these formal meetings and one-on-one sessions, he had them to his home for dinner to build a personal bond, and they reciprocated. He spoke at events held at contractor plants and made speeches in their communities. Many of his public addresses came at their request (or from members of Congress). He considered the executives groups to be effective communication tools, and he thought none of the CEOs took advantage of what they learned, although he pointed

---

[7]   Mueller Interviews: NASM 5/2/88 and Putnam, 6/27/67; Mueller, IATA, Amsterdam, 10/23/69.

out that they "got to see what the other guys were doing and that helped in terms of keeping their attention on what needed to be done." However, "no one wanted to be the guy that ended up stopping the program because he didn't do what needed to be done." Consequently, the executives group became a driver in supporting Apollo. The first executives meeting took place less than two months after he joined NASA, and they initially met quarterly, then several times a year. He held meetings at contractor plants and NASA centers, exposing the CEOs to each other's problems, and it soon became an effective tool in getting top management attention. He later said it is "amazing if you can get the CEO to come and see what the total program is and what his group's problems are, how rapidly those problems get addressed and solved."[8]

Originally, Webb thought that Mueller should concentrate on running human spaceflight while he took care of Congress. His reluctance came from Holmes' "end run through Congress and the White House," Mueller said. The administrator had become sensitive to anyone establishing external relations, although after Mueller's performance at the first budget authorization hearings, he gained Webb's confidence. Mueller used Freitag to manage OMSF's relationship with Congress. And as Freitag recalled, Mueller "wanted me to be his personal congressional liaison. This was the antithesis of the way you usually do things. Ordinarily, an administrator wants all congressional liaison funneled through his office. However, Webb was willing to allow Mueller to do this," although he told him to "keep it under control."[9]

II

To address Teague's complaint about the lack of information, Mueller proposed monthly briefings for his subcommittee, an idea that the representative immediately latched on to. These meetings provided information about OMSF, their problems, and what their future prospects looked like. Mueller called these briefings "fairly innovative," and before the annual budget authorization hearings Teague would take the subcommittee to visit NASA centers and contractor plants. Mueller judged this arrangement to be constructive because, he said, the subcommittee "got to see and understand the complexity of the program," and after a few months from what had been an adversarial relationship they began to work well together. The briefings built credibility because Mueller made the effort to establish real communications, which took time and cultivation. While at times he might have told the members more than they wanted to know, he said he wanted them kept up to date because they needed the information to make the correct decisions. It was "imperative for us

---

[8]   Freitag Interview, Lambright, 6/27/91; Disher Interview, Ertel, 1/27/67; Mueller Interviews: Slotkin, 9/10/09 and 2/22/11, McCurdy, 6/22/88, JSC, 1/20/99 and Putnam, 6/27/67; Correspondence, 1965-1969, GEM-39-16.

[9]   Mueller Interview, NASM, 6/22/88; Freitag Interview, Lambright, 6/27/91.

to get them involved so they understood what was important and what the problems were and why we were making the decisions we were, so they'd have some confidence that we were doing the right thing," he said.[10]

Teague played a key role in the House. In addition to chairing the NASA Authorization Subcommittee he led the NASA Oversight Committee, which gave him additional influence. Nonetheless, support from Teague "was not blind support. The committee probed, asked questions, and tested the reasonableness of our approaches," Mueller wrote. They had a window into the program from the monthly briefings covering both success and failures. The "program was almost continually under fire from critics who doubted we would ever reach the Moon and who felt the funds should be spent on other national needs," he said. But Teague believed in the program, and worked tirelessly on its behalf. These committees reported to the House Committee on Astronautics and Science (the "House space committee"), chaired by George P. Miller of California. Miller attended a few of Mueller's monthly briefings. And together with several space committee members, he attended Teague's budget authorization hearings, although their main contribution involved putting questions into the *Congressional Record*, mainly for show. (Because George Miller and George Mueller's names were both pronounced "Miller", it became common practice in Washington to call the chairman "Miller-Miller" and the head of human spaceflight "Miller-Mueller," using the original German pronunciation for Mueller.)[11]

While Teague remained Mueller's key contact in the House, Senate committees initially worked with Webb. The administrator also worked with the appropriations committees, while Mueller mainly dealt with the authorization committees. But building and maintaining agency credibility with Congress remained a top concern throughout Mueller's tenure. In addition to the meetings, hearings and reviews, he regularly visited and corresponded with the members, always congratulated them on their re-election, and contributed to the Democratic Party at Teague's suggestion (despite being a registered Republican). As he did with other constituencies, he made Congress part of the solution, and by "the end of 1964, our credibility with Congress had been restored," he said.[12]

Dealing with the White House also remained very important. So Mueller said he "spent a fair amount of time working on that side of the equation," but was not as effective because the Bureau of the Budget was a difficult agency to deal with. And Webb "was far more knowledgeable in that arena than anybody else," having served as budget director in the Truman administration. When Congress created NASA, they established the National Aeronautics and Space Council to plan space policy. Mueller established a good relationship with NASC executive director Edward C. Welsh and helped him to understand the complexity of the lunar landing mission. Consequently, Welsh aided in garnering support for NASA during White House

---

[10] Mueller Interviews: Slotkin, 9/9/09 and 2/22/11 and NASM, 6/22/88.

[11] Mueller to Teague, 7/20/78, Committee on Science and Technology, GEM-161-2; Meeting Notes, Teague, 9/17/63, GEM-79-12; Mueller Interview, Slotkin, 9/9/09 (second session).

[12] Mueller Interview, Lambright, 9/21/90.

budget discussions. As Mueller recalled, "in order to get through the Bureau of the Budget you had to have some external influences within the White House," and having Welsh as an ally proved useful. He spent time with the NASC director, regularly invited him home for dinner and kept him informed, especially about the post-Apollo program.[13]

Mueller respected the administrator's political acumen, although he did not particularly enjoy his company. He said, "I stayed as far away as I could because he's the kind of guy that talks to somebody and . . . that's the way it's going and then he talks to somebody else and he changes direction." Webb, a true politician, "had a political way of looking at things . . . so there wasn't any fundamental direction he would take." He just did whatever it took to get a "particular problem solved" for the moment. Nonetheless, Mueller felt Webb did a number of important things at NASA. Yet while a consummate politician, he was not the kind of person to run a program because he did not like to get involved in the technical details, which Mueller saw as, "a good thing" because it left *him* free to implement the program.[14]

Mueller met the president a second time about six weeks after the meeting in the White House hallway. He briefed Kennedy during his visit to the Launch Operations Center on November 15. The president spent a little more than two hours at Cape Canaveral, and sandwiched between sightseeing and witnessing a Polaris launch he received two NASA briefings, one by Mueller and the other from von Braun. Mueller spoke for twenty minutes, showing Kennedy models of the hardware, all built to the same scale, and briefly discussed the overall schedule for the Saturn I/IB Program. According to Seamans, "The President seemed quite interested in what George Mueller had to say," although Mueller's recollection is that Kennedy mainly "smiled and said thank you." Von Braun's briefing took place in front of the Saturn I on its launch pad, where he described the fifth Saturn I mission (SA-5) scheduled for January 1964. The president expressed considerable interest because it would carry the heaviest payload ever launched into space by any nation, and told Seamans to brief the White House press corps about it. Unfortunately, Kennedy did not live to witness the launch because a gunman assassinated him the following week; and on November 22, Lyndon B. Johnson became President of the United States. "Johnson was one of the more astute political maneuverers in our history," Mueller recalled. However, following the president's death "any chance for the United States 'backing off' of Apollo that Kennedy had initiated vanished," wrote space historian John M. Logsdon, and "the program became in a sense a memorial to the fallen president." And the assassination made the landing on the Moon in the decade of the 1960s a matter of national pride.[15]

---

[13] Mueller Interviews: NASM, 6/22/88 and Slotkin, 2/24/10 and 6/9/10; McDougall, *The Heavens and the Earth,* 173, 175, 309 and 400.

[14] Mueller Interview, Slotkin, 2/24/10.

[15] "Outline of Lunar Program Briefing for President Kennedy..." 11/15/63, GEM-100-17; Seamans, *Aiming at Targets,* 113; Mueller Interviews: Slotkin, 6/10/10 and 2/22/11; Logsdon, *Exploring the Unknown*, 419.

2-1   George E. Mueller presents Apollo to John F. Kennedy, November 15, 1963. (NASA photo)

Mueller called Johnson "a space buff" who, like Kennedy, used the space program as a weapon in the Cold War. While the new president supported Apollo, his first State of the Union Address, delivered six weeks after he took office, contained only one sentence devoted to NASA in its thirteen pages. Listed fourth among the ten ways to achieve "a world without war," Johnson, said, "we must assure our pre-eminence in the peaceful exploration of outer space, focusing on an expedition to the Moon in this decade in cooperation with other powers if possible, alone if necessary." Mueller said Johnson and Webb were "politically very similar and with pretty much the same kind of a background." They did not trust each other but respected each other. "So that was a fortuitous arrangement," he added.[16]

Webb did not interfere with the all-up testing decision, though the triad watched the debate "with interest," Mueller said. They were more curious about it than to try to second guess him. They had "a keen awareness something had to be done," he

---

[16]   Mueller Interview, NASM, 5/1/89; "State of the Union Message..." The White House, 1/8/64, GEM-69-12; Mueller Interview, Slotkin, 6/8/10; Webb to the President, 12/20/63, Budget, GEM-66-16; Lambright, *Powering Apollo,* 133.

2-2   John F. Kennedy and NASA officials intently listen to George E. Mueller's Apollo briefing, November 15, 1963. (NASA photo)

recalled. He used trend charts to show the impact of switching to all-up testing, which made a dramatic difference, and the idea appealed to them. They recognized the necessity to take some risk to achieve the lunar landing goal, and "to get moving at the earliest possible date." They supported Mueller, and the administrator wanted unanimity because he wanted to give the new man a chance. And after winning their confidence they went along with the decision, although Dryden had reservations and cautioned Mueller not to act too precipitously. But as Seamans said, "George created quite a stir with his revised program. Words like *impossible, reckless, incredulous, harebrained,* and *nonsense* could be heard behind the scenes. After announcing the plan ... George followed up immediately with detailed schedules. George didn't sell; he dictated – and without his direction, Apollo would not have succeeded."[17]

The initial reaction to the all-up testing decision was "almost incredulous," Mueller recalled. Von Braun's team was horrified. However, with the backing of the

---

[17]   Mueller Interviews: Lambright, 9/21/90 and Van Ripper, 12/7/66; Seamans Interviews: JSC 2, 11/20/98 and Van Ripper, 12/6/66; Seamans, *Project Apollo*, 51.

triad, the center directors had no way to appeal the decision. Mueller considered Seamans' support particularly important, and later called the general manager "more of a supporter than an overseer." Recalling an early visit to Huntsville after announcing the decision, Mueller described how Arthur L. H. Rudolph, the Saturn V program manager, tried to convince Seamans that all-up testing did not make sense. "Arthur Rudolph cornered Bob by the models of the Saturn V ... and asked him how one could possibly test all the huge stages of Saturn V at the first firing! Bob said 'Talk to George.' Arthur went on to try the same argument on me. I simply said, 'Why not?' and backed that up with thorough and logical details of exactly how we would proceed. With Bob on board, all three centers embraced the concept and, best of all, it worked!" Mueller said the size of the launch vehicle did not affect all-up testing. "The arguments are the same no matter how complicated it is and it's either you can do it or you can't do it. So you might as well – and it saves designing three or four or five different vehicles in order to carry out the sequential testing. So you're way ahead of the game deciding you want to do it right." When he first discussed all-up testing, he said "there were enough astonished looks," and no one "jumped on board at the first reading." He later noted, engineers "want to test it and test it and test it. And I want to test it and test it and test it at a subsystem level. However, at a system level you're much better off testing the system because in the end that system has to work. And then the only way you find out is if you test it as a system."[18]

Von Braun sent Mueller a message on November 6 with his initial reaction to the all-up testing decision. While not objecting, he raised a number of detailed technical issues, said that he needed more money, and argued that it had no long term schedule impact while extending the near term schedule. He concluded by reminding Mueller that the Saturn V remained on the critical path. After thinking further about it, von Braun sent a more detailed rebuttal on November 12, which began, "We believe the philosophy of flying live all stages, modules, and systems, beginning with the first R&D launching to be a worthy objective," and there is no "fundamental reason why we cannot fly 'all-up' on the first flight." He then laid out a dozen pages of conditions, including "critical technical, scheduling [and] funding" considerations, and asked for an additional $138 million during FY 1964 and $314 million in FY 1965. He tried to back away by raising objections, but because of the program office structure Mueller had a direct line of communications into the centers, and "there wasn't any way [for him] to back away without [me] finding out about it." In addition to objections in Huntsville, Houston raised some issues, although Mueller contained them; and while there were many reasons given for why it would not work, he refused to change, believing it essential to achieving the schedule.[19]

If Mueller followed MSFC's original schedule they could not land on the Moon

---

[18]   Mueller Interviews: McCurdy, 6/22/88 and Slotkin, 6/8/09, 9/10/09 and 6/15/10; Hodge Interview, Lambright, 10/5/91; "Bob Seamans at the Helm!" Mueller, MIT, Cambridge, MA, 6/10/09.

[19]   Von Braun to Mueller, 11/6/63 and 11/8/63, GEM-91-10; Mueller Interview, Slotkin, 9/9/09.

that decade. Almost everyone in Huntsville agreed with this, but not the decision to use all-up testing to fix it. While Gilruth and von Braun initially opposed the idea, von Braun would become more vocal in his opposition. However, the second stage presented a bigger problem than the first stage, because in the traditional step-by-step approach more time is spent testing the first stage. And it made no sense to fly the first stage, then the second stage before combining the two. Mueller reasoned, "If you lost a vehicle, you were likely to lose it at any stage so you might as well go as far as you can and find out where the problems are." When he first talked about all-up testing, the center directors "looked askance, and said, 'You couldn't possibly think of anything so silly,'" as he recalled. To convince von Braun, Mueller held a meeting in Huntsville and concluded by saying "the only way I can get to the Moon in this decade is through this program." Von Braun finally said, "Well it's risky, I agree. I support that idea." Nonetheless, support at MSFC remained far from unanimous, and most of von Braun's people, including his deputy, Rees, remained unconvinced. "But it was a decision that I did make," Mueller said.[20]

Konrad Dannenberg, deputy Saturn V project manager, remembered, "von Braun was first very reluctant ... [but] finally agreed, yes if we go my way, we can't do it on time, and he finally agreed." Adding, "it was basically a little bit against his philosophy ... It was a great risk. And if George Mueller had not had the luck that it worked, he would have looked very bad. But he was lucky." However, Mueller later insisted that luck had nothing to do with it; it was "reasonable and logical." Having accepted all-up testing, von Braun did all he could to make the upper stages work properly. The logic behind the concept of all-up testing was "sufficiently good," Mueller said, so "eventually you had to accept it as being a reasonable thing to do." In the worst case, "you might lose an upper stage or two," but he insisted that it did not change the risk profile. "The risk was there in any event, and the progress was certainly a lot faster than it otherwise could possibly be." You "don't decrease the risks by testing sequentially; you only spread the risk out." You do not gain anything by sequential testing, "so logic prevailed," as he termed it.[21]

MSC leadership viewed spacecraft development from an aircraft development perspective. Gilruth said Mueller "had very strong ideas, and he was not well versed in aviation and aerodynamics. He was much more on the electronic side." Although careful not to publicly criticize him, Gilruth left no doubt that he thought Mueller did not understand the most important aspects of human spaceflight. For his part, Mueller characterized aircraft development as "an art form ... that carried over into the early parts of Apollo." Spaceflight represented "an entirely new world ... And the thing that really kills programs is the changing requirements." That is why the incremental development of aircraft did not work for spacecraft. They were different, and it became necessary to introduce more discipline into the process, in

[20] Mueller Interviews: Lambright, 9/21/90 and JSC, 8/27/98.
[21] Dannenberg Interview, Launius, 7/25/00; Mueller Interviews: Slotkin, 2/22/11 and McCurdy, 6/22/88.

particular system engineering, program control, and configuration management. Mueller said, "Gilruth really was not equipped mentally for that complex a program. He was back in an era when he could design an airplane and fly it. And he had his students around him and they organized it the way he wanted to." However, Shea and Low helped convince Gilruth to go with the all-up testing decision. As Mueller described it, Gilruth took "time to absorb it before he committed himself. But the folks at Manned Spacecraft Center were not all that enthusiastic I must say."[22]

III

Before Mueller arrived at NASA, Webb appointed a panel of consultants under the leadership of Bell Labs' former president Mervin J. Kelly to assess the management of the human spaceflight program. (Kelly, like Mueller, a graduate of the Missouri School of Mines and Metallurgy, served as the Labs' research director when Mueller worked there during World War II.) The Kelly committee identified non-competitive government salaries as a key issue at NASA. Unable to change civil service salaries, they suggested the agency use a contractor like Space Technology Laboratories to provide additional technical support. He criticized MSC management and Gilruth in particular, writing that MSC "resented any external influence" and resisted support contractors. He told Webb it was desirable to strengthen Houston's management, and named MSC as the Achilles' heel of the human spaceflight program. And Kelly was "particularly ... emphatic in believing that Dr. Gilruth was not properly equipped to manage this Center," according to Mueller.[23]

Kelly also recommended that NASA strengthen its overall ability to manage large programs, because agency personnel lacked the proper skills and background. People with the requisite experience existed in the military and in the aerospace industry, although attracting qualified civilians to the agency would be difficult. Following a meeting with Kelly, Mueller wrote a memorandum to Webb saying that after a "rather brief review" of NASA operations, he thought "the national interest would be best served" if they brought in people with program management experience and the skills developed in defense programs. He noted the heavy participation of major contractors possessing such experience already working in important roles in the lunar landing program. Webb asked Robert S. McNamara to assign military program managers to NASA, and the defense secretary agreed. He also suggested that Mueller appoint an air force officer as his deputy. But Mueller first had to mend fences with the air force, so he enlisted Teague's help with Schriever, both graduates of Texas A&M. The representative then told the general,

---

[22]   Gilruth Interview, NASM, 3/2/87; Mueller Interviews: Slotkin, 6/15/09, 2/24/10 and 2/22/11; Seamans Interview, Van Ripper, 12/6/66.

[23]   Mueller Interviews: NASM, 2/15/88 and Sherrod, 3/20/73; Eichel to Mettler, Mueller, Solomon, 5/31/63, GEM 27-6; Seamans, *Apollo*, 34; Meeting Notes, Nick Golovin, 9/11/63, GEM-79-12.

"we have a civilian space program [and] I want you to help it, not hinder it." Schriever originally wanted the air force to run the human spaceflight program, although he supported NASA after the agency got the assignment in 1958. Mueller asked Schriever for some air force program managers, and after "some reasoning" with the air force hierarchy, they concurred because they too could appreciate the program "from a national point of view," Mueller explained. Two of Mueller's Space Technology Laboratories colleagues, James R. Burnett and Robert A. Bennett who worked directly with Colonel Samuel C. Phillips, Minuteman program director, recommended him to Mueller. And he asked Schriever to assign the colonel to NASA, calling Phillips "a perfect fit" for Apollo. Mueller did not know him on a personal level, but knew him "well enough to know he was an exceptional individual."[24]

Schriever agreed to let Phillips go, but the first word the colonel received about his impending transfer came in a telephone call from Mueller telling him the Apollo Program needed him. Phillips initially did not want the job because he saw himself in line to become commander of the air force Ballistic Missile Division, and believed it better for his career, not considering "how exciting it might be to be in charge of a program to get people on the Moon," he recalled. As Schriever remembered, when Mueller first asked for Phillips, he called it like "pulling teeth." "George Mueller came over to ask me could he have Sam Phillips," but he ran "the most important missile program we had, as far as the Air Force was concerned." But Schriever told Phillips, "I'll not stand in your way. I think it's a good opportunity for you from a career stand point. If you want the job, I'll make you available." Schriever then got a commitment from the air force chief of staff not to forget Phillips at promotion time before turning him over to NASA effective December 30, as deputy director of the Apollo Program. And the air force promoted him to brigadier general effective with his transfer. Mueller later called this a turning point for Apollo, because Phillips helped him to put in place the infrastructure and management practices "essential for a system-wide approach," and installed "the kind of insight and controls ... needed to carry out the program." And by letting Mueller have Phillips, Schriever said, "you know we were giving him quality."[25]

Mueller held regular off-site meetings to get the OMSF management working together. As he recalled, "these off-sites played a real role because ... they weren't in a formal setting where they had to take positions. It's when you got them off to the side and said 'well look, here's the problem and how are we going to solve it?' And I'd get them all talking and get them to reach a common conclusion." He wanted his people to interact outside the formal briefing room, and invited the senior staff, a group of "very opinionated people" who had to be convinced to follow his direction, to discuss his proposals and if they had better ideas, he would change. "But until

---

[24] Schriever Interviews: JSC, 4/16/99 and NASM, 9/8/89; Skaggs to Slotkin e-mail, 6/3/11.
[25] Mueller to Webb, 9/27/63, SCP-40-1; Mueller Interviews: Lambright, 9/21/90, Slotkin, 9/9/09, JSC, 1/20/99 and Putnam, 6/27/67.

then we would have to go in one direction," he said. Each senior staff member was expert in a particular area, and had to be convinced to pull together to achieve a common goal. To break down walls they would talk about problems because, he said, "If you can define a problem people can really work on it." He considered it important to establish relationships and make sure they were good and remained open. He used these off-sites to facilitate interaction, and he included support people who did not usually attend such meetings even though they played important roles in the program. He also included people below the center and program director levels because they had regular contact with the contractors. He considered communications the secret to program management, "communications on a level that is free and easy and not constrained by the fact that you're the boss." So he held these off-sites, he said, "to create an environment where you could have a free exchange and understanding and then begin to build those interpersonal relationships which are so important."[26]

Mueller introduced Phillips at such an off-site meeting in January 1964. Initially, there was "concern and wonderment" about what Phillips would be doing, and how he would do it. So, to avoid conflict or turf battles, as well as to give everybody a chance to get to know each other, he initially named Phillips as his deputy program director; although he assigned him day-to-day responsibility for Apollo as fast as he could accept it. Mueller said that von Braun, who was used to working with military officers, felt that "Phillips was great, but Gilruth wondered about him." However, everyone soon learned that they would much rather have Phillips as program director than Mueller, who officially held the position. It took a little while to get the general established, Mueller said, because as an outsider they were "kind of suspicious of him. However, he was a terrific guy, and within about six months everybody thought he was wonderful and they were encouraging me to get out of it." From his first days at NASA, Phillips and Mueller worked well together, and when the general later became director of the Apollo Program little changed. The role of Space Technology Laboratories in the ballistic missile program involved technical direction of the contractors while the air force performed program control and had overall program management. To a large extent Phillips and Mueller established a similar working relationship at NASA. Phillips involved Mueller in key decisions, "But there was never any argument when there was a technical question and I got involved," Mueller said, while program decisions were Phillips' to deal with. He or his people chaired the various reviews, which Mueller usually attended as an active participant. And Mueller considered Phillips "a great guy" who "did the right things." He could not remember any major disagreements because they "thought alike," he said. And while they had different ways of operating, their differences did not prevent them from working well together.[27]

---

[26]   Ibid.
[27]   Mueller Interview, Slotkin, 6/15/09, 9/9/09 and 2/22/11, and Putnam, 6/27/67; Phillips Interview, NASM, 9/8/89; NASA News Release, No. 63-287, "NASA Appoints Phillips," 12/31/63, NASA Archives.

Phillips recalled that NASA had "a program structure involving headquarters and centers and contractors, built into or overlaid on the line structure of headquarters and centers." While "the basic requirements and skeleton of the organizational structure ... existed," he had to complete the design, staff it, and make it work. He called the organization that Mueller had established "totally consistent" with his experience in the air force ballistic missile program, and "totally in accord" with how it should be structured and managed. However, whilst Mueller had introduced the concepts, the final design of the organization and bringing in a significant number of new people to flesh it out became Phillips' top priority. Phillips asked Schriever to assign experienced officers to NASA, and got "good people, starting with the top jobs," he said.[28]

When first formed, NASA only had a few active duty military officers in its management, although the astronauts mainly came from the armed forces. Several army officers joined the agency with von Braun, people like Rocco A. Petrone and Lee B. James. Petrone was a lieutenant colonel and West Point graduate, and served as director of launch operations at the Cape. Mueller considered him a "typical army engineer, very competent, very dedicated and did a remarkable job managing the development of the Cape." Furthermore, Petrone and Debus "made a tremendous combination." James, another military academy alumnus, retired as a colonel and served as director of program management at MSFC. The navy also assigned a small number of officers to NASA early on, people who later retired and stayed with the agency such as Freitag and John K. Holcomb. Freitag, as previously noted, played a prominent role in assisting Mueller with Congress, while Holcomb worked as OMSF director of operations. Drawing from the air force, Phillips asked to have fifty-five nominated officers assigned to NASA. Writing to Schriever, he outlined his needs, showing that the agency had 256 military officers on detached duty in March 1964, and requested the assignment of fifty-five additional air force officers to the Apollo Program. Mueller called them "quite important because there weren't hundreds of other people we could bring in who had that kind of a background." Phillips needed "people with big program experience," to work in key positions. And working with Shea and Gilruth at MSC, Petrone at the Cape and von Braun at MSFC, he placed these officers in key positions. Because MSFC had a long history of working with the army, von Braun and Debus were more open to the assignments than Gilruth. Phillips said, "I'm sure that my assignment was met with a lack of enthusiasm by a lot of people within NASA," and there "was less than hearty enthusiasm about my bringing in more" air force officers. However, von Braun asked for help, while "Gilruth, Shea and company didn't ... [and] Petrone had an open mind." Consequently, he placed a number of people with "worthwhile experience" throughout the human spaceflight program, he said.[29]

---

[28]   Phillips Interview, NASM, 9/29/89.
[29]   Ibid., 9/8/89; Personnel – 55 position package request, 1964, SCP-43-15; NASA Military Requirements, 3/29/64, Personnel – Military – Fifty five positions..., SCP-44-1-4; Mueller Interview, Slotkin, 2/24/10.

First Phillips filled key positions in the Apollo Program Office. Next, he worked with von Braun to superimpose a program management organization on the existing Huntsville structure. MSFC had already reorganized into two entities – Technical Operations, consisting of the engineering laboratories, and Industrial Operations, which included program management and the management of contractors. Phillips brought in Edmond F. O'Connor, an air force brigadier general to lead Industrial Operations with three project managers working for him: Rudolph led the Saturn V project, James managed the Saturn I/IB projects, and Lee Beaulieu, a civilian, led the engine development projects. Dannenberg, Rudolph's deputy, called Industrial Operations "necessary to really manage and organize this very major network of contractor activities and our own in-house activities." Although MSFC's laboratory directors were initially unhappy about this because it gave some of their authority to the project managers, after some "arm-twisting" they accepted the change and fell in line.[30]

Phillips recruited David M. Jones, another air force brigadier general who became OMSF's deputy associate administrator for programs. And he remembered "several colonels" joining his staff, one as chief of program control and another in charge of configuration management and system engineering. In the end, Schriever transferred all fifty-five requested officers, ranging from captains to several one-star generals. They took assignments based upon their experience, and contributed much to the success of the Apollo Program. At the program's peak NASA had over four hundred active duty military officers on detached duty, a very important factor, because they brought skills in disciplines which the agency did not have in abundance. They became the "backbone of the rigorous controls" that the agency implemented, and without them Apollo would have cost a lot more. Phillips also brought some civilians to the program, and one of his early acquisitions as head of program planning and control in the Apollo Program Office was Jerald R. Kubat from Boeing, who also brought "a young man" named James B. Skaggs to the agency. Mueller called Kubat, "the best program control guy at Boeing" and Skaggs was of the same caliber. Hiring people from industry to work for the government was always difficult because of the salary issue, but because of Apollo's prestige some joined to support the program "in the national interest," Mueller said.[31]

Phillips faced little overt resistance when he first arrived at NASA, although he thought there "had to be, at least in the back of the minds of many people, some resentment of outsiders coming into their program." It did not create obstacles, though he said there "clearly was a certain amount of a wait-and-see kind of attitude" that imposed extra demands on him and the other officers that he brought to the agency. These people had to "earn a place in the scheme of things," and had to make others comfortable with them. When it came to program planning and control, system

---

[30]   Phillips Interview, NASM, 9/29/89; Dannenberg Interview, Ray, 4/6/73.

[31]   Phillips Interview, NASM, 9/8/89; Adjustment to Marshall Organization, Organizational Announcement (Lee James), 12/5/68, MSFC, Huntsville, AL; Mueller Interview, Slotkin, 9/9/09.

engineering, and configuration management he did not find resistance or resentment, but he ran into problems with flight operations because human spaceflight operations experience resided with just a few people working on Mercury and Gemini. So he found "resistance and reluctance and even maybe elements of resentment to intruders coming in from outside," primarily at MSC. But Houston was having other problems with people at headquarters and at the other centers as well.[32]

Phillips found the organizational interfaces "fairly difficult" at first, both between headquarters and the centers, and among the centers. He characterized the agency "as a group of feudal baronies," which resulted from the origins and evolution of the organization. Some of it had to do with a desire for control, and "how much of the total pie" each center would receive. And some people displayed a considerable amount of resistance to outsiders in key positions, outsiders such as Mueller and himself. However, most people recognized the need to create a team that included NASA civilians, the military and industry. And Mueller later attributed much of the success of Apollo to the program management organization they established, with its lines of communications between people at headquarters, the centers and contractors. With Phillips' help, he said, "by the time we got into our actual flying program" they had "a very strong, well-working team," though they devoted a great amount of time to establishing communications and building teamwork to overcome parochialism. It took time to get communications flowing, he said "because it was antithetical to what they were brought up to think about." The centers thought they were self-sufficient, and did not need outside help. "It took a long while to become a team," but starting in early 1964 the Apollo program management structure got progressively stronger.[33]

But as Mueller readily admitted, "we [initially] had a very great problem because we were dealing with three centers ... We had the whole set of ... very important people that had to be corralled and [set] going in the same direction and then we had multiple layers of contractors and support troops." They all had to understand their roles and responsibilities, and this created a "mammoth communications problem." And until he got the necessary communications in place, they had difficulty getting things done. But building this team took more than the formal organization and meetings; Mueller had to establish personal relationships, which took an effort by everybody. To begin with people remained at arm's length, but once they began to understand what he had in mind they relaxed. He had to get people to understand what they needed to do, and to make sure they did it; calling that "the secret of the success of the program, because so many programs fail because everybody doesn't know what it is they are supposed to do." He concentrated on making sure people understood their roles, knew what was important, and did their jobs correctly. He started selling his ideas before he joined NASA, and continued selling after he moved to Washington. He got people to tell him their ideas, and then to bounce them

---

[32]   Phillips Interview, NASM, 9/29/89; Phillips to Shea, 2/23/64, Systems Management, SCP-46-17.

[33]   Ibid.; Mueller Interview, Slotkin, 9/9/09 (second session).

against his own in order to arrive at common ground; and when he discovered better ones, he adopted them. Proceeding in a logical, organized manner, over time he got to the point where he understood enough about what people needed or felt they needed, and then provided direction in ways they would accept. Yet, while some people never changed, they all "eventually accepted the fact that we had to do this together or we weren't going to get it done," he said.[34]

IV

The original estimate that Apollo would cost $20 billion came from a combination of bottom-up estimating and the addition of a "fudge-factor" which Webb called an "administrator's discount." Dryden took the estimates from the centers and more than doubled them before Webb added his discount. Because of changing requirements, the program offices worked with contractors to rewrite specifications and renegotiate contracts. And in February 1964, Mueller said, "We have every confidence that, with the level of funding requested, we can accomplish the beginning of manned lunar exploration within this decade and within the overall cost estimate of $20 billion." Then in a report to Congress in March, he testified Apollo would cost $19.5 billion, a figure which included Mercury and Gemini through the end of FY 1969. By adding the Ranger, Surveyor and the Lunar Orbiter programs, the projected cost increased to $23.4 billion. Based on these estimates, he told Webb the lunar landing program as organized and staffed would carry out the mission, it had the proper controls in place, and the first landing would take place within the decade at a cost of less than $20 billion. He accepted this baseline as an amount that he could live with, and used it in subsequent speeches and congressional testimony.[35]

However, obtaining funding for human spaceflight became problematic because of the rising cost of the Vietnam War and the Johnson administration's Great Society programs, which ate into the federal budget. But as Mueller later observed, "To some extent the Apollo Program held the country together in spite of the Vietnam War because every time we had a success the whole public became enthusiastic about it and that was a positive thing." Nonetheless, the White House told NASA to reduce its spending, and Webb asked Seamans to find ways to eliminate expenses. As a result, the administrator's annual review in December 1963 was devoted almost exclusively to program costs. Mueller instructed the center directors to provide greater confidence that the agency could accomplish the goal within its cost and schedule predictions, and kicked off the review by declaring, "We can live within the $20 billion run-out cost and accomplish the lunar landing mission

---

[34]   Ibid., 2/24/10.
[35]   Bizony, *The Man Who Ran the Moon,* 41; Mueller Interviews: Slotkin, 9/9/09 and Lambright, 9/21/90; Mueller, Caltech Associates, Los Angeles, CA, 2/6/64; Viewgraphs for Webb, 11/24/64, Notes & Transcripts, GEM-80-3; Draft 11/16/64, Mueller Summary for Webb on Cost and Schedules, SCP-40-10.

before the end of 1969 if NASA general management decides they want us to." However, he recommended expanding the space program to develop a national capability for space exploration. He wanted to extend human spaceflight beyond the initial lunar landings with a robust post-Apollo program, initially called the Apollo Extension Systems. At the time, NASA management thought Apollo would only be the first step – with Apollo-one followed by Apollo-two, leading to a mission to Mars in the 1980s. Yet despite Mueller's confidence about program cost, behind the scenes the Apollo program control staff called these estimates overly optimistic. In a memorandum to Phillips they wrote, "Our estimates have, in fact, significantly increased for the same scope of work" as a result of acquired experience. Thus, while Mueller assured every one of the validity of the $20 billion cost estimate, the Apollo program control staff claimed it would cost more. Nonetheless, Mueller's position became the party line, although the staff continued to caution Phillips to "at least recognize those parts which can be substantiated and those parts which cannot."[36]

NASA required a large number of engineers, scientists and technicians, and by early 1964 some 35,000 engineers and scientists toiled among the estimated 200,000 persons working on human spaceflight. The agency expected to add another 100,000 contractors by 1966, before reducing staff toward the end of the decade. Based on estimates by the National Science Foundation, the US space program employed about 2.8 percent of the nation's cadre of engineers and scientists. "This is a substantial number," Mueller admitted at the time, but contended "it will not strain the national supply." But, he cautioned, the cost of the space program could only be guaranteed if the agency maintained its existing schedule. If Congress provided less money or it stretched out the schedule, the cost would increase because the "time required to do any portion of the research often expands in relation to the time available." And creating a corollary to Parkinson's Law, he observed that "development time extends to meet any delivery milestone." Nonetheless, he always argued that Apollo was not a crash program, because unlike crash programs in which cost was not a major consideration, Apollo funding was limited. Therefore, he concluded, "there is a sound economic reason and justification for maintaining our lunar program on a schedule which includes landing a man and returning him to Earth in this decade. To extend this date would cost appreciably more money and unnecessarily delay this first major step in space exploration." Expanding on this argument he claimed that the requirement for scientists was exaggerated, because "the nation can well afford the effort being made in the manned space program." Nonetheless, criticism of the human resources drawn upon by NASA continued.[37]

---

[36] Dick, *Space Age*, 362; Mueller Interviews: Slotkin, 9/9/09 (second session) and 2/24/10; Briefings for Webb, 1964, SCP-40-10; Newman to Phillip, 11/25/64, SCP-41-2; Levine, *Managing NASA*, 185-188; Van Nimmen, et al., *NASA Historical Data Book*, 1958-1968, 115.

[37] Mueller, Caltech Associates, Los Angeles, CA, 2/6/64; "Parkinson's Second Law," Stuart Crainer,193; "...Manpower Requirements in the Manned Space Program," Mueller, AIAA, Los Angeles, CA, 2/3/64.

When Mueller testified before Teague's subcommittee in support of the OMSF budget on February 17, he said that since the previous May, when the final Mercury mission was flown, "manned space flight has been in a period that probably appears to the general public as a relatively quiet one." With another nine months remaining before the scheduled first piloted Gemini flight, NASA kept busy "moving from design to manufacture to actual delivery of hardware." He pointed out that 1965 would be a year of test and validation, and that 1966 would begin the operations phase of Apollo. Gemini would expose astronauts to long duration flights, and to new operational techniques, including rendezvous and docking. And because the air force Manned Orbiting Laboratory Program would also use the Gemini spacecraft, this spacecraft had become an important part of the nation's defense. Bridging the gap between Mercury and Apollo, Gemini was expected to resume human spaceflight in the last quarter of 1964, increasing the operational proficiency and knowledge of the technology necessary for human space flight, while extending journeys into space to up to two weeks, all to the benefit of Apollo. After detailing what he expected each Gemini mission to achieve, he said the agency was focused on completing production and system testing of the Gemini spacecraft. Yet, although 1963 had been a year of development headaches, "The program is now largely over the development hump" and the Gemini-Titan system would qualify for flight in 1964. He called Gemini "far more advanced, complex and versatile" than Mercury, and said McDonnell Aircraft planned to deliver twenty-two operational and test versions of the spacecraft. The air force was modifying and rating the Titan II for human spaceflight by introducing redundant backup systems, adding an abort capability, replacing the inertial guidance with ground radar guidance, and deleting unnecessary systems. And in addition to the twenty-eight Titan II flight tests already conducted, further tests were planned.[38]

Mueller predicted that Apollo would open the door to exploration of space and the solar system. After the Gemini Program, human spaceflight would resume aboard the Saturn IB and Saturn V in Earth and lunar orbit to perfect operational techniques and test spacecraft systems. He described all-up testing, telling the subcommittee that all flights would be scheduled to use complete space vehicles with live stages, and all systems would be operational. He reminded the committee the air force had tested Minuteman in an all-up mode, which he said "brought us to the point where the technology and experience is such that we can now drop the step-by-step procedure." Then, in a strong defense of all-up testing, he argued, "It will permit us to land an American astronaut on the Moon, and return him safely to Earth, in accordance with our schedule, even though we are operating this year on a reduced budget." Using all-up testing placed NASA in position to "capitalize on successful flights," to gather a large amount of flight data early in the test program, and obtain information that was needed by design engineers. And by testing essentially complete spacecraft earlier, and by acquiring additional data per flight,

---

[38] Mueller's Statement before the House Subcommittee on Manned Space Flight, 2/17/64, GEM-44-13; Mueller Interview, Slotkin, 9/10/09.

the probability of meeting the scheduled landing date improved. However, thinking beyond the initial missions to the Moon, Mueller observed "at some time in the distant future Apollo will be only a historical milestone." Thus, "we must continue to plan ... the follow-on steps in the evolution of the national manned space program." He predicted, "[by] the time the Apollo lunar landing has been accomplished, the Nation will have developed the scientific and engineering skills and industrial capacity to forge ahead into other space endeavors." He discussed three scenarios: expanded lunar exploration, Earth orbiting laboratories, and missions to the planets. And he said NASA would study "the potentialities of the present Saturn series of boosters, the feasibility of reusable boosters, and larger launch vehicles." Wrapping up, he told the committee that the buildup of the Gemini and Apollo programs was progressing in an orderly manner, and that although accelerating the program might decrease costs, slowing and stretching it out would only increase them. He argued for an orderly program, before outlining how much money OMSF would need in FY 1965 and reminding the members that the space program supported "plants, laboratories and test facilities in almost all of our fifty states." Congress later appropriated more than 96 percent of the administration's FY 1965 request of $5.25 billion.[39]

In his congressional testimony in 1964, Mueller gave Gemini a positive spin, but when he spoke about it just before the final Gemini mission in November 1966 he gave a less positive view of where the agency had stood before the first flight of that program. "At that time we still had a schedule problem, we still had a dollar problem, and we had just changed managers ... so there was a fair amount of uncertainty involved." And during an early visit to MSC he announced, "We had to finish Gemini in order to get it out of the way of Apollo," making his priority clear. He later claimed "one of the things that I did do with the Gemini management group [was] to explain that we had to get this done or we were going to have to cancel it." At the time, he called Gemini "a stopgap program," yet they made it as germane to Apollo as possible. He said he "probably would" have canceled it, "but Bob Gilruth made a cogent argument ... there would be a two-year gap between when our last manned [Mercury] flight and the first flight we could reasonably expect to have on the Saturn vehicles, and that would not be good." Gilruth said they were well into Gemini development, and it made sense to finish its flight test program, and "maybe we can learn something in the process?" This convinced Mueller. Nonetheless, at that time, he recalled, the "public was fixated on the Moon. Anything that didn't get us to the Moon didn't get much attention."[40]

Mueller's statements about canceling Gemini engendered a call from James S. McDonnell, Jr., the president of McDonnell Aircraft who was known as "Mr. Mac." Asked, "Are you really serious about that?" Mueller replied, yes, "dead serious." He did not want Gemini to overlap Apollo, and as the Gemini flight schedule continued to slip he established a cutoff date of January 1967. To control cost, he knew the

---

[39] Ibid.
[40] Mueller Interviews: Putnam, 10/4/66, Slotkin, 6/15/09, 9/10/09 and JSC, 1/20/99.

agency had to control the schedule, and early in 1964 he spent time "introducing the idea of getting a day's work done each day and getting people imbued with the idea that they had to figure out what that day's work was in order to be sure that they could get it done," emphasizing this theme in internal discussions as well as public speeches. It took more than a year before planning approached the point where they could define what required to be done each day, although in early 1964 he said the continuing Gemini slippage was "characteristic of a program when it isn't under control."[41]

V

Implementing configuration management in the Apollo Program remained a key objective. Phillips sent agency personnel and encouraged contractors to attend a joint air force/aerospace industry configuration management symposium in Los Angeles in late February 1964. Afterwards, a local newspaper wrote, "It doesn't take gypsy tea leaves or a crystal ball to forecast that air force management policies are going to have a big effect on [the] National Aeronautics and Space Administration ... The effect on NASA of streamlined Air Force management is as yet invisible, a few years from now it may well be said, NASA learned from the Air Force." Phillips planned to impose portions of the air force configuration management manual on the Apollo Program, noting that it was not too late because the program still had about seven years remaining. And with the multiplicity of contracts and constantly changing scope, it required configuration management. Phillips told the newspaper that the two most important factors influencing the success of Apollo were R&QA and configuration management. "By emphasizing these two factors, we may eliminate hardware teardowns and rebuilds," he said. A well-defined program "will provide us with the tools we require to fulfill the Apollo Program objectives." He planned to make it a requirement in all Apollo contracts, and by the fall he expected the centers to begin implementing configuration management in the Apollo Program. However, its use was spotty and remained that way for quite a while.[42]

Mueller's calendar was filled with travel, meetings, presentations, speeches, and luncheon and dinner meetings. The agency tracked his movement in 1964, estimating that he worked twelve hours per day, five days per week, plus about eight hours on Saturdays. During the six month period from April 1 to September 30, 1964, he spent forty-six percent of his time in meetings with outside personnel, forty-four percent with OMSF staff, five percent with one or more of the triad, and another five percent with other NASA personnel. He traveled away from Washington half of the

[41]   Mueller Interviews: JSC, 1/20/99, Putnam, 10/4/66 and Slotkin 9/9/09 (second session).
[42]   Management Skill Studied," *San Bernardino Star-Telegram*, 3/8/64, CM [configuration management] Symposium, SCP-41-6; Wong to Distribution, Phillips' Staff Meeting on CM [configuration management], 2/27/64, SCP-41-2; Johnson, *The Secret of Apollo*, 138-139.

time, split between visits to NASA centers, meetings with contractors and making public appearances. His 1965, 1966 and 1967 calendars show similar schedules, and on average he worked seventy-two hours per week between 1964 and 1967, excluding work at home during off hours, although he worked at home most Sundays as well.[43]

Mueller's travels frequently took him to Los Angeles, where he spent time with NAA. Rocketdyne generally made good progress on their Saturn engine projects, although he became increasingly concerned about progress with their F-1 engine, as well as the space division's two projects. The agency pressured Harrison Storms, the space division head, to replace the Apollo spacecraft project manager, leading to the appointment of Dale D. Myers as project director. However, problems at the space division continued as costs escalated and schedules slipped. NAA "was simply over its head," Mueller recalled. They had a poor management structure, and he called Storms "a manifestation of the problem," believing that his background as the X-15 program manager did not qualify him to build a complex spacecraft like Apollo. Unfortunately, Mueller said, "Stormy was probably the kind of guy that is a good one for leading one of those breakthrough programs [but] he certainly was not a guy that could get down to the nitty-gritty details of a program like Apollo. And he tended to be more emotional about them than analytical." NAA also had significant problems with the S-II stage. In designing it, engineers tried to use proven technology to avoid problems and delays, but the company did not escape them.[44]

The F-1 engine had previously experienced instability problems, called "pops;" in reality fairly substantial explosions that blew up part of the engine. NASA wanted an engine that was within the state of the art, for improved reliability, so Rocketdyne scaled up the Saturn I's H-1 when designing the F-1. Joseph P. McNamara, head of Rocketdyne's liquid rocket division called the F-1 "a big dumb engine" which used kerosene and liquid oxygen. But its sheer size created problems not experienced with the H-1, making it susceptible to combustion instability. The F-1 problems surfaced well before Mueller arrived at NASA, but engine research in the United States and Europe led to the development of new instruments that Rocketdyne used to diagnose the problems and evaluate changes to correct them. Minor modifications in the thrust chamber led to significant improvements; though NASA did not officially declare the problems resolved until January 1965. Indeed, it was possible to detonate explosives within the combustion chamber of the final engine while it was running without inducing instability. "Looking back there were a myriad of those problems," Mueller recalled, and added, there "wasn't any lack of challenge."[45]

There were also bureaucratic constraints within the contractor organizations which impacted the human spaceflight programs. As Mueller observed, "In any corporation there's a whole set of procedures that they've developed, which may or

---

[43] Schedules, 9/64-4/67, GEM-92-6; E-mail, Mueller to Slotkin, 8/12/09.

[44] Memorandum for Mr. Webb, 4/16/67, NASA Archives; Mueller Interviews: Lambright, 9/21/90 and Slotkin, 2/24/10; Bilstein, *Stages to Saturn*, 191.

[45] Mueller Interviews: NASM, 2/15/88 and 5/2/88; Bilstein, *Stages to Saturn*, 107 and 112-115.

may not make any sense in your particular project, which are almost impossible to change." In dealing with corporate bureaucracy, he had to either work around the procedures or get them changed, which in some cases meant changing people. He also found each system and subsystem had its own set of people problems. With a decentralized set of contractors and subcontractors, when problems occurred the agency had to rapidly surface them in order to insure that enough resources were applied. In most cases he found it better to leave contractor practices alone, so NASA did not arbitrarily change procedures. However, contractor procedures represented constraints, and at times the agency had to change them. But while NASA had the right to change non-performing contractors, they preferred not to do that because "changing a major contractor is something that one doesn't do very readily," he explained. Instead, the agency usually requested the contractor to reassign personnel. And besides the practical issues that would result from changing contractors, there would be "a hue and cry in Congress" to deal with, and even if the change was carried through, the agency would have to start over with the new contractor and build a new system.[46]

The first Gemini-Titan flew a successful unpiloted three orbit mission on April 8, proving the structural integrity of the space vehicle. Addressing a technical meeting at Patrick Air Force Base, Florida the following day, Mueller honored President Kennedy's role in establishing the Apollo Program and remembered his visit to the Cape, recently renamed the Kennedy Space Center, where the president had stood in front of a Saturn I and called the sight "astonishing." Reviewing accomplishments, Mueller said Mercury "established that man was capable of doing important work in space for periods as long as a day and a half" and Gemini would extend that time to two weeks. While the US led the Soviet Union in some areas, the Soviets led in others, and the two were equal in still others. Some people did not believe the US led in spaceflight, but Mueller, quoting Teague, noted, "By putting Americans on the Moon first, we will wipe out the last doubts that the United States has achieved pre-eminence in space flight." He said Apollo had already achieved several important objectives by establishing the industrial base with which it would "develop, test and manufacture equipment that depends on very advanced technology," and NASA had assembled the team required to build the Apollo infrastructure.[47]

Mueller said that the human spaceflight program was "even more vast" than the Manhattan Project or the ballistic missile program, and claimed the "total capability being developed is of definite military significance . . . [but] research and development must first be completed before applications are possible." In comparing NACA's contributions to aviation with NASA's to spaceflight, he pointed out that NACA did not work for the military, but the armed forces benefited from its R&D. Similarly, NASA did not work for the military, though DOD benefited from its efforts. Then

---

[46] Ibid.
[47] "GT-1," NASA, ksc.nasa; "Manned Space Flight Objectives and Accomplishments," Mueller, American Ordinance Association, Patrick AFB, FL, 4/9/64.

he argued that if the US did not have the Apollo Program, the nation "would not permit the Soviet Union to go it alone in manned space flight." Thus ninety percent of the OMSF budget would be spent even without Apollo. In conclusion, he compared landing on the Moon in 1969 with Lindbergh's landing in Paris in 1927. The man who as an eleven year old schoolboy in St. Louis saw the first solo flight across the Atlantic as one of the great achievements of his lifetime, said, "His goal was not Paris but the development and demonstration of the capability, equipment and know-how for the transatlantic flight." And as Mueller would later repeat, "the Moon is our Paris."[48]

---

[48] Ibid.

3

Getting ready

> "Give us the tools, and we will finish the job."
> *Mueller quoting Winston S. Churchill during the*
> *Battle of Britain, February 1, 1965*

Just as the explorations of the fifteenth and sixteenth centuries preceded the scientific and industrial revolutions, Mueller said, "we are moving into a new era of history, in which civilization itself will be transformed in ways that we cannot possibly foresee." Later he noted "we are writing important history." And when Kennedy decided to go to the Moon and the timescale to get there, he overcame the greatest obstacle. After three years, tangible evidence of progress in building a national space power existed. Yet while Mueller made major decisions about the organization and management of human spaceflight during his first few months at NASA, it took time for them to take hold. But while focusing on Apollo, he did not ignore Gemini as the agency prepared for the resumption of human spaceflight after the last Mercury flight in May 1963. As the press and the public turned their attention to Gemini and Apollo, Mueller urged everyone not to forget the need to extend Apollo beyond the initial lunar landings. However, as Congress and the administration moved on to address other national priorities, they became increasingly reluctant to spend more money on space. In addition to its cost, the psychological impact of the Vietnam War and the resulting unrest at home played a role in how Americans viewed space exploration. Nevertheless, Apollo remained popular as Gemini got underway.[1]

Mueller had sympathy for the requirements of the scientific community and he did his best to support space science with a coherent scientific program. He needed the support of scientists to assure continued funding for human spaceflight, and thought that without it NASA would not succeed. However, proponents of space science failed to understand how much of their funding came from Apollo. Rather

---

[1] Mueller, Missouri School of Mines and Metallurgy, Rolla, MO, 5/31/64; Mueller, Aviation/Space Writers Association, Miami, FL, 5/28/64; Mueller Interview, 9/9/09 (second session).

than detracting from scientific research money, Apollo was augmenting it. Nonetheless, opposition to Apollo by the scientific community created problems, and these were exacerbated by criticism from members of PSAC and the National Academy of Sciences' Space Science Board. This led Mueller to establish the Science and Technology Advisory Committee for Manned Space Flight with the assistance of Townes. STAC increased the credibility of human spaceflight within the scientific community, although Myers, the man who succeeded Mueller as associate administrator for manned space flight later said, "George was the symbol the scientists were against, because of his dedication to go to the Moon." Nonetheless, Mueller supported the inclusion of science in the human spaceflight program, despite opposition from within OMSF. "There was a great deal of reluctance on the part of Gilruth to get any science in the program," he remembered. The MSC director "wanted to do engineering and get it there and back like you did with the test aircraft." And his view of space science can best be summed up in a remark made in 1987, when he said, "Once you had a Moon rock, you pretty much had them all." Houston was an engineering center which regarded science as an unnecessary burden to be put up with. However, there were space scientists who thought anything spent on human spaceflight wasted money which should have been devoted to scientific research.[2]

To Mueller, Newell, head of the Office of Space Science and Applications, was a bureaucrat, not a scientist, who wanted to run the space program and actually fueled discontent in the scientific community. Newell thought that money spent on human spaceflight detracted from space science, considered spending on space to be a zero sum game, and fought to increase the space science budget. However, he failed to recognize that the Apollo Program supported his budget, and that it tied directly to the overall agency budget which was also dependent on Apollo. Congress and the administration tied NASA's budget to human spaceflight, something that many in the scientific community could not relate with. Yet, while many scientists thought human spaceflight detracted from funding for science, Mueller pointed out "we probably spent more money on science out of the Apollo Program than the rest of the money that space science was getting." And he noted "scientists are not very objective or realistic" when dealing with questions outside of pure science.[3]

Nevertheless, Newell laid the lack of funding for space science at Mueller's feet, and claimed that Seamans decided OSSA would be responsible for all space science including science conducted on piloted missions. He said Seamans put money for science into the OMSF budget as a gimmick and did not mean for Mueller to be in charge of it, an argument which Mueller could not understand. Newell argued that OMSF exercised control over science performed on Gemini and Apollo which should have been assigned to OSSA, and that Mueller appeared lavish in allocating

---

[2]   Mueller Interviews: NASM, 11/8/88, 6/22/88, Sherrod, 3/31/70 and Slotkin, 6/15/09, 9/10/09; Gilruth Interview, NASM, 3/2/87.

[3]   Mueller Interviews: NASM, 11/8/88 and Slotkin, 6/8-9/10 and 2/22/11; Compton, *Where No Man Has Gone Before*, 89.

funds to engineering aspects of human spaceflight but became very cost conscious when it came to supporting space science. However, Mueller claimed he had agreement with Newell for them to jointly organize scientific experiments, though OMSF had sole responsibility for conducting them. Nonetheless, he explained, "no matter what you said, they still felt that they weren't getting their fair share of the resources ... [although] they've gotten their share of the resources all along, in fact, more than their share."[4]

STAC was effective in dealing with PSAC and the Space Science Board because it provided Mueller with an external view of what science ought to be performed, and as independent observers they let their peers in the scientific community know what NASA did. Mueller enlisted Townes to chair STAC, who in turn recruited colleagues that outshone the members of PSAC and the Space Science Board – according to Mueller. STAC, he said, had "outstanding people looking at what we were doing and agreeing that this was the best thing to do." And he called them, "Outstanding guys in every case." STAC won over some members of PSAC, becoming a key to support within the administration for the lunar landing program. Nonetheless, Mueller said that Wiesner continued to oppose lunar orbit rendezvous, fearing that "we were going to kill all these people because we'd put them out in space without being able to demonstrate that they could survive." Mueller countered, "I don't know how you quite show that you can survive unless you try it, but there was a fair part of the physiological community that felt that the weightless environment would destroy the ability of the human mechanisms to work. So we had to answer all of those questions, and we needed to do it in a way that was unbiased, and STAC helped in that regard."[5]

John E. Naugle, who later replaced Newell as head of OSSA, called this "a continuing battle" between Newell and Mueller. "George wanted to get his hands on everything he could. Or, if he couldn't have his own hands on it, then he would like to have his own space science program." However, Mueller took over science for Apollo while Newell, he claimed, did little more than talk about it. "Homer Newell was really the one that felt like he should run everything ... I took over the science in Apollo so it worked very well." Adding, Newell "tried to create the illusion that he was the chief scientist [of NASA] but actually he wasn't all that great." Mueller called STAC "one of the most effective science committees I've ever seen. Homer Newell never managed to put together anything quite like that."[6]

Webb announced the establishment of STAC on March 30, 1964, a few days before the first Gemini flight. A NASA press release said that the committee would advise Mueller about the "scientific and technological content of the manned space program," and he would use the committee to oversee the whole Apollo Program. At

---

[4] Newell, *Beyond the Atmosphere,* 291; Mueller Interviews: Slotkin, 6/15/09 and NASM, 5/1/89.

[5] Mueller Interviews: NASM, 11/8-6/22/88, McCurdy, 6/22/88 and Slotkin, 6/15/09, 2/24/10 and 2/22/11.

[6] Mueller Interview, Slotkin, 6/15/09

their first meeting, held at KSC, the STAC members received a series of briefings about space science projects and then reviewed the agency's advanced technology and research programs. Mueller asked for their help in understanding the composition of the lunar surface, the impact of radiation on spacecraft, and the physiological impact of spaceflight on humans. He also sought assistance in several technological areas – thrusters, engine stability, reliability, redundancy and sensors. Over the years, STAC would advise on important questions concerning science and technology; and as an indication of their interest, during Mueller's tenure at NASA the committee met two to four times each year, and with most of its original members remaining active throughout.[7]

Mueller said that he pushed space science into a prominent position in the Apollo Program because there was no point going to the Moon without doing scientific investigation. In that sense he supported Newell "to the maximum extent" possible, he said. But he had to forcibly inject science into the program because of opposition from within OMSF, especially at MSC, although STAC helped to convince Gilruth. STAC also aided Mueller in his dealings with the scientific community, and tried to deflect criticism from the PSAC, which persistently argued that "the Apollo Program didn't do any science." Nevertheless, opposition from scientists plagued the Apollo Program throughout.[8]

Addressing a conference devoted to the peaceful uses of space in late April, Mueller said the agency would fill the pipeline for the Apollo effort, and announced that recent reviews found "no technological problems of such a major nature that they would interfere with the accomplishment of the program schedule." But, he said, "the most challenging technical task before us is the integration of all systems and subsystems." NASA had already determined that radiation would not pose a threat to astronauts on the Moon, but many unknowns remained. He compared NASA to its predecessor, the NACA, which had concentrated on R&D, pointing out that some of the "most significant advances in ... aviation resulted from fundamentals of flight developed by the NACA," and similarly, NASA was developing the fundamentals of spaceflight. He then predicted that this competence would serve the nation long after the end of the Apollo Program. Keynoting another human spaceflight conference the following week, he called the first Gemini test a success, and observed that NASA produced launch vehicles "far larger than those required at present for military purposes, which are making this country second to none in the power to lift objects into space." He also pointed out that one-third of the money to pay for landing on the Moon had already been appropriated, and FY 1965 funding would bring the program half-way to the Moon.[9]

During the spring of 1964 Mueller received three honorary degrees, the first of

---

[7]   "Manned Space Flight Advisory Committee Meets," NASA Press Release, 3/20/64; STAC, KSC, FL, 3/27-28/64.

[8]   Ibid.; Naugle Interview, Lambright, 11/7/91; Mueller Interview, Slotkin, 6/15/09.

[9]   Mueller, Conference on the PUOS, Boston, MA, 4/29/64; "The Manned Space Flight Program," Mueller, AAS, NY, NY, 5/5/64.

seven that he would eventually receive. Giving the commencement address at Wayne State University in Detroit, he said it was "essential that universities take a leading role in the development and dissemination of information needed for the exploitation of scientific and technological developments." At times in the past, the US had failed to capitalize on the technology that it developed. The first powered flight by the Wright brothers was in 1903, while Europeans moved ahead in aeronautics and Germany became the first to use the airplane in war. Similarly, although Robert H. Goddard invented the liquid fueled rocket in New Mexico, the Germans exploited rocketry during the Second World War and developed the first operational jet aircraft. Then when it came to ballistic missiles, he noted, the US "became so preoccupied" with it they failed to pioneer space exploration. However, after the Moon is conquered, "this versatile capability remains for other manned spaceflight applications." And as he would argue repeatedly over the next five years, "The time is approaching when we must give more thought to the exploitation of the capabilities provided through these efforts," because the nation must plan for the future to avoid the mistakes of the past.[10]

Continuing his efforts to garner support for human spaceflight, Mueller spoke to a group of bankers the following week and addressed the space program's economic impact. Highlighting the infrastructure built, the people employed, and the contracts awarded "within the framework of the free enterprise system," he called Apollo the "largest job of research, of development, and of manufacture ever attempted." NASA worked through the private sector, and while the agency spent ninety-five percent of its money with contractors, its main economic value derived from the knowledge obtained. And, he added, "the new knowledge acquired in space exceeds by far the value of funds so far spent. For knowledge, more than guns and butter, is the true power of modern states." Expanding on this theme later, he said, "the space effort will continue to reveal processes and materials which offer a potential for new industries and new products, and thus for new job opportunities." And the flow of history "indicates that a small spring of basic research is often the source of a mighty river of technology-based industry." Thus, "it is most imperative to augment our reservoir of knowledge with new ideas that can come only from basic knowledge."[11]

On May 28, Mueller witnessed the launch of the sixth Saturn I (SA-6) carrying a boilerplate Apollo capsule to evaluate the structural integrity of the space vehicle. He went to this launch, as he would all Gemini and Apollo launches; a practice that he continued after leaving NASA. Following this launch, his hometown paper, the *St. Louis Post-Dispatch* quoted him calling the Saturn I rocket "the most powerful in the world." Yet while SA-6 experienced a problem at liftoff when one of its eight H-1

---

[10] Space Research – the Implications for the Earth," Mueller, Wayne State University, Detroit, MI, 5/20/64; "Space Research: Directions for the Future," SSB, 4/68; Freitag Interview, Lambright, 6/27/91; Cushman to distribution, University Affairs, 7/23/68, GEM-108-7.

[11] "The Space Program and the National Economy," Mueller, PA Bankers Association, Atlantic City, NJ, 5/27/64; Mueller, Technology Club of Syracuse, Syracuse, NY, 2/1/65.

3-1 George E. Mueller, Wernher von Braun, Eberhard F. M. Rees and Rocco A. Petrone at SA-6 launch, May 28, 1964. (NASA photo)

engines shut down prematurely, the booster operated in an engine-out mode as designed, and performed as planned.[12]

Mueller addressed a meeting of space journalists a few days later, acknowledging that the "people you write for and the people I work for – the people of the United States … are engaged in a tremendous national effort to achieve leadership in the vital area of manned space flight." He cited "highly tangible evidence" of progress, and said that because "crew safety is the first consideration of the US manned flight programs, the buildup in flight duration in Gemini will be gradual," going from three orbits later that year to a flight of up to two weeks on the seventh mission. NASA would gain two thousand hours of human spaceflight experience on Gemini, and he insisted that the agency was taking no short cuts. NASA, working with the air force, modified the Titan II before rating it for use in Gemini, and he highlighted the fact that the malfunction detection system had no counterpart on the Titan missile.

----

[12] *St. Louis Post-Dispatch,* AP, *Washington Post*, 5/29/64, GEM-83-4; Bilstein, *Orders of Magnitude,* 104.

This system would provide advanced warning of potentially dangerous launch malfunctions, allowing the astronauts to abort if necessary, making it the key to crew safety during powered flight. And because redundancy would improve safety, each component and system critical to crew safety had backup features.[13]

Returning to his alma mater to receive his second honorary doctorate on the silver anniversary of receiving his bachelor's degree in electrical engineering from the Missouri School of Mines and Metallurgy in 1939, family and friends heard Mueller pay homage to the pioneering history of Missouri, and acknowledge the contributions of St. Louis-based McDonnell Aircraft. The local press hailed him as a returning hero, with headlines and articles. In his commencement address, he asserted that the space program was contributing directly to the strength of the country, and there were many precedents for investing in this kind of technology. "Through their roads, the Romans controlled their world," he said. British sea power was another beneficial national investment. Then he claimed "the world looks at our space program ... as an index of our scientific and technological vitality." Quoting the president, he posited, "A nation's prestige is measured by a new yardstick: its achievements, or lack thereof in space," and knowledge gained from human spaceflight would provide immediate returns. He spoke of the experiments planned for Gemini, and pointed out, "One of the most exciting aspects of any scientific investigation is that we never know just what we will learn." Then quoting STAC member and once and future presidential science advisor Lee A. DuBridge, he concluded, "A thousand mysteries surround the Moon. Does it hold the key to the history of the Earth and the solar system? Quite possibly."[14]

II

By the end of May, a team of NASA and McDonnell engineers in St. Louis began system testing of the Gemini spacecraft for the first piloted flight, and the same team re-inspected the spacecraft upon its arrival at the KSC, minimizing some of the duplication experienced during the Mercury Program. That summer also saw the beginning of efforts to convert McDonnell's contract from cost plus fixed fee to cost plus incentive fee. However, negotiations dragged on into the fall, and the two sides remained wide apart on contract costs even while preparations for the mission were underway.[15]

As NASA looked towards the post-Apollo period, Webb established a Future Programs Task Group which recommended additional studies of extended Apollo missions beginning in 1968, and called for space stations and human mission to Mars in the 1970s. Mueller published an article in *Astronautics & Aeronautics* magazine

---

[13] Mueller, Aviation/Space Writers Association, Miami, FL, 5/28/64.
[14] Mueller, Missouri School of Mines and Metallurgy, Rolla, MO, 5/31/64.
[15] Hacker & Grimwood, *On the Shoulders of Titans*, sections 11-1and 10-3.

called "Apollo Capabilities," and after describing the "basic" Apollo Program, he addressed "extended" capabilities, writing, "Apollo spacecraft can be adapted with minor modifications to longer and more sophisticated missions." He said an extended Apollo spacecraft launched on a Saturn IB could operate in low Earth orbit for more than sixty days. Or NASA could launch a research laboratory into polar orbit on a Saturn V. And extended lunar missions would be possible using equipment launched aboard Saturn rockets at a cost ranging up to $2 billion. Webb supported plans for extended missions, although outside of the agency some people criticized such plans as too vague or overly ambitious. Mueller had high hopes for the Apollo Extension Systems, and authorized Bellcomm to prepare further studies.[16]

Mueller's third commencement address (and third honorary degree) of the season took him to New Mexico State University during the first week of June. With New Mexico's Clinton P. Anderson, chair of the Senate space committee in attendance, he discussed the importance of continuing the quest for scientific knowledge and the need to support human space exploration. To emphasize the role of humans in space he said, "The most dramatic demonstration came when [L.] Gordon … Cooper, Jr.'s automatic controls failed" on the final Mercury flight and he "oriented the spacecraft manually by observing the lights of Shanghai, then fired his retro-rockets for return to Earth when John H. Glenn, Jr. gave him a radio signal from a ship off the coast of Japan." He pointed out that the space program had an unbroken string of successes in sixteen major launches, including Mercury, Saturn I, and a Gemini flight from the Cape; and three Little Joe II-Apollo flight tests at White Sands in New Mexico. Adding, "We fly a spacecraft, manned or unmanned; only after a comprehensive set of ground and flight tests have been completed and every deviation … explained."[17]

While opposition to human spaceflight continued in the scientific community, it also came from the political right. Former President Dwight D. Eisenhower published an article which said the race to the Moon "diverted a disproportionate share of our brain-power and research facilities from equally significant problems, including education and automation." Rebutting Eisenhower, a *Houston Post* editorial quoted Mueller's recent remarks at New Mexico State University to show that Apollo was not a crash program. Adding their two cents to the argument, the *Miami Herald* editorialized that the successful launch of SA-6 "showed how far we have come in little more than a thousand days" since the start of Apollo. However, agreeing with Eisenhower, the Miami newspaper called for a "more orderly and moderately paced" space program. Adding, "This is too vast an undertaking, with too great a meaning for all mankind, to be pressed with a full-speed-ahead-damn-the torpedoes urgency."

The question of whether Apollo was or was not a crash program would not go

---

[16]   Hitt, et al., *Homesteading Space*, 7; "Apollo Capabilities," AIAA, 6/64, GEM-110-3.
[17]   Mueller, New Mexico State University, University Park, NM, 6/6/64.

away, and Mueller addressed it again at an Apollo executives meeting in June. The agenda called for an exchange of views about Apollo Program problems, with OMSF officials and the contractor CEOs discussing key issues. One session focused on communications with the public, and NASA's Disher prepared a position paper about public communications showing that the time span of Apollo greatly exceeded other national R&D programs. Borrowing Mueller's arguments to show it was not a crash program, he told them that the problem existed because the public thought of it as a crash program, thus creating an unfavorable impression of the whole space program. To change this perception, he said, industry should explain "in a low key tone" that Apollo was not a crash program.[18]

Phillips briefed the CEOs about NASA's plans to implement configuration management and the center directors – washing their dirty linen in public – argued against it. Gilruth opposed configuration management because it required too many people; and based on what he heard, the expense would be too great. Phillips pointed out that it took Douglas Aircraft four months to confirm the configuration for the S-IV stage on SA-6, while it only took Martin Marietta only two days to confirm the configuration for the modified Titan II used for the first Gemini test flight. Von Braun also objected, because he said it removed decision making from working engineers. Boeing's Allen called configuration management a "fundamental of good management." But von Braun continued to object because he wanted flexibility in R&D programs which he claimed configuration management would remove. Allen countered that it worked equally well in R&D and production programs. Mueller then pointed out that everything was on cost and schedule with the Titan III, which was the first air force program to use configuration management from the start, and said that using configuration management "doesn't mean you can't change it. It doesn't mean you have to define the final configuration in the first instance before you know that the end item is going to work .., It means you define at each stage of the game what you think the design is going to be within your present ability. The difference is after you describe it, you let everybody know what it is when you change it. That's about all this thing is trying to do." Nonetheless, the center directors continued to resist configuration management.[19]

Following this meeting, Phillips wrote to Mueller saying that the MSFC project offices were "ineffective in view of apparent strength and competition from R&D laboratories." Von Braun managed programs much as he did before the establishment of the Apollo Program Office, and Huntsville meetings were "attended by masses with no apparent leader," leaving decision making authority diffused, while decisions and directions were passed on to contractors "without compliance with normal contractual procedures." Specifications were out of date, contractors delivered their hardware without proper identification, and specification trees were non-existent. Thus, almost a year after Mueller had introduced the air force

---

[18] AEG, 6/17-19/64, SCP-42-7.
[19] Ibid.; Johnson, *The Secret of Apollo*, 139-140.

Table 3-1: Apollo Program Office GEM boxes, air force System Program Offices, and GEM boxes as modified by Phillips

	GEM Boxes	AF System Program Office	GEM boxes as modified by Phillips
1	System Engineering (SE) (CM a subset of SE)	Engineering	SE (Test initially in SE)
2	Program Control (PC)	PC	Program Planning & Control (PC)
		CM	(CM a subset of PC)
3	Test	Test & Deployment	Test
4	R&QA	R&QA in Engineering	R&QA in SE
5	Flight Operations (FO) (Procurement at centers)	Procurement & Production (FO not part of SPO)	Operations (Procurement at centers)

approach to program management, there remained confusion about how to implement the GEM boxes. Configuration management, which was part of system engineering at MSFC, came under program control at MSC. While MSFC considered it to be a staff function, it operated as a line organization at MSC. Consequently, Phillips reintroduced program management in the Apollo Program, organizing it around three elements – program planning and control, system engineering, and operations – with testing and R&QA now operating as a part of system engineering (Table 3-1). He said these functional groups had "to work extensively together, because all aspects of all their jobs require a great deal of integration with other elements, and a great deal of iteration."[20]

After Shea went to Houston to manage the Apollo Spacecraft Project Office in October 1963, he said Washington did not need its own system engineering function or additional people to do system engineering at headquarters, arguing instead that Washington should use the centers to perform this function. Others at the centers agreed, saying, as Phillips recalled, "we don't need Washington's help to tell us how to build a spacecraft," or "we don't need people in Washington to tell us how to build a launch vehicle or ultimately a Saturn V." However, while Mueller and Phillips agreed that Washington did not need to design or direct the design of hardware, they were required to provide overall direction. Headquarters, they said, should provide "overview or oversight in inter-center matters, whether the centers wanted it or not," and they used interface panels together with a panel review board as the mechanism to provide that oversight. While the Apollo Program Office had a full complement of support from Bellcomm, Phillips did not think the company could carry out all aspects of the system engineering function as he saw it. Consequently, he changed the GEM boxes and "hybridized" the air force SPO

---

[20]   "Management Techniques..." 6/8/64, MSFC, GEM-76-3; Cotton to Phillips, Mueller "mirror image" boxes, 8/19/64, SCP-40-7; Phillips Interview, NASM, 9/29/89.

structure by putting configuration management in program planning and control, rather than leaving it in system engineering. He also established a separate test directorate reporting to him, which received support from both Bellcomm and the program planning and control group. This "bastardization" of the air force approach to testing, as he called it, "worked quite well," he admitted, and by using a standalone test organization his program management model nearly matched Mueller's GEM boxes. Phillips' key role as Apollo program director involved integrating and managing the work of the centers and contractors. He used the design reviews leading up to the flight readiness review to manage the process, chairing the reviews and the panel review board, which he called "the ultimate decider of interface issues." He also participated in important center-led activities, including project reviews, and often people from the Apollo Program Office – either Bellcomm or NASA staff – participated in project office reviews.[21]

By early July, all-up flight tests of the Saturn IB were less than eighteen months away, and Saturn V all-up flight testing would begin in 1967. FY 1965 marked the peak of Apollo efforts, but in looking beyond the initial lunar landings, Mueller said, "we are approaching the planning of new programs along a philosophy of fully capitalizing on the capabilities in the Apollo Program" to fly missions to extend time in orbit and on the Moon, carrying larger crews, and increasing the lift capacity. As the operational phase of Apollo approached, the agency devoted increased time to planning for lunar exploration. Mueller also wanted to use scientist-astronauts on future flights. It was logical that scientists should explore the Moon, and he saw scientist-astronauts as useful in persuading the scientific community to accept and support follow-on lunar exploration. As he later explained, "I was trying to get essentially the Apollo Program extended for ten years or so to really explore the Moon."[22]

By August, concerns about NAA's performance led to a meeting between Mueller and Atwood to review progress. In notes prepared before the meeting, Mueller wrote: "remove work from N.A.A.," and consider competing one or more of the company's contracts. And when they met, he told the CEO about plans to conduct an assessment of Storms and his management team, and expressed reluctance to allow NAA to bid on any new NASA business while problems persisted on their existing contracts. When they departed, Atwood gave his assurance that performance would improve, but Mueller remained dubious.[23]

The agency flight tested another Saturn I (SA-7) with a boilerplate Apollo capsule on September 18. The spacecraft carried ballast to simulate the size, weight, structure and center of gravity of a piloted command module. This was a suborbital flight to test the spacecraft structure in the Earth's atmosphere, and its ability to jettison the

---

[21] Ibid.

[22] Mueller, NASA-University Conference, Washington, DC, 7/7/64; Mueller Interview, Slotkin, 2/24/10.

[23] Notes on Review with Lee Atwood, 8/29/64, NAA, GEM-84-3.

launch escape system. With the cancellation of the Saturn I program, the three remaining vehicles would carry Pegasus meteoroid detection satellites to test for space dust traveling at cosmic velocities which could pose an impact hazard to spacecraft. With ninety foot winged panels, Pegasus would record the impact of space dust on SA-8, 9 and 10. After the success of SA-7, NASA declared the Saturn I operational, although the agency had no plans to use it beyond these three robotic missions.[24]

NASA tried to use the Program Evaluation Review Technique as the standard reporting system in the Apollo Program, because each contractor had its own way of keeping tabs on progress. The agency wanted a single system, but implementing PERT created problems because of the "not invented here" syndrome, as Mueller described it. Never a strong advocate of PERT, he regarded it as "a questionable invention," and felt that most contractors did not use it effectively. In the end, NASA never succeeded in getting it fully adopted. "Everyone gave lip service to it and tied it in to their existing systems," he said, but they did not always use it as intended; nevertheless, PERT remained the standard reporting system.[25]

Mueller frequently visited contractor facilities. He liked to walk around the shop floor and talk to the workers, and had fairly unlimited access before the contractors figured out what he was doing. Recalling, he said, "eventually management caught on and we had escorts." He looked at work being done while assessing progress, and chatted with workers to find out what they thought about the progress being made. This gave him a feel for morale at the firm, because "in the long run the guys down on the line really determine whether this thing gets built ... right or not," he said. Because of these visits, he picked up things that needed to be done differently. And after talking with the workers, he would meet with their managers to see what they knew. He liked to tell the story of one plant visit, where he saw a PERT diagram "up there with all those boxes and things." Because he did not believe the schedule, as he related the story, he took "the guy aside who's really doing the work," and asked, "'how do you really keep track of it?' And he has a desk drawer he pulls out, and there's his schedule." It may or may not have had any relationship to the schedule shown to NASA. Leading him to say, "PERT isn't of any use, nor is any other system, unless the people that are actually doing the work use it." But PERT was not the only way to keep track of a program, there were many ways to look at progress and find the critical path. The trick, he said, "is to keep what's going on in the program reported," and not just "a form up there that keeps people off your back." A reporting system such as PERT only identifies problems. What is important is to forecast what will happen and fix it before it becomes a major problem, which was the role of program control.[26]

---

[24]   Press Kit, Seventh Saturn I, 9/13/64, Saturn Program, GEM-91-15; Bilstein, *Stages to Saturn*, 329.

[25]   Mueller Interviews: NASM, 5/2/88 and Slotkin, 9/9/09; NASA Authorization for FY 1966, GEM-45-3.

[26]   Mueller Interviews: Slotkin, 6/10/10, and NASM, 2/15 and 6/22/88.

III

After a little more than nine months as Mueller's deputy, Phillips became Apollo program director – the general had been operating in this role for months and on October 27 Mueller made it official. Just before this, Phillips sent recommendations for a speech that Mueller planned to give to the senior staff at MSC. The general identified parochialism as the most serious problem at the centers, said it built barriers between them and with headquarters, and insisted that these walls had to come down. The centers refused to bring good people from the outside into NASA, and he called Apollo "too big and its objectives too important … to permit 'family' loyalties to militate against bringing in really good people." Despite it being the third year of the Apollo Program, he wrote, "Hectic beginnings … led to procrastination on firming up requirements and making certain decisions for months and even years." However, with program definition over, the time had arrived to complete the design. But NASA tended to make things too complex, and room existed for simplification, although he cautioned that there was no such thing as a simple change. They could not be avoided, he explained, so "we need a well-disciplined system to be able to accommodate changes." And he said it was time to "make a clear division between the operational configuration and R&D appendages." The program's engineers needed to learn how to react expeditiously and to regard time as an important factor. They needed a sense of urgency to do things in a timelier manner. He suggested that the centers bring their contractors into the "family" and give them more meaningful responsibilities. Furthermore, "We must discipline all parts and all levels of our organization to plan and pursue specific actions when things are not going according to plan." And he recommended the centers take immediate action when a problem was discovered, rather than leaving things to work themselves out or let nature take its course, as they were wont to do. Mueller welcomed Phillips' suggestions, and when he spoke at MSC he told them they had done an outstanding job building the center from the ground up, while flying Mercury, conducting the Gemini Program "at top speed," and building "a full head of steam" on Apollo – all under the intensive scrutiny of the press, public, and Congress. He said that a fundamental reason for the reorganization of OMSF was decentralization of program authority to the centers. But as OMSF strengthened its capabilities, it delegated responsibility for managing portions of the program with the Apollo Program Office providing the necessary guidance.[27]

However, as Mueller explained, in the early days of Apollo cost did not become an issue. Launching Mercury occurred in an atmosphere of a crash program because the public demanded something be done to catch up to the Soviet Union. It had been a time when Congress did not debate cost, authorizing funds to begin the program in a mood similar to a declaration of war. However, the mood changed when NASA

---

[27] Phillips to Mueller, 10/3/64, GEM-43-4; Mueller Interview, Slotkin, 2/24/10; "Presentation by Mueller before the Senior Management of the MSC," Houston, TX, 10/5/64.

succeeded in building launch vehicles with a lift capability greater than anything claimed by the USSR. And, he said, while America's adversaries still led in human spaceflight, "every day that passes increases our chances of taking a long stride ahead of them with the first manned Gemini flight." He claimed success in making it clear to Congress that NASA's objective was not a race to the Moon, but to develop the national capability for human spaceflight with lunar exploration as only one step in the process, "neither the first nor the last." However, he added, "Under the present conditions, there is a great deal more emphasis on holding costs." The agency found itself heavily engaged in internal reviews of the FY 1966 budget, and it needed to justify all funding requests. And Congress looked to NASA to meet its commitments, although the agency was failing to do that by missing the launch dates for the initial Gemini missions. But the administration and Congress measured success by the ability to meet commitments, which, he said, "leads to the conclusion that key events generally slip at a relatively constant rate throughout the life of a program." Applying a similar analysis to Apollo milestones leads to questions about the agency's ability to meet the lunar landing goal by the end of 1969; although "it does not follow that we cannot improve our schedule performance," because others have met theirs and NASA met key milestones in developing major subsystems. Yet progress in R&D could not easily be predicted, which made it necessary to plan work in shorter intervals – the next six months – and not overly focus on the long term. Congress and the administration judged NASA "on the basis of how well we get our work done in the next six months ... [and i]f we cannot accomplish the things we have scheduled in that time, there will be a tendency to suspect that we cannot accomplish in the next five years the things we have scheduled in that period."[28]

Mueller identified three problem areas – attitude, communications, and how the agency organized its work. To fix attitudes, NASA had to look internally and believe that schedules could be met, but the organization must realize schedule slippage was "at least as serious to the program as a poorly performing system." R&D required trading off cost, time and performance, whereas at NASA engineers tended to give performance the top priority at the expense of the other factors. Nonetheless, he said, "we cannot and will not cut corners with respect to performance, reliability or testing that will endanger the success of a mission." However, at times the agency focused on increasing the performance of "nice to have" things, ignoring cost and schedule, although it would only take a subtle shift in attitude and a willingness to establish day-by-day work plans in order to achieve the desired change. He then observed that "we can expand and improve our system for monitoring between all levels," and we must "insist from every level that each day's work be accomplished that day." If "we could operate in this fashion for six months we could substantially improve our schedule performance. And if we could do it for six months, we would find it easy to continue for a year and then two and then five." To continue to receive support from the administration and Congress, he argued, "We must accept in our

---

28   Ibid.

minds and hearts that schedules can be met and that it is vital and important to meet them."[29]

Mueller did not get involved in the day-to-day direction of Gemini, focusing instead on Apollo during his first year at NASA. Mathews ran the Gemini Project Office in Houston, and recalled, "I was left pretty much to run Gemini on my own devices," because it basically involved MSC and the air force. Mueller considered Mathews a good manager, a good engineer, and one of his favorite people at NASA, though his skills were not in the same ballpark as Shea or Low, "but almost," he said. But as Mueller pointed out, Gemini was "simply a takeoff on Mercury." Its main value lay in the fact that it filled the gap between Mercury and Apollo. Low had carried "the burden" of Gemini in Washington while Mueller was busy with larger issues, but when Low was appointed as Gilruth's deputy in May 1964, Mueller took over as acting Gemini program director. Nonetheless, Mathews remained in charge, and, both Mueller and Low gave him full credit for its ultimate success.[30]

After becoming Gilruth's deputy, Low spoke with Mathews daily. They both were "early birds," and saw each other first thing every morning. According to Low, "Chuck felt that he ought to communicate, and he did." Low kept MSC's director informed, while Shea, "would occasionally come up and tell us something ... But it really was sort of a minimum – when he felt like it – and anything beyond that was somewhat like pulling teeth." Yet, "At the same time, he had this hot line to George Mueller ... where he did by-pass ... Bob Gilruth and me ... and worked out his problems on the telephone, directly with George Mueller." Nonetheless, Low said, "We all had good relationships with Joe ... Except that under Joe Shea the Apollo project was run as a closed corporation." Mueller later insisted that Shea had kept everyone informed, and that the complaints he heard of concerned Shea communicating too much and people did not like his frankness. Mueller said Shea "kept Bob Gilruth as fully informed as he could. But Bob believes in having faith in human beings; he gives free reign to everybody ... In many instances, I remember the things [that] Bob later complained about [were] reviewed in management council. What can you say if Bob wasn't listening?" But after Phillips came on board, Low developed a good working relationship with him, and they spoke daily, although, as noted earlier, he never developed the same relationship with Mueller.[31]

In the scores of interviews, both on and off the record before and after Mueller left the agency, Low frequently expressed his feelings about Mueller and Shea, and their style of management. Sherrod interviewed Low dozens of times, and in 1972 told Low in an off-the-record interview, "Joe says that he and George Mueller thought exactly alike, as engineers." Low responded, "let me tell you what they missed, both of them. They missed the fact that when you have a user of your wares,

---

[29]   Ibid.

[30]   Mathews Interview, Lambright, 10/4/92; Mueller Interviews: Putnam, 10/4/66 and Slotkin, 6/15/09 and 2/25/10; Lambright, *Powering Apollo*, 134.

[31]   Low Interview, Sherrod, 11/7/69; Mueller Interviews: Slotkin, 9/9/09 (second session) and Sherrod, 8/19/71.

you've got to listen to them, and you've got to make certain that he will accept and understand your decisions." Low defined the "users" as Houston's operations people and particularly the astronauts. Adding, "I learned a lot from the users ... And there was no point in my forcing anything down the astronauts' throats ... Same with the operations people ... That means that the best ... analytical technical decision might not be the best overall decision. It may well be that Joe and George always jointly thought of the best analytical technical decision, but it might not have been the best overall decision."[32]

Low wanted to satisfy the *users*, while Mueller answered to the *customer* – the American people and their elected leaders in Washington. With widely different perspectives, Low failed to understand Mueller's point of view, and Mueller did not agree with Low's. Mueller considered the flight controllers and astronauts as highly trained people. However, they were no different than thousands of others working on the human spaceflight program, despite the lavish attention paid to them by the press and the public. He acknowledged the astronauts' skill as pilots, their bravery and willingness to take personal risk. Their health and safety remained of major concern, but he did not let their personal foibles, their wants or desires drive the design and development of the space vehicle that they would fly. They and the flight controllers were a part of the program, but they did not run it; he did. As he later said "nothing that I've seen addresses the question of how much influence did the astronauts have on the actual detailed design. And it turns out they had more than they should have because no one wanted to argue with them ... and you tried to figure out how to fix it so they could have what they wanted. But ... it would have been better if there had been a better level of engineering review for those things." Satisfying the pilots remained part of MSC's culture, because "the pilot was always the guy that had to fly the thing, so he had a very large voice in how to design it. And that's probably why we had so many accidents with aircraft because the pilots didn't really understand the system," he said. Adding, "astronauts are good for the unusual things that come up ... [Though] if you look at where we had problems in flight for example, some guy brushed a switch that shouldn't have been brushed and caused a whole host of problems until we figured out what switch was in the wrong position. Humans are prone to make mistakes and that's true of all humans. So you have to be awfully careful that you don't put them in a position where they can make mistakes that are dangerous for themselves and for the system as a whole."[33]

After Mueller got more involved with Gemini, he began to introduce the air force program management methodology. However, Gemini was so far along that he could not implement the program plan in the same manner as it was being rolled out for Apollo. Nonetheless, he assigned people with the right skills to the Gemini Project Office. He implemented the GEM boxes, and attributed some of Gemini's success to it. Another key to Gemini's ultimate success came from establishing work package managers for each subsystem, "not only for it's getting built and working," but getting

---

[32] Low Interview, Sherrod, 6/21/72.
[33] Mueller Interview, Slotkin, 2/24/10.

"it done on time and within budget." And "shortly after the sub-system manager concept went into work, the stuff began to get delivered from the sub-contractors and began to work," he said. McDonnell Aircraft and Martin Marietta both had many problems in 1963; "they were in terrible shape," he recalled, slipping well behind schedule. It took significant action to get improvement, but by the next spring the Titan II was no longer on the critical path and by year-end it was working "reasonably well," although some problems remained. Whereas Bellcomm provided system engineering support for the Apollo Program Office, NASA was mainly using government employees on Gemini simply because the agency did not have sufficient Bellcomm resources. Because Gemini was a follow-on to Mercury, and as such was thought to be a modification to existing systems, it suffered different problems than Apollo and also had fewer contractors involved. And Mueller noted, "You need to tailor your management structure to the program you're managing ... Although we did try to make sure that these various disciplines were represented in the planning for Gemini and we did have people working on those disciplines in the organization."[34]

At a STAC meeting at the end of October, John F. Clark, who was the director of space sciences in OSSA, discussed the role of astronauts as sensors, manipulators, evaluators and investigators – a characterization that Mueller adopted and used in subsequent speeches. Clark said the space science program would not come to full fruition until scientists stepped on the Moon or other planets. Like Mueller, he spoke of space exploration leading to human planetary travel in the 1980s. Then as he would repeat at future STAC meetings, Mueller described how the Apollo Extension Systems would use surplus Apollo equipment. And with production of Apollo related hardware exceeding the needs of the "basic" Apollo Program, STAC supported the proposal for a post-Apollo program to use that excess equipment. However, STAC member Harry H. Hess, who was chairman of the Space Science Board, said the science board had considered the question of what NASA should do after Apollo and had concluded that scientific planetary investigation by automated probes ought to become the primary goal; and only later should the agency use astronauts for space exploration. Acknowledging that NASA could not change objectives for the Apollo Program, Hess argued that lunar exploration should continue after the initial Moon landings, although only as a secondary goal. Mueller tried to convince Hess of the value of human spaceflight, but he remained unconvinced and opposition from the Space Science Board would continue to pose an obstacle to Mueller in his quest to develop post-Apollo plans.[35]

IV

Mueller set out to improve the relationship between NASA and its contractors in order to overcome the adversarial feelings that had built up because contracts were

---

[34] Mueller Interviews: Putnam, 10/4/66 and Slotkin, 2/25/10.
[35] Minutes of the Third Meeting of STAC, JPL, Pasadena, CA, 10/29-30/64.

running late and over budget. Government contracting officers tried to enforce the letter of contracts, while the contractors argued they did not know what the agency wanted. NASA's procurement people blamed the contractors for overrunning, but Mueller blamed the overruns on lack of definition before the contracting process began. He considered this a joint responsibility which had to be addressed together, first by defining the program more clearly, then by renegotiating the contracts and establishing proper communications between the contractor's top management and NASA management. He found "a hodgepodge" of different kinds of contracts. The agency had poorly estimated contract costs, and some contractors complicated this by submitting low-ball bids in the expectation of getting well through changes. Mueller wanted to restructure the contracts on a rational basis, and change to incentive type contracts in order to better align the interests of the agency with the contractors. And it took months to rewrite specifications and work out the details of the new contracts. As a first step, NASA established incentives based on schedule and cost – the two things that both sides agreed on while preparing new specifications. However, a program the size of Apollo required the ability to change and the agency provided for that in the new contracts. Mueller considered this important because, he said, you "can't make a contract at the beginning when [for example] you aren't sure what kind of insulation you're going to use in the second stage [of the launch vehicle] that defines exactly what that's going to cost and how you're going to go about doing it. So you need to preserve flexibility." He introduced incentive fee contracts, calling them "effective in almost every instance because it got management's attention." He considered cost plus incentive fee contracting to be the most effective way to procure R&D. However, he had to keep incentives simple and precise to insure the agency could measure and incentivize the proper objectives. Using these contracts as a means of communication not only focused the contractor's attention, it made NASA focus on what it wanted to incentivize.[36]

As Gemini remained behind schedule, Mueller called it "the usual kind of a schedule that NASA had been used to." And when it looked like Gemini might overlap the Saturn test schedule, he decided he would not to permit it. At a Gemini meeting he announced he would not extend Gemini past the first Saturn IB launch date, saying "however many flights we got off is how many flights there would be in the program." Then in November he called Mr. Mac to tell him that NASA and his company were wide apart in contract negotiation. He laid it on the line, and said, "We can't spend any more money on this program and that comes from the Bureau of the Budget. Basically, all we can do is chop off the program earlier and earlier so that we live within the total money made available to us. We have $1.3 billion ... to carry it out. The problem is that our original estimation was $540 million." When McDonnell asked how much money would be for his company, Mueller said "$770 million – somewhere around there," everything included. McDonnell responded, "I can tell you that we will take that figure and do our best to work something out as

36   Mueller Interviews: NASM, 5/2/88, 6/22/88; FAR, subpart 16.4; "Apollo Program Oral History," 7/21/89, NASA Archives, RN 18924.

regards our work to get it done within that." They discussed canceling hardware and eliminating duplication at the Cape. The CEO said that his company could save money by reducing spare parts and Mueller replied, "I think it will take creative work on both our parts." Although he would not eliminate hardware without further study, he did say "I would be enthusiastic in terms of cutting down spares ... I'm in favor of hardware and in favor of testing at the factory rather than at the Cape." McDonnell asked if he could bid on Apollo work, but Mueller responded "I have been all up and down the line on that, including, Webb, and that answer is 'no.'" He said that Low would visit St. Louis to conclude the negotiations, and then pointed out, "I place first emphasis on the schedule because I think that's the only way you save on cost on a program like this. Whatever we do I would like to find a way of creating the situation where we can in fact make it worthwhile to McDonnell in terms of profits to better this cost ceiling ... I am for the least total cost. The more profit you make [the better] so long as we can get the total costs down." McDonnell then observed, "In the free enterprise system I find people are still receptive to this kind of thing." Mueller responded, "I will back this 100 percent." In other words, McDonnell concluded, you are giving us "more profit or earnings for reduced overall cost to NASA." Before the call, Mueller had met with the Apollo program control staff, who told him, "We feel MAC's estimate is greatly inflated probably because of contingencies and also for negotiating purposes ... We feel that $740 M[illion] is ample to do the program and besides that's all there is." The contract was finalized on December 18 with a cost base of $712 million and an incentive fee of four to eight percent. It was approved by Mueller on January 28, 1965. And after converting the Gemini contracts to cost plus incentive fee, he said "the schedule [is] accelerating and costs are down."[37]

During a two week period in the fall of 1964, Mueller made a series of speeches defending the space program and discussing the future of space exploration. He said that landing on the Moon was not NASA's first priority. "There is a considerable degree of misunderstanding of the overall purposes of our national space program ... Many people seem to believe that a landing on the Moon is our paramount objective. This is not so ... Our principal goal is to make the United States first in space." The Apollo mission milestones, he explained, provided focus for the development of a capability to enable the US to compete in the space race, and provide knowledge to manage the largest R&D program ever undertaken. He pointed out that more than 250,000 people worked in the US space program, and said the knowledge they gained would be applicable elsewhere. Direct benefit of this investment included the Saturn IB, with the power to place a basic space station in low Earth orbit, and by adding a third stage it could land instruments on the Moon, or send a scientific spacecraft to Mars. The Saturn V provided "a quantum jump in

---

[37] Mueller Interview, JSC, 1/20/99; Phone conversation with McDonnell and GEM, 11/27/64 4:04 pm and Memorandum from J. Fried to Mueller, 11/27/64 12:45 pm, Notes & Transcriptions, GEM-79-12; Hacker & Grimwood, *On the Shoulders of Titans*, 10-3; Mueller, AASA, Atlantic City, NJ, 2/12/66.

launch vehicle capability," and could put a large space station in orbit or launch a space probe "to the depths of the solar system." He called Apollo the precursor to a true space station, and the lunar excursion module a test bed which could be used to "evaluate the feasibility, and demonstrate the solution of most of the operations required for space exploration of very long duration." But because of the space program's long term nature, he added, it is "difficult to maintain in the public mind the perspective that was established with the national decision in 1961, to undertake Project Apollo." Then, referring to the flight of a crew of three cosmonauts on a one day mission in the first Voskhod spacecraft in October, he pointed out, because of the time between Mercury and Gemini, new Soviet space feats clouded "the national perspective with envy for the most recent Russian accomplishment." He expressed concern about what the USSR might do next, believing that they could certainly attempt a circumlunar mission "because they had the capability of doing that."[38]

This would not be the last time that Mueller spoke of extended planetary missions. According to OMSF staffer Jay Holmes, "Mueller was a protagonist of going to Mars; always had been … In the course of our thinking about what the post-Apollo might be, Mueller would say, 'Mars is where we want to go,' [and] if we do a good job on this lunar program, maybe we [will] get a chance to go to Mars afterwards." His "eyes would light up with enthusiasm," Holmes recalled. And when someone suggested a piloted Mars flyby for the 1970s, Mueller had Bellcomm study it; and his staff churned out hundreds of possible post-Apollo scenarios.[39]

When Mueller spoke to several groups in early November, he observed that the US space program would provide the "capability required for exploration of space … [but] it must be recognized that there is no such thing as <u>unmanned</u> exploration of space. The only question regarding man is his location." He called an astronaut a "light weight, versatile and mobile computer with intelligent judgment," with the ability to anticipate events and make decisions. Humans could translate data into information, and correlate, systematize, and recognize patterns. They could also determine the information needed; their ability to evaluate was indispensable, and they could deal with the unexpected. At the highest level, humans could act as scientific investigators, responding "creatively to unexpected situations, postulating theories and hypotheses, and devising and initiating systematic measurements." Astronauts, he said, "can operate in an un-programmed situation and reap full benefits of the true objective of manned operation – that is exploration of the unknown." Thus, he posited, "As a sensor, unattended instruments are satisfactory for most purposes. For manipulation, automatic machines are difficult but not impossible to construct. For evaluation, man is essential. For investigation, he must be a scientist." However, scientific investigation would not be the only reason for human space exploration. Humans could assist with practical applications, where

---

[38]   "Man's Role in Man-Machine System in Space," Mueller, AIAA Manned Space Flight Meeting, Houston, TX, 11/4/64; Mueller, NAREB, Los Angeles, CA, 11/12/64; Mueller Interview, Slotkin, 2/25/10; Voskhod 1, NSSDC, NASA.

[39]   Holmes (Jay) Interview, Sherrod, 6/28/72.

their ability allowed for the evaluation and investigation of equipment malfunctions. And anticipating some of the arguments that would surround future satellite repair missions, he raised the question of whether the repair would be worth the expense of transporting astronauts to perform the work, because placing humans in space came at a high price. In near space, machines could do sensing and manipulation, but as exploration ranged deeper into the solar system the speed of light placed limitations on the ability to manipulate machines on a real time basis by remote control. Then citing a Space Science Board study, he declared "the very technical problems of control at very great distances, involving substantial time delays in command signal reception, may make perfection of planetary experiments impossible without manned controls on the vehicles."[40]

Mueller said "accomplishment of manned lunar flight and safe return within this decade is feasible for the United States and there is a good chance that, with sufficient funds, this can be accomplished prior to its accomplishment by the Soviets." The basic Apollo mission would last for two weeks, though he claimed it would be easy to extend these missions by months, and it would "not be difficult to modify the LEM to provide for the possibility of extending man's stay time on the Moon beyond the one to two days contemplated in early missions." He rejected the argument that human spaceflight took engineers and scientist away from more productive endeavors, claiming instead that it absorbed excess workers from declining missile development projects. So the space program did not conflict with other efforts that required engineers and scientists. To the contrary, he said, "it contributes to the fundamental solution of these problems by bringing about giant steps in economic and technological development, giving people the opportunity to help themselves through new economic activities."[41]

In late 1964, Mueller continued to emphasize the importance of crew safety, which he said was evident by the effort undertaken to rate Gemini's launch vehicle for human flight at a cost about four times that of the production Titan II. The Apollo spacecraft, the Saturn IB and the Saturn V were in testing and production, and the design of the lunar excursion module was essentially complete and production tests were underway. NASA had twenty-eight astronauts in training, and together with the National Academy of Sciences, was planning to select a class of scientist-astronauts. He continued to argue that although "the pace of the manned flight program is brisk, it is not a 'crash' program," and insisted that the Apollo development schedule was conservative compared to other large national R&D programs. Yet, despite fears in the medical community, he said there was no evidence to show that well-conditioned and well trained astronauts would suffer ill effects from flights of up to two weeks. Nonetheless, he cautioned, "we must face even the sad prospect of possible fatalities in the manned flight program ... [And

---

[40] Mueller, National Editorial Writers Conference, Cocoa Beach, FL, 11/13/64; "Mans Role in Man-Machine System in Space," Mueller, AIAA Manned Space Flight Meeting, Houston, TX, 11/4/64; Compton and Benson, *Living and Working in Space,* 68.

[41] Mueller before the NAREB, Los Angeles, CA, 11/12/64.

b]ecause our entire program is conducted in the open and is reported day-by-day, development-by-development, our difficulties and failures receive as much or more attention than our successes."[42]

NASA struggled to control cost at the end of 1964, while the administration asked for additional budget cuts in order to pursue other national priorities. As he planned the FY 1966 budget, Webb assumed that the agency would receive between $150 and $250 million less than in 1965. Mueller considered several radical options, including canceling Gemini after the first piloted flight, eliminating the Saturn IB or even the lunar excursion module, and also looked at the impact of rescheduling the first lunar landing. He considered these options drastic, but planning for radical contingencies became a regular exercise at NASA. As the end of 1964 approached, it looked like 1965 would not go well for the US space program. There were "all sorts of things going around Washington at that time," Mueller remembered, "and the whole Apollo Program was up for grabs ... it was a favorite whipping boy." However, after NASA began to show the ability to control its budget and make progress, some of this uncertainty began to fade away. As he recalled, "the bean counters were always willing to delay or otherwise change things but fortunately we had a dead president with a commitment and when the chips were down why they finally decided we're 'gonna' do it." In his quest for support, Mueller used Kennedy's lunar landing goal to focus attention on human spaceflight, taking advantage of the "definite end date and in our case a definite number of dollars we had available to ... work with."[43]

V

Following weather delays and mechanical problems, GT-2 stood ready for launch on December 8. The countdown went smoothly, and just before noon the first stage engines ignited – only to shut down one second later, leaving the fully armed and fueled booster standing dangerously on the launch pad. However, the launch crew carefully made it safe. An investigation discovered that the malfunction detection system had stopped the engines following a drop in hydraulic pressure caused by the failure of what turned out to be an unneeded improvement. During development, the walls of the engine servo-valve housing had been thinned in order to achieve a minor weight saving, and this led to the engine failure. The production version of the Titan II experienced overpressures without incident. This proved one of the points made in Mueller's address to MSC's senior management about "nice to have" changes. The agency then set about fixing the problem that had caused the overpressure.[44]

---

[42]   Mueller, Colloquium, University of California, Berkley, CA, 3/16/65; Mueller, National Editorial Writers Conference, Cocoa Beach, FL, 11/13/64.

[43]   Webb to the President, 11/30/64, and Phillips to Seamans, 12/7/64, SCP-41-2; Mueller Interview, Slotkin, 9/9/09 (second session), 2/24-25/10.

[44]   Hacker & Grimwood, *On the Shoulders of Titans*, 9-5.

   Mueller frequently spoke about the role human spaceflight played in the nation's defense, although never as extensively as in early 1965 when he said, "Our national security, our ability to lead the Free World, and our future economic growth are bound upon the efforts which we expend to excel in the acquisition of new scientific knowledge and the development of new technology." He said that NASA's appropriations were a part of a national security package. Along with the Atomic Energy Commission and DOD, the civilian space agency's efforts provided essential technology to defend the nation. However, while money for defense related hardware and maintenance of forces decreased, defense R&D funds increased, which included NASA's budget. "Viewed in these terms," he claimed, investment in space R&D "is not an increase in federal expenditures, but rather a reallocation of expenditures within the national security package." Again appealing for support for the agency budget, he quoted Winston S. Churchill, who said during the Battle of Britain, "Give us the tools, and we will finish the job." Then in the spring of 1965 after a bruising budget battle, Congress appropriated $5.175 billion for NASA in FY 1966, about 1.6 percent less than the administration's request and only 1.4 percent below the 1965 appropriation. The budget reduction came to $75 million, which had little impact on the Apollo Program.[45]

---

[45]   Mueller, Technology Club of Syracuse, Syracuse, NY, 2/1/65; Dick, *Space Age*, 363.

4

Resuming flight

Kennedy Space Center: the "new port of destiny."
Mueller, March 12, 1965

As Apollo-Saturn development continued, human spaceflight resumed aboard the Gemini-Titan. During the first nine months of 1965 the agency flew three piloted Gemini missions, demonstrating spacecraft maneuverability, and Edward H. White II became the first American to perform an extravehicular activity in space. However, the Soviet space program conducted unannounced flights just before scheduled NASA missions, and behind the scenes the Apollo Program experienced significant schedule delays and cost overruns. Nonetheless, Mueller continued to promote the post-Apollo program as a means of maintaining the technical capability that was being built for Apollo.

Discussing his testing philosophy for Apollo in January 1965, Mueller said with larger space vehicles, test costs increased to the point where NASA had to reduce launches to save money. Augmenting flight testing with ground testing had been his philosophy since he managed the Able spacecraft program (1958-1960) at Space Technology Laboratories, where he learned the value of thorough ground testing prior to flight. He planned to combine increased ground testing with all-up flight testing in order to maximize the amount of information obtained from each flight. Beginning with individual parts, testing moved to components, subsystems and complete systems, then ground testing the entire vehicle to the maximum extent possible, with R&QA extending throughout. NASA ground tested Apollo subsystems in 1964, and planned to conduct complete system tests in 1965 and flight tests in 1966. The schedule called for human spaceflight aboard the Saturn IB to begin in 1967, followed by piloted flights on the Saturn V commencing in 1968. With the support of Congress, he said, "these schedules will be met and our goal of a manned lunar landing in this decade can be accomplished."[1]

Meanwhile, NAA's space division continued to suffer problems in building the

---

[1]   Mueller, Joint Technical Society Meeting, Columbus, OH, 1/21/65.

command and service modules of the spacecraft and the S-II stage of the launch vehicle, while Rocketdyne made good progress with the Saturn's engines. Although the F-1 engine schedule remained tight, development continued without significant issues after overcoming earlier combustion instability problems. Some of the engines that Rocketdyne built for the Apollo Program experienced schedule slips, but these were considered minor compared to the issues experienced by the space division, as Mueller prepared to meet with Atwood again in January 1965 to discuss why the spacecraft development costs were escalating. Shea identified problem areas, though he said cooperation between NASA and NAA remained good. Myers had a "good management hold" on the spacecraft, but testing still left much to be desired, and schedule problems raised questions about the space division's ability to handle the workload.[2]

Apollo flight tests had not begun, though Mueller was already looking beyond them and the subsequent lunar landings to the post-Apollo period. Of like mind with Webb and Seamans when it came to the Apollo follow-on program, the agency still had to convince Congress, the administration, and the scientific community of its importance at a time when other urgent priorities demanded attention. This would not be easy, but Mueller argued that Apollo and Saturn could be used to expand scientific exploration of the Moon and the planets without significant further development cost, thus increasing the return on the existing investment in human spaceflight. However, even some scientists supporting human spaceflight criticized the Apollo Extension Systems as unimaginative. They thought serious scientific work required scientist-astronauts, not pilots with rudimentary scientific training. Nonetheless, Mueller had high hopes for the post-Apollo program, and considered both Gemini and Apollo to be building blocks for an orbiting research laboratory, a precursor to an orbital workshop; although further steps in space exploration depended on what the agency learned from them. He planned future spaceflights, including additional missions to explore the Moon, extended operations in low Earth orbit, "or off to the planets," as he put it. The Apollo Extension Systems would exploit existing hardware, and studies showed that a number of different missions could be accomplished using equipment developed for the Apollo Program. Extending lunar exploration could include lunar orbital flights, and NASA had a long list of experiments to carry out in lunar orbit. Bellcomm looked at possible lunar surface investigations, using the Moon as a base for studies of the Earth, and confirmed the feasibility of reaping "additional rewards" from Apollo, according to Mueller.[3]

With its engines repaired, on January 14, GT-2 completed its preflight tests and was prepared for another liftoff attempt. This time nothing interfered, and on the morning of January 19 it was launched after an almost flawless countdown. It was a

---

[2] "Unnamed" and "Summary of NAA Schedule Trends," SCP-49-1; Shea to Mueller, 1/9/65, SCP-49-1.

[3] Lambright, *Powering Apollo*, 139; Compton, *Where No Man Has Gone Before*, 65; Mueller, Colloquium, UC, Berkeley, CA, 3/16/65.

suborbital mission to test the spacecraft structure and heat shield during re-entry over the South Atlantic. Based on this success, NASA declared Gemini operational, and announced that the next flight, scheduled in the spring, would carry astronauts, and that a longer duration mission would follow in the summer. With GT-2 finally out of the way, the launch crew completed testing the Gemini 3 spacecraft and mated it to its Titan booster; and by mid-February GT-3 stood erect, ready for final testing.[4]

In March, as NASA prepared to resume human spaceflight, Mueller said the "eyes of the world" would focus on the Cape "to watch the beginning of new voyages of space exploration." Over the next two years, the agency was planning to launch ten Gemini spacecraft, at the rate of one every two or three months. NASA scheduled the next mission to last about a week, to test rendezvous and assess the astronauts' ability to perform useful work outside the spacecraft. Between the government, industry and universities, about 300,000 people worked in the US space program, with most of them involved with Apollo. And Mueller claimed that "coordinating all this activity and translating it into finished space vehicles on the launch pad, ready to take explorers to the Moon, is probably the greatest single engineering and management task the human race has ever undertaken." But half a decade remained to achieve the goal. He said, "It will be an exciting thing to watch this great engineering enterprise move forward, not in secret but before the eyes of the world." Then he called KSC the "new port of destiny," which "must be an open port – a port that adequately tells the story of a free people's desire to enter and use space ... 'as an avenue toward peace.' "[5]

When Mueller testified before the Senate space committee later that month, Senator Margaret Chase Smith asked him about the differences between the Manned Orbiting Laboratory and Apollo Extension Systems. He explained that they differed mainly in the types of experiments they would conduct: DOD focused on military applications, while NASA had more interest in science and technology. Although the civilian space agency closely coordinated and shared information with DOD, they had very different objectives. He considered the MOL important, and supported it as a national program, not as competition, while the press questioned whether it would duplicate parts of NASA's post-Apollo program. On the other hand, National Reconnaissance Committee chair Edwin H. Land believed, "you can't do anything from MOL that you can't do from a satellite." Mueller in turn pointed out that "you can't do anything from a satellite that you haven't already thought through and developed. So the MOL had the advantage ... of free will, which would have made it in the long run a substantial difference in how you approached the reconnaissance that you were trying to set up with the space platforms." He made similar arguments

---

[4]   Hacker & Grimwood, *On the Shoulders of Titans*, sections 9-5 and 10-3; Mueller, Technology Club of Syracuse, Syracuse, NY, 2/1/65. The first three Gemini missions were assigned Arabic numerals, but thereafter Roman numerals were used. The GT numbers were not affected, so GT-4 was the Gemini IV mission.

[5]   Mueller, Greater Titusville Chamber of Commerce, Cocoa Beach, FL, 3/12/65.

about piloted versus robotic spacecraft, recalling that Newell had said "you can do anything with our satellites including landing on the Moon and doing everything you needed to do there." In one sense, Mueller said, "he was right and in another sense he was wrong ... you can do anything you thought about doing but you can't do anything you haven't thought about doing, whereas the manned area allows you to do the unexpected and exploit the things that you really hadn't thought about before."[6]

Meanwhile, Gemini 3 went through final ground simulation tests in mid-March, and after a flight readiness review Mathews committed GT-3 to launch. Then after fixing some minor technical problems, the Titan II lifted astronauts Virgil I. Grissom and John W. Young into orbit on March 23. Following the practice in the Mercury Program, Grissom named the spacecraft *Molly Brown*. And while the astronauts considered this a cute name, it did not amuse Mueller, who ordered future Gemini flights to be designated only by their numbers in the sequence. (During Apollo the astronauts returned to naming the spacecraft if there was a requirement for individual radio call signs.) Once in orbit, Grissom tried to perform a biological experiment, but it failed when he twisted the handle too hard and broke it off. Young successfully performed another test. Then they conducted the first in-space maneuver to change their orbit, demonstrating the ability to make small, precise changes in speed such as would be required for rendezvous and docking in orbit.[7]

After three orbits the spacecraft landed short of the intended splashdown point because its theoretical lift characteristics during re-entry proved inaccurate. Shortly after landing, Mueller said that the "technical factors which caused the spacecraft to land short of the carrier have been identified and will be corrected." The agency called the mission a success, and declared Gemini-Titan operational. Mueller wrote, "in spite of the aerodynamic anomalies which caused an up-range landing, the GT-3 spacecraft was sighted by recovery aircraft only nine minutes after splashing into the Atlantic." And as he later explained, "All major objectives of the first three Gemini flights having been met, the Gemini system was determined to be qualified for the advancement of US manned space flight." He called the human factor in spaceflight of great import, and the most problematical, because "human response to prolonged space flight is unknown." Determining whether astronauts could endure two weeks in space was a key objective, and proving the ability to rendezvous and dock continued to be necessary in order to validate the lunar orbit rendezvous mode of landing on the Moon.[8]

Ever the spoilers, the USSR launched the second Voskhod spacecraft five days

---

[6]   NASA Authorization for FY 1966, GEM-45-3; Mueller Interview, Slotkin, 2/24/10.
[7]   Logsdon, *Exploring the Unknown, Volume VII*, 207; Hacker & Grimwood, *On the Shoulders of Titans*, 10-5; Mueller, School of Electrical Engineering, Purdue University, Lafayette, IN, 4/22/65.
[8]   Ibid.; Hacker & Grimwood, *On the Shoulders of Titans*, 10-5; "Gemini Comes of Age," Writings, 1964-1969, GEM-110-9; Mueller, Lecture II, University of Sydney, Australia, 1/10/67.

before Gemini 3, and Aleksey A. Leonov became the first human to perform an extravehicular activity in space. "It was an amazing feat," MSC's Christopher C. Kraft, Jr. recalled, and the "Press and media and US politicians reacted with shock and dismay that the Russians were once again proving to be superior in space." Because of Leonov's EVA the media complained that the US remained behind in the race to the Moon. Like the initial reaction to Sputnik, NASA officials could not understand the public reaction; although even they were taken by surprise when the cosmonaut performed his EVA. However, as Jay Holmes said, "those close to the NASA program considered EVA a rather modest extension of capability and therefore of modest significance." Nonetheless, the press and public considered the Soviet space feat to be a major concern. Mueller later said that while the Soviet Union always seemed to come up with a surprise right before a NASA mission, "We anticipated that and we had enough knowledge to know that they could, but not enough knowledge to be willing to explain that we did know. At that time this whole surveillance satellite system was very, very closely held, so not very many people knew it existed, much less what it was producing. And we did know at NASA ... so we watched them building their complexes and could pretty well guess what they were going to be able to do." However, Mueller said, "They took more risks ... And that paid off" in propaganda value for them. But he added "we took a fair amount of risks too."[9]

Mueller addressed a space congress at Cocoa Beach in early April, calling Gemini 3 a success and saying that the space program had "entered a period in which successful performance ... is expected as a matter of course ... We have every reason to believe that the United States maintains its lead in space science ... " It was clearly in the lead in applications. He observed that the Saturn I had a lift capability of about twice the Voskhod launch vehicle. Nonetheless, the Soviets led in human spaceflight operations by a considerable margin. Yet Gemini 3 performed the first spacecraft maneuvers, showing the ability to change orbits, a key program objective and one that was essential for Apollo. The Soviets may have additional surprises, but the US would show leadership in spaceflight with the Apollo Program. And Apollo would not be the end of human spaceflight; it was an "intermediate objective," he said. At the press conference following this speech, he said the nation should "be prepared to lose a pair of US astronauts on a space mission because of accidents." Interestingly, the media did not latch on to this statement, as they later exploited a similar remark he made about space rescue.[10]

Discussing GT-3 at another press conference, a journalist asked about problems with the experiment that Grissom had broken, and Mueller pointed to the need for better review and integration of experiments before flight. "The lesson to be drawn,"

[9] Seamans, *Project Apollo*, 59-60; Kraft, *Flight*, 214; Hacker & Grimwood, *On the Shoulders of Titans*, 11-1; Holmes to file, Arthur C. Clarke, 6/29/65, GEM-67-4; Mueller Interview, Slotkin, 2/24/10.

[10] Mueller, Space Congress, Cocoa Beach, FL, 4/6/65; *Daytona Beach Journal*, 4/5/65, GEM-83-4.

he said, "is that equivalent emphasis must be put on the review and design of experimental equipment and its compatibility with the spacecraft and conditions of space flight," as is put on any other aspect. And NASA should define experiments and the required equipment earlier than in the past. He said the schedule called for GT-4 to fly in the third quarter of the year, adding "we are working awfully hard to see if we can advance that date." Asked about plans for an EVA on Gemini IV, he said that NASA planned one on Gemini VI, but added "[we] are looking into the possibility of accelerating the application of these experiments." With the possibility of an EVA, journalists asked numerous questions about what the astronauts might do. When he was asked about the possibility of a rendezvous on GT-4, Mueller stated an emphatic no.[11]

Looking back across almost fifty years, Mueller said "the press by and large spends all of its time trying to unearth problems in any of the programs, so they have something to write about ... [T]he press was mostly concerned about what we were doing wrong and hardly ever spent any time ... [with] what we were doing right." And "it was not a very happy relationship." Reporters focused on problems and tried to entrap him into saying something that would make headlines. He described press relations as adversarial, and "We've never quite ... built the kind of cooperative relationship with the media that would be most productive" and "maybe it isn't possible." Consequently, "in public office your relationship with the media is one of essentially survival." You need to "stay away from them as much as possible and keep them as far away from you as you can." Julian W. Scheer, who ran NASA's Public Information Office, assigned Alfred P. Alibrando to OMSF to interface with the news media. Mueller felt comfortable with this arrangement because, he said, "the less we had ... to do with the media, the less problems we got into." NASA ran Apollo as an open program, and the press had access to everything. The agency had few classified papers, and NASA management frequently talked to reporters. Mueller met with editorial boards, gave interviews and spoke at press conferences; though he claimed he did not seek personal publicity. Nonetheless, except for Webb and some of the better known astronauts, he had the highest number of mentions in the print media of all agency officials between 1963 and 1969.[12]

II

Mueller spoke to the scientists attending NASA's Lunar Exploration Symposium in late April. The agency asked for recommendations about establishing a long-range scientific program for human exploration of the Moon, and it established working

---

[11]   Mueller, News Conference, MSC, Houston, TX, 4/19/65; Mueller, Lunar Exploration Symposium, MSFC, Huntsville, AL, 4/26/65.

[12]   Mueller Interviews: Slotkin, 9/9/09 (second session) and 2/22/11, and NASM, 6/22/88; Google News Archives, 1963-1969, news.google.com.

groups to operate throughout the Apollo Program. These groups would advise the agency about issues such as the optimum geological traverse on the lunar surface, how and where returned samples should be analyzed, and which scientists should participate in the analyses. Following the initial meeting, the working groups held summer workshops at Woods Hole, Massachusetts under the sponsorship of the Space Science Board. While Mueller spoke about the initial exploration of the Moon at this first meeting, he looked beyond Apollo, as he always did, and predicted the momentum of advanced missions would soon match on-going programs. And he said studies pointed to the Apollo Extension Systems "as the logical means of extending the capabilities inherent in the Apollo Program," noting that the FY 1966 budget contained $48 million for advanced studies. With equipment designed for the Apollo Program, he added, "it appears quite possible to carry out many types of experiments in missions lasting months instead of weeks."[13]

Around this time, Mueller established the Manned Space Flight Experiments Board to coordinate experiments with the twin goals of selecting the most meaningful and feasible ones. After the appropriate project office had reviewed an experiment's feasibility, the new board would prepare its recommendations. And once selected, a principal scientist would develop the experiment, qualify it for integration into the spacecraft, and train the astronauts. The board would concern itself with operational feasibility, leaving it to an experiment's sponsors to determine its scientific merit. As chair of the experiments board, Mueller only decided which experiment flew in the case of a disagreement. However, space historian W. David Compton wrote, "Not everyone was happy with the new experiments board; some space science officials protested that Mueller usurped their prerogatives to select and evaluate experiments," although he saw his role as insuring compatibility of experiments with the spacecraft and the operation of the mission.[14]

A few days after the lunar science meeting, Mueller spoke to the space medicine branch of the Aerospace Medical Association, where he raised questions about a human's ability to survive and operate in space. He asked rhetorically if it would be possible to maintain and enhance the astronaut's contribution to the "man-machine system," and assured them he was "optimistic on all counts." As the pace of human spaceflight increased, experimentation during Gemini would be increased to include preliminary evaluation of the most critical medical areas involving the cardiovascular and musculoskeletal systems. Three of eleven experiments on Gemini IV would measure the impact on the human heart, muscles and bones, and NASA planned to repeat these experiments on subsequent flights to gain a broader base of data. The Manned Space Flight Experiments Board had already approved sixteen medical experiments designed to learn more about how extended periods of weightlessness

---

[13]   Mueller, Lunar Exploration Symposium, MSFC, Huntsville, AL, 4/26/65.

[14]   Compton, *Where No Man Has Gone Before*, 25; "Earth Orbital Mission Program," undated, GEM-72-4-5.

affected the human body, and to assess measures to overcome its effects. Both NASA and the air force were investigating ways to extend human spaceflight, and many of these studies were biomedical or behavioral in nature. In conclusion, he told the physicians, "Space research has already provided positive results of value to medicine on Earth and there is clear evidence of rapid increase in the volume and significance of such results."[15]

In mid-May, Phillips sent a member of his staff to review the implementation of configuration management at the centers. And in a detailed report, he learned that MSC still did not have a single fully qualified person, although several people were assigned to configuration management on a part time basis. Specifications were not under configuration management, contractors received waivers from filing plans, and MSC had no plans to hire any full time configuration management personnel. MSFC had similar issues, though the report said von Braun at least recognized the problem, had asked for help, and was planning to hire some qualified people. KSC had not yet implemented configuration management, one partially qualified person was working on it in a staff position, and Debus did not want any help from headquarters in this area. Because the centers did not take configuration management seriously, neither did their contractors. And the report predicted, "It will take approximately six months to implement CM across the Apollo Program and then an additional 12 to 18 months for its full benefits and results to become evident." In other words, configuration management would not be fully implemented in the Apollo Program until the end of 1965, and perhaps not until mid-1967 would its full impact be felt.[16]

Meanwhile, a debate about performing an EVA on Gemini IV took place. When Mueller learned of MSC's plans for a possible EVA he gave the idea a lukewarm response and offered no encouragement, but neither did he stop planning or training for it. Then in mid-May, Gilruth demonstrated EVA for Seamans, and Mathews flew to headquarters to convince Mueller to add it to the mission. Mueller wanted to know how they could move the EVA officially scheduled for Gemini VI forward to Gemini IV. But Mathews told him they were ready, and "all EVA gear had been qualified, or nearly so, and the crew trained." With no apparent reason not to, Mueller and Seamans approved it, and after much internal debate the administrator also approved, though Dryden saw it as merely a stunt to match what the Soviets had done.[17]

As preparation for GT-4 proceeded, fuel cell development problems plagued the program. Until this new electrical power system became available the duration of a mission would be limited by the capacity of chemical storage batteries, which was about four days. Originally scheduled for the third quarter, Mueller announced that

[15]   Mueller, Space Medicine Branch, Aerospace Medical Association, NY, NY, 4/28/65.

[16]   "Trip Report – CM [configuration management] Assessment and Recommendations, Lt. Col. Taylor, 5/14/65, SCP-50-1.

[17]   Mueller Interview, Slotkin, 9/10/09; Hacker & Grimwood, *On the Shoulders of Titans*, 11-1; Seamans *Apollo*, 60.

because of the successful flight of Gemini 3 and progress in ground testing there was "every indication that we have slipped ahead of schedule on the launch of Gemini IV." He considered rendezvous and docking to be more important than EVA, because if the astronauts could not do the former there would be no chance to do the latter. Nonetheless, his philosophy remained that they should do as much as possible as soon as possible, and since they were not ready to rendezvous, it made sense to try EVA. Then on May 25, NASA announced, "White would step into space on the next Gemini flight and use a 'zip gun' to propel himself." And as Dryden expected, the press and public assumed the agency had added the EVA merely to keep up with the Soviets.[18]

When speaking before a group of materials engineers at the end of May, Mueller said that during 1965 the test program would include about 4,300 "vigorous" qualification tests of major flight components, each of which had to be successfully completed before the space vehicle could move to system testing. NASA planned about 1,600 qualification tests of Saturn components, and 1,000 for the Apollo spacecraft. And, "Failure anywhere along the line has the potential of causing a schedule delay." Thus, the timetable depended on the successful completion of literally thousands of individual tests and measurements of materials, components and assemblies leading to qualification testing of the subsystems and ground support equipment.[19]

Then on the morning of June 3, millions of people around the world watched GT-4 rise from its launch pad at KSC, and for the first time live TV covered the event via the Early Bird satellite that had been placed into geostationary orbit over the Atlantic a few weeks earlier. Because of the prospect of an EVA, the press and public showed greater interest in this mission, and the number of reporters registered to cover the flight from MSC's newly opened news center exceeded its capacity. Once in orbit, the astronauts tried to maneuver Gemini IV closer to the Titan booster's second stage to simulate the final phase of a rendezvous. But the astronauts did not have a good feel for the orbital mechanics involved, despite extensive training and simulations on the ground. Using thrusters to "catch" the booster had the counterintuitive effect of opening rather than closing the separation because the additional thrust placed the spacecraft into a higher and therefore slower orbit. After using half their fuel they gave up on this station-keeping effort. And as Mueller later said, "when you have a new vehicle the response is different and you have to learn how to calibrate the responses ... And that took a bit of doing."[20]

---

[18]   Mueller, Aerospace Fluid Power Systems & Equipment Conference, Los Angeles, CA, 5/19/65; Mueller, Conference on PUOS, St. Louis, MO, 5/26/65; Mueller Interview, Slotkin, 9/9/09 (second session); Hacker & Grimwood, *On the Shoulders of Titans*, 11-1.

[19]   Mueller, Society of Aerospace Material and Process Engineers, San Francisco, CA, 5/27/65.

[20]   Hacker & Grimwood, *On the Shoulders of Titans*, 11-2; Mueller Interview, Slotkin, 9/10/09.

4-1   Edward H. White II performing America's first EVA, Gemini IV, June 3, 1965. (NASA photo)

Next, the crew prepared for the EVA, opened the hatch, and White ventured out. While floating freely about fifteen feet from the spacecraft, he paid no attention to time as he spoke about the spectacular view while restrained by a twenty-five foot tether with a hose supplying oxygen. Because of the design of the communications system, when his partner James A. McDivitt spoke to White, Houston could not. It was sixteen minutes before McDivitt checked with mission control and they could tell him to get White back inside the capsule, and another seven minutes before White complied. Without specific tasks to perform, moving about in zero-gravity appeared easy. When attempting to close the hatch, the astronauts found it hard to shut, and after a struggle White settled physically exhausted in his seat. NASA later concluded that gases escaping from materials inside the cabin had created a pressure which made the hatch difficult to close, and resolved the problem by developing an elaborate two hatch system for Apollo.[21]

Because of an onboard computer failure, the spacecraft re-entered the Earth's atmosphere manually, splashing down after a flight lasting almost ninety-eight hours. Medical examinations showed no serious after effects of weightlessness, although both astronauts lost weight, bone mass, and blood plasma. At the post-flight press conference, Mueller called Gemini IV "this country's most successful mission," and the *St. Louis Globe-Democrat* wrote, "Dr. Mueller closed the happy meeting with an announcement he still believed the United States could land a man on the Moon by 1969 ... [although w]hether we get there first or not depends on what other people are doing." Following Gemini IV, Webb sent a "Dear George" letter which said, "As we see the increasing evidence of the correctness of your decisions with respect to Gemini and particularly GT-4, this is just to let you know how much Dryden, Seamans and I appreciate your thoroughness, your dedication, and the strength you add to this team." The administrator was satisfied with his decision to hire Mueller.[22]

Returning to Washington, Mueller told the National Space Club that it would take "a great deal of effort and a number of years for the United States to achieve first place in space," and Gemini IV did not overcome the Soviet lead of several years. Despite this success, it remained too early to move the date of the lunar landing forward. However, the *New York Times* quoted Shea saying, "a lunar landing could happen as early as mid-1968 if every ground test and flight test between now and then went perfect – something experience has taught the agency not to expect." Nevertheless, speculation increased that NASA might speed up the lunar landing schedule.[23]

Mueller held an off-site meeting at Myrtle Beach Air Force Base, South Carolina, where Phillips discussed the latest Soviet accomplishments in space. The CIA, using its still secret reconnaissance satellites, provided NASA with updates about what the Soviets were doing by watching as they built infrastructure and launched rockets. Phillips said, the "Russians have consistently up-staged the US in their timing of space firsts and it appears that they will continue to do so." He then recommended reexamining programs to insure that the agency took advantage of its success, and suggested NASA seek the advice of an internationally recognized expert like Arthur

[21]   Hacker & Grimwood, *On the Shoulders of Titans*, 11-2; Lecture II, Mueller, University of Sydney, Australia, 1/10/67; Kraft, *Flight*, 221-222; Bilstein, *Orders of Magnitude*, Ch. 5.

[22]   Hacker & Grimwood, *On the Shoulders of Titans*, 11-2; "NASA Officials Cite Feats of Gemini Flight," *St. Louis Globe-Democrat*, 6/8/65, GEM-83-4; Webb to Mueller, 6/10/65, GEM-44-7.

[23]   Mueller, American Trial Lawyers Association, Miami Beach, FL, 7/27/65; Mueller, Bannow Memorial Address, Bridgeport, CT, 6/28/65; "Rights Leaders Threaten to Mar Astronauts Parade," *St. Louis Globe-Democrat*, 6/14/65 and "Space Chief Says US is Far Behind," *NYT*, 6/15/65, GEM-83-4; Hacker & Grimwood, *On the Shoulders of Titans*, 11-2.

C. Clarke to help win the public relations battle. Acting on this suggestion, Mueller invited Clarke to Washington to discuss how to generate wider public support for the US space program. The author cautioned him to avoid creating the impression that "a space project or undertaking is being carried out for the propose of generating popular enthusiasm." A landing on land instead of in the ocean would be interpreted as a major accomplishment, as would developing a reusable booster, a permanent orbiting laboratory, or a base on the Moon. A long range flyby/capture mission or landing on one of the moons of Mars as a prelude to landing astronauts on the Red Planet would also be popular. They discussed how to improve the public's reaction to the space program, such as beaming live TV pictures from space. Mueller later called Clarke "more than a science fiction writer . . . He visualized what could happen and what should happen long before you had the necessary engineering to make it happen – a futurist." Mueller engaged Clarke as a consultant, and over the years they developed a friendship that lasted until Clarke's death in 2008.[24]

After converting all Gemini contracts to cost plus incentive fee, Mueller sent a letter to the Apollo contractors about converting their contracts as well. Discussing this with the center directors and their procurement people at the end of June, he spoke of the problems encountered in converting the Gemini contracts. "You can't incentivize a contract unless you understand, in some depth and detail, what it is that you are trying to do," he said. He wanted all major fixed fee contracts converted to incentive fee by year-end, and said that if they had not defined the end items, "we should have them defined." And, like the Gemini contracts, he planned to incentivize schedule and cost performance. He observed, "we find improvements continually being developed and the improvements often cost more in terms of time (and, actually, when you are finished with it, in terms of performance) than having it work right in the first place." Though the problem, he said, was not getting good performing subsystems, "It is getting subsystems put together to make systems . . . That is why the emphasis on schedule."[25]

III

During the summer of 1965, Mueller established a new organization to manage post-Apollo programs, tapping Disher to organize the Saturn/Apollo Applications Office as its deputy director, a job that he held for the next eight years. Mueller planned to appoint Mathews director after completion of Gemini, and in the interim appointed his deputy for programs, General Jones, as acting director. The office officially opened on August 6, a minor milestone, although behind it lay years of space station

---

[24]   Resume of Discussion in the OMSF, Myrtle Beach AFB, SC, 5/13-18/65, SCP-49-6; Arthur C. Clarke, clarkefoundation.org; Mueller Interview, Slotkin, 2/24/10; Holmes to file, Arthur C. Clarke, 6/29/65, GEM-67-4.

[25]   Mueller to CEOs, 6/30/65, 5/7/65, SCP-48-5.

studies and post-Apollo planning. Many issues faced Disher and Jones while they organized the new office; and because of the importance that he placed in the Apollo Extension Systems, Mueller promoted the new office despite a degree of uncertainty about what it could accomplish. Its establishment began an intense period of "trying to find a theme that we could sell," according to Disher, and "we must have gone down 250 different roads on Apollo Applications." Within NASA, everyone agreed there should be an Apollo follow-on program, but few supported a bold new goal such as a human mission to Mars, so they looked for something in-between. And by the fall of 1965 Disher and Jones developed a plan they thought Webb and Seamans would support. They asked Mueller for $260 million to fund it, but only received $40 million, which was just enough to keep it alive for another year. With money from the previous budget they defined the program, and planned to use FY 1967 appropriations to continue definition and conduct some experiments. This led to the development of an airlock device to enable astronauts to move between Apollo and an empty Saturn fuel tank which they planned to use as an orbital workshop. They also funded a major experiment called the Apollo Telescope Mount.[26]

In late June, Mueller told a group of educators "the most important achievement of GT-4 was the splendid condition of Astronauts McDivitt and White" following four days in space; they experienced no long term ill effects from their time in zero-gravity. He announced that Gemini V would rendezvous with a target pod about the size of a briefcase. After ejecting the pod, the astronauts would let it drift about sixty miles away, and then rendezvous with it using onboard radar. To rendezvous, he said, "requires an exceptional degree of readiness for precision launching of the spacecraft booster, a very accurate guidance and navigation system and responsive and effective systems for maneuvering in orbit" by exceptionally well trained astronauts. And while NASA launched GT-3 and GT-4 two quarters behind the original schedule, plans for GT-5 called for a launch in the third quarter, meaning that the agency shaved a full quarter off the delayed schedule. Consequently, it had a good chance to gain additional time and complete all ten piloted Gemini flights by early 1967, close to the original schedule.[27]

Mueller gave a more extensive report about the GT-4 mission in an address at the end of June, calling it "a turning point in many respects in the American space program." Because the astronauts did not suffer adverse effects from four days in zero-gravity, he expressed confidence the human body would be able to withstand longer periods of weightlessness without ill effects. And Gemini IV demonstrated the ability of an astronaut to maneuver outside a spacecraft using a handheld propulsion unit, proving the effectiveness of the spacesuit as protection from environmental dangers. White had suffered no disorientation and had performed his assigned tasks. Gemini IV marked a transition from Mercury, where astronauts flew as passengers,

---

[26] Disher Interview, Ertel, 1/27/67; Compton and Benson, *Living and Working in Space,* 20-21; Schneider to Mueller, History of AAP, 12/5/69, GEM-50-1; Hitt, et al, *Homesteading Space,* 7.

[27] Mueller, National Aerospace Education Council, Washington, DC, 6/24/65.

to what Mueller described as "a new generation of maneuverable vehicles, from which man can emerge into space and do useful work." Gemini was fully operational and ready for longer missions. Noting that Gemini V would be an eight day mission, which was the approximate time to fly to the Moon and back without spending any time in lunar orbit, he said that this would demonstrate the ability of astronauts to withstand prolonged weightlessness and perform their duties effectively.[28]

In his public remarks, Mueller gave many reasons for why he thought it important for the US to be pre-eminent in space, calling it imperative "for reasons of national security, national pride of achievement, and international prestige." Expected spinoff benefits included the discovery of new scientific knowledge, economic stimulation, social progress, development of new technology, and the "compelling urge of man to explore and discover." NASA worked closely with DOD, to make its capabilities available to the military. And taken together, he said, the civilian space program provided "a national resource of enduring value," allowing the country to operate in space and engage in "the wide variety of missions that may be required by the national interest."[29]

Although successful, Gemini IV left two important questions unanswered: the ability to perform station-keeping in the final phase of a rendezvous and why the onboard computer failed towards the end of the flight. Seamans asked the Langley Research Center to study the orbital mechanics of the mission, "especially the complex decisions on attitude and velocity changes and probable fuel usage both with and without computers." After reviewing the flight data, Langley engineers concluded that Gemini IV had enough fuel to perform the station-keeping alongside the Titan stage but the crew had not been adequately trained for the job. One of the engineers wrote, "no one was 'adequately trained', in that the differences between motions on Earth and motions in orbit were not intuitively realized or 'second-nature' to anyone." After returning the capsule to St. Louis, engineers removed the computer and it worked perfectly in hundreds of tests. IBM, the manufacturer, was unable to duplicate the failure on the ground. Unable to diagnose the exact problem, they added a manual switch to bypass memory areas believed to have caused the failure.[30]

Since the agency had no reason to repeat the EVA performed on Gemini IV, Mueller scratched them from the next three flights, allowing the astronauts to concentrate on other mission objectives. NASA named L. Gordon Cooper, Jr. and Charles "Pete" Conrad, Jr. as the crew for GT-5, which would fly a mission to evaluate the rendezvous guidance and navigation system, demonstrate the ability to remain in orbit for eight days, and evaluate the effects of weightlessness. The agency set the launch date for August 9, but because this did not leave enough training time,

---

[28]   Mueller, Bannow Memorial Address, Bridgeport, CT, 6/28/65; Mueller, American Trial Lawyers Association, Miami Beach, FL, 7/27/65.

[29]   Ibid.

[30]   Hacker & Grimwood, *On the Shoulders of Titans*, 11-3; Seamans, *Apollo*, 60.

chief astronaut Donald K. "Deke" Slayton flew to Washington and convinced Mueller to "reluctantly" give them ten additional days.[31]

From the day that he arrived at the agency, Mueller pushed hard to sell a post-Apollo program. He commissioned studies and spoke about an expansive follow-on program, although Congress remained reluctant to fund it. Nonetheless, the Apollo Extension Systems approved by Webb in August contained a total of twenty-nine missions between 1968 and 1971. Operating in low Earth orbit and on the lunar surface, the agency planned launches at the rate of about eight per year, with two-thirds of them carrying astronauts. However, as Compton wrote, "A manned program of that magnitude seemed to have little chance of becoming reality, but until circumstances forced him to back down from it, Mueller kept the pressure on the centers to plan for big things." In late June, NASA revealed the names of the first six scientist-astronauts, including two medical doctors and four others with doctorates in physics, engineering, and geology. One man dropped out shortly after beginning astronaut training for personal reasons. The geologist, Harrison H. Schmitt, would go on to become the only scientist to set foot on the Moon. However, to undertake the planned number of Apollo flights, NASA would need more astronauts, and Mueller authorized the recruitment of a new class in September, and in April 1966 the agency announced the names of nineteen new pilot-astronauts.[32]

While MSC remained deeply involved in spacecraft development, engineering work on the Saturn V approached its conclusion at MSFC. With the end in sight, Mueller and von Braun had concerns that the expertise built up for Saturn would dissipate unless something came along to preserve it. They spent time discussing post-Apollo programs capable of maximizing MSFC's talent, and hence building an orbiting space station, which was one of von Braun's earliest dreams, became part of their planning. Mueller saw this first space station as a step along the way toward human missions to Mars, viewing the Apollo Extension Systems as a means of developing the expertise for planetary exploration and thus maintain the industrial base. Michael I. Yarymovych, who would become technical director of the MOL, but worked for NASA at that time, called Mueller the driving force for the post-Apollo program "because he saw that from the day after Apollo lands on the Moon he will be asked what to do for an encore and we were beginning to realize ... landing a man on the Moon was a great but one-time event with very few follow-ups." However, the Bureau of the Budget objected to the Apollo Extension Systems, considering it poorly conceived, overly ambitious, and not justified by the experiments planned. Yarymovych said the budget bureau asked "if we couldn't find enough missions to justify it, how could we justify large space stations." Despite that, Mueller wanted to continue to fly experiments after Apollo to keep the scientific community interested in human spaceflight. In an article published in *Astronautics & Aeronautics* that August, he wrote that "extending the capabilities of

[31] Hacker & Grimwood, *On the Shoulders of Titans*, 11-3.
[32] Hitt et al, *Homesteading in Space*, 41-42; Compton, W*here No Man Has Gone Before*, 66-67.

Apollo opens up the Earth-Moon system to intensive exploration by man through a variety of scientific, biomedical, technological, and operational experiments." Appealing to scientists, he said NASA had identified nearly two hundred "meaningful" experiments, and "as our knowledge and interest in space grow, the number of experiments will continue to grow also to take advantage of our increasing payload capability." Both lunar orbit and lunar surface exploration planned for the post-Apollo period would uncover new scientific knowledge. But he would fight a losing battle, as Congress remained reluctant to spend large sums on human spaceflight after Apollo. Then in late August, testifying before the House space committee he described plans for the Apollo Extension Systems, emphasizing the scientific rather than the engineering reasons for this investment. He described the planned missions, displaying considerable enthusiasm for a continuation of human spaceflight after Apollo. Webb and Seamans also testified at these hearings, but downplayed the cost involved. And the House staff report published after the hearing noted that NASA's testimony did not state the expected cost for the program.[33]

On the morning of August 21, Conrad and Cooper rose into orbit on Gemini V in an attempt to fly an eight day mission. The fuel cells caused early concern before a specific problem developed. After about two hours in space Cooper ejected the target test pod and turned on the rendezvous radar. Because of another fuel cell problem, Houston ordered Cooper to put off the initial rendezvous and to power down all non-essential equipment. Failing to fix the faulty fuel cell, the astronauts rested while engineers on the ground tried to figure out what to do next. Unable to do anything about the fuel cells, NASA canceled the rendezvous, so the astronauts worked on experiments until flight planners came up with the idea of conducting a "phantom rendezvous" that would fly Gemini V to a moving point in space in order to test the spacecraft maneuvering system. This worked out well, and they performed four burns for an imaginary rendezvous. And as Gemini V circled the Earth, the faulty fuel cells produced enough power to allow the mission to establish a record for long duration spaceflight of eight days in zero-gravity. After they were back on Earth, Charles A. Berry, MSC's medical director, worried about the effect of weightlessness. However, after two days the astronauts' bodies returned to normal and concern about spending time in weightlessness subsided. As Mueller said at the time, the mission "provided a great deal of valuable information on the physiological effects on man of long-duration flight" demonstrating that zero-gravity would not be a threat to the health of astronauts on a lunar landing mission.[34]

The president sent Cooper and Conrad on a goodwill tour to six countries in

[33]   Hitt, et al., *Homesteading in Space*, 8; Yarymovych Interview, Benson, 2/2/76; "Beyond Apollo," *Astronautics &Aeronautics*, 8/65, Writings, GEM-110-3; Compton and Benson, *Living and Working In Space*, 42.

[34]   Hacker & Grimwood, *On the Shoulders of Titans*, 11-3; Kraft, *Flight*, 232-234; Mueller, Hartford Rotary Club, Hartford, CT, 10/4/65.

Europe and Africa, accompanied part of the way by Mueller, who called their warm receptions "an indication of the world-wide excitement generated by this country's space accomplishments." In Athens, he reported, "the [Greek] royal family, scientific and technical communities, and the man in the street joined in a warm and enthusiastic welcome, eager for every detail of the Gemini V flight." Thus, "Our accomplishments ... are inherently important from the aspect of international affairs, and may be considered as a measure of our ability to compete with a formidable rival ... [and] as a criterion of our ability to maintain technological eminence worthy of emulation." In Athens, the astronauts attended the International Astronautical Congress, where they met the crew of the second Voskhod mission.[35]

In his most extensive public discussion of the Apollo Extension Systems to date, Mueller addressed the Athens space congress saying that by 1970 the US would have three boosters rated for human spaceflight. The Saturn V could launch a crew of three into low Earth orbit for ten to fourteen days; orbit the Moon for four to six days; or land two astronauts on the Moon for twenty-four to thirty-six hours. He spoke of extending lunar orbit missions to twenty-eight days, and of lunar surface visits of up to two weeks – the length of one period of daylight. The basic Apollo mission afforded limited time for exploration, but follow-on missions would provide longer periods and help to answer questions about the Moon's origin, evolution, and properties. Orbital missions could survey the Moon using remote sensing techniques or send probes to the surface. And he said that future lunar surface missions were a logical extension of the basic Apollo mission because they could increase surface time, use more scientific instrumentation, and permit increased astronaut mobility. However, he said, the "foundation of any long-range and broad national space program is a strong manned Earth-orbital program." Post-Apollo missions could "demonstrate man's ability to perform useful tasks in space," and he expressed the need to focus on making the spacecraft more compatible with extended periods of flight. Then he predicted, "Man's greatest contributions in space will come when he can bring his intelligence to bear on the spot. He is needed as explorer, scientist, and creative man of judgment, as well as sensor, filter, data processor, pattern recognizer, and manipulator." In conclusion, "The application of Apollo and Saturn equipment provides the means ... to accomplish this prolonged exposure and experience of man in the space environment as well as the early exploration of the Moon."[36]

IV

With Mueller away in Europe, William B. Taylor, assistant director for engineering studies in OMSF, briefed the administrator and general manager about the Apollo

---

[35] Hacker & Grimwood, *On the Shoulders of Titans*, 11-3 and 11-4; Mueller, Hartford Rotary Club, Hartford, CT, 10/4/65.

[36] "Some Applications of Apollo," Mueller, XVI IAC, Athens, Greece, 9/14/65.

Extension Systems. Seamans supported Taylor, but Webb objected to his referring to the Apollo Program as the "first lunar landings." The administrator did not want the post-Apollo program tied to any particular date or to the lunar landings because of a "far-fetched possibility that there may never be a lunar landing, or at least in the now planned time frame" he explained. While supporting the Apollo Extension Systems, Webb nevertheless speculated that by 1968 the national objective could be changed from landing on the Moon to, for example, putting a synchronous satellite over Vietnam. But Seamans objected to this kind of "schedule flexibility," as he called it. The administrator suggested taking the cost of the post-Apollo program out of the Apollo budget and showing it separately, even proposing that it not use Apollo in its name because as a former budget director he knew "if it had the word Apollo connected to it the Bureau of the Budget bookkeepers would just add the cost of the extension program to the present Apollo Program and would then say it cost a lot more than had been predicted." Taylor presented a $408 million budget, which Webb and Seamans considered too high. They did not have a particular amount in mind, though they also felt that $100 million would be too low. And following this session, the administrator approved a $250 million Apollo Extension Systems budget for FY 1967. But the budget bureau thought $250 million excessive and countered with an offer of $100 million, although they indicated willingness to listen to the rationale for a higher amount. When he returned from Europe, Mueller thought neither $100 nor $250 million adequate, and argued for the original request. Without full funding, he claimed, the US would fall back into second place in the space race, and $250 million did not cover half of the experiments identified for the post-Apollo period. Moreover, without more funding, he predicted that morale would decline, and that the lower figure represented poor economic policy leading to increased unemployment and would leave the investment in Apollo unused between 1968 and 1971. Failure to fully fund the program also represented poor political strategy, and he speculated that it could become a political issue in the 1968 presidential election. These arguments did not convince Seamans or Webb, and they left the budget request at $250 million; though as Mueller later pointed out, the real problem was usually with the Bureau of the Budget; they had "a very simple minded view of life, and that is, how do we get within our budget this year, and where can we cut?" They had so much money to spend and limited places to cut, and the space program remained high on their list of things to cut back. In "almost in every instance I can remember," Mueller said, "the Administrator had to go to the President and finally cut a deal with him" which exceeded the budget bureau recommendation. Adding, "how much of that was a charade and how much was real, I don't know, but it happened every time I was involved."[37]

By the fall of 1965, Gemini flew on a regular schedule. With successful missions in March, June and August, human spaceflight returned to the front pages of the nation's newspapers, and NASA planned two more flights that year. Five missions in

---

[37] Ingram to Mueller, AES Presentation to Mr. Webb, 9/21/65, SCP-48-4; Compton and Benson, *Living and Working in Space*, 42-43; Mueller Interview, NASM, 5/1/89.

nine months, more than one every two months, with each extending the time spent in space or practicing maneuvers needed for Apollo. Astronauts who later became well known received assignments to fly on Gemini VI and VII, while NASA shuffled flight plans trying to figure out which mission to focus on which objective, as slow progress fixing balky equipment and fuel cell problems affected schedules. Perhaps the greatest problem at the time involved the Agena target vehicle, required for the rendezvous and docking tests. Though usually a reliable missile, the Agena suffered development problems long before Mathews took over the Gemini Project Office, and he worked to resolve them to keep up with the flight schedule. As NASA trained the crew of Gemini VI to perform rendezvous maneuvers, Edwin E. "Buzz" Aldrin Jr., who earned a doctorate in aerospace engineering at MIT prior to being selected as an astronaut, writing his dissertation on the pilot's role in orbital rendezvous, teamed up with Dean F. Grimm of MSC's flight crew support division to develop the trajectories and maneuvers to permit the astronauts to "catch" a target vehicle. Walter M. Schirra and Thomas P. Stafford, who would fly aboard Gemini VI, practiced rendezvous simulations and worked with Aldrin and Grimm to determine the best procedures to use on GT-6, which was planned for launch on October 25. A two day mission, Mueller assigned seven scientific experiments to it, but reflecting Houston's negative view of space science, Schirra said, "On my mission, we couldn't afford to play with experiments," considering rendezvous to be sufficient, though in the end he did conduct several radiation and photography experiments in orbit.[38]

On the eighth anniversary of Sputnik, October 4, Mueller spoke about the "compelling urge of man to explore and to discover," again dwelling on the role that the civilian space program played in national defense. He said he and Schriever had worked out details of NASA's contribution to the MOL program. Looking to the future and giving purpose to the nation's efforts in space, he argued that, "the exploration of space, the Moon and the planets will answer vital questions about the origins, early history and evolution of the solar system and the universe as a whole; it will enable us to investigate for life on other planets; and potentially ... give us an understanding of the origin of life itself." If science and the discovering the origins of life did not provide sufficient motivation to support the space program, he turned to the earthly spinoffs. Of equal importance, he added "are the improved methods of production and quality control that have been developed as part of the space effort." He spoke about Gemini VI as the first mission to attempt to rendezvous and dock with a target vehicle, and said it would be "one of the most important and significant milestones in our space program." Following that flight, Gemini VII with Frank F. Borman and James A. Lovell, Jr., planned a mission of up to fourteen days. Neil A. Armstrong would conduct another rendezvous and docking test on Gemini VIII in 1966, and David R. Scott would perform an EVA for a complete orbit lasting about ninety-five minutes. As Apollo moved rapidly ahead, he called 1965 a crucial year and said the "prospects look good." He predicted that extended lunar exploration

---

[38]   Hacker & Grimwood, *On the Shoulders of Titans*, 12-1; National Space Data Center, *Gemini 6A*.

could lead to "continuously supported outposts on the Moon," similar to those in the Antarctic. Other possibilities for the post-Apollo period included "manned Earth-orbital space stations of indefinite lifetime . . . [and] exploratory flights in deep space, leading ultimately to manned planetary exploration, with Mars as the first likely target."[39]

The design of the Earth orbiting workshop went through a number of configurations beginning with a "spent stage laboratory" based on studies conducted in Huntsville going back to 1959. It would use a Saturn S-IVB upper stage and would be called the orbital workshop. MSFC received the lead role in developing it, an important factor in gaining von Braun's support. However, giving Marshall direction of the project created conflict with Houston, and confused the previous division of responsibilities between the centers. In the past, MSC developed human spacecraft, while MSFC handled launch vehicles and engines. Assigning the orbital workshop to Marshall made them competitors with Houston. Douglas Aircraft, the developer of the S-IVB stage, promoted the idea of using it for the orbital workshop well before NASA made the final decision. And Mueller recalled Douglas pushing the idea very hard. "They had the S-IVB stage, and they were trying to figure out how to use it in the future." Von Braun, a longtime supporter of the spent stage concept, agreed with Douglas, and saw the value of using the Saturn upper stage because of its size, and the ease of launching it into orbit.[40]

However, while this went on, not all was well in the basic Apollo Program. The second stage of the Saturn V reached a critical point at the end of September, when NAA produced the first flight-weight version and prepared a static test to simulate flight loads. NASA design criteria called for the S-II to withstand 150 percent of the maximum calculated flight load, and required testing to that level or beyond. During an overnight static test, the S-II suffered what NASA called a "catastrophic failure," after reaching only 144 percent of the maximum flight load. Whilst the expression "catastrophic failure" may imply to a laymen a disaster or calamity, to a structural engineer it meant that the failure occurred all at once, without warning, destroying the stage. A catastrophic failure is binary, either it fails or it does not, and this is the definition the agency used. Perhaps it was also a disaster or calamity, but that was not what structural engineers understood the term to mean. Recalling these events, Mueller called the attitude exhibited at NAA "amazing," and he argued that the tests ought to have been continued to at least 150 percent, because "the next one might fail at 110 percent." Nonetheless, NAA management felt the S-II stage failed at a level which was never expected to be reached in flight, so it did not represent a big deal to them. On the other hand, as Mueller later pointed out, during the second Saturn V flight test the launch vehicle experienced oscillations well above the design

---

[39]　Mueller, Hartford Rotary Club, Hartford, CT, 10/4/65.

[40]　Compton and Benson, *Living and Working in Space,* 22-23 and 33; Hitt, et al., *Homesteading in Space*, 10 and 18.

limit and it held together; only MSFC's conservative design philosophy kept it from being a catastrophic failure.[41]

Atwood said schedule pressure got tense after the S-II failed, and "I think really made enemies of the company and NASA." The agency criticized NAA because of problems encountered in manufacturing the S-II, and everyone worried that it "might be the Achilles' heel of the whole program," according to Atwood. He called the agency's design philosophy very conservative, and did not think it essential to test to 150 percent of the expected flight load, because "the S-II was never projected for that kind of a margin of strength and safety and weight." After this test failure, MSFC engineers pinpointed the fault in the manufacturing process, and blamed NAA for shoddy workmanship. Von Braun assigned Rees, to monitor progress, and Rees and Storms promptly went "at each other's throat." They disagreed about personnel and technological issues, and according to Joe Goss, an NAA engineer on the project, "The Germans wore their design arrogance ... They knew all about booster-vehicle design," though when the company "introduced aircraft design elements ... they didn't think we knew what we were doing." Then following a project review in Huntsville on October 4, where NAA unveiled dramatically increased costs and further schedule delays, von Braun formed a review team to meet at the space division headquarters in Downey, California at the end of October to determine what to do next. He met with Atwood, who flew to Huntsville on October 14 in anticipation of the on-site review, but the S-II situation remained critical. MSFC's O'Connor asked Mueller to write to Atwood, telling him "reorganizations to correct the ... situation are mandatory," and he sent a letter to Storms on October 18 informing him, "The continued inability or failure of [the space division] to project with any reasonable accuracy their resource requirements, their inability to identify in a timely manner impending problems, and their inability to assess and relate resource requirements and problem areas to schedule impact, can lead me to only one conclusion – that management does not have control of the Saturn S-II program." Rees wrote to Phillips on October 21 and informed him that, based on his personal observations, the management of the space division exhibited bad behavior and had attitude problems, with internal personnel issues contributing to significant cost and schedule issues. He said the firm had an inadequate cost estimating system, and their management was bloated and inefficient. And they did not manage subcontractors well, failed to perform per the contract, and had serious quality, planning, control and reporting problems. All of this was further compounded by internal management issues.[42]

Mueller said NAA had "problems in production and delivery" of all of the pieces

---

[41] Catastrophic failure causes a complete cessation of a function. Such a failure occurs in a binary state, i.e., a function is either a success or a failure. Yang, *Lifecycle Reliability Engineering*, 10; Mueller Interview, Slotkin, 2/24/10; "Apollo: Looking Back," Caltech, Pasadena, CA, 5/18/71, SDC Speeches.

[42] Gray, *Angle of Attack*, 196-198 and 160; Atwood Interview, NASM, 8/25/89; Johnson, *Apollo*, 143; Rees to Phillips, 10/25/65, NAA, SCP-50-5.

of equipment that they were associated with. Moreover, "They simply didn't have adequate control of their production process, and they were meeting schedules by not completing work" but by delivering the hardware to KSC with the intention of completing the work there; which was a poor way of doing things. Adding, "We were having problems with the S-II stage, and we were having problems with the engines, as well as with the command module," and quality remained a significant issue. NASA had problems with other contractors, but because NAA had so many Apollo contracts, they became a bottleneck for the whole program. And "for some reason," Mueller thought NAA management were not getting across to their workforce the importance of doing things right, and on time. They worked in an "airplane mode", instead of a "Space mode" which had more rigorous requirements than those in an experimental aircraft program. Other Apollo contractors had similar difficulties in meeting schedules, but "to some extent they had simpler problems," he said.[43]

Complaints about NAA led Mueller to write to Atwood on October 27, telling him he had become "increasingly concerned that the rate of progress in both the S-II and CSM projects does not support the requirements of the Apollo Program ... Recent delays and major rescheduling in both projects, together with the tremendous cost increase in the S-II project recently proposed by NAA and continued excessive cost rates in the CSM project, have convinced me that we will not meet the objectives of the Apollo Program unless substantial improvement in your operations can be made." Referencing previous conversations, he added, "I know you are also concerned ... While I am optimistic that converting our contracts to an incentive base as quickly as possible is a big step in the right direction, I have no basis for confidence that this action is all we should take." He informed the CEO that Phillips would lead a group to review relations with NAA. In his reply, Atwood agreed to support the review, and said he awaited Phillips' correspondence. The general then sent a letter saying he would lead a tiger team, including Shea and Rees, to review the company's performance on all Apollo contracts.[44]

Phillips led "a major task force" of approximately 150 NASA and air force personnel in a detailed examination of the NAA space division's projects, identifying problems and determining what needed to be done to correct them. The company got deeply involved in the review, and while Atwood cooperated, Storms "was probably the largest part of the problem ... since he was highly success-oriented, and unwilling to admit that there was any possibility that something more could be done to do the program properly." As Mueller described it, "In a sense, any of these programs could be run perfectly if management did exactly the right things at the right time, but human beings being what they are ... you have to have some external forces occasionally to cause the internal management review to take place." Because some of the same problems occurred in every project managed by the space division, Mueller thought they had management problems. And he pointed out that once their

---

[43]   Mueller Interview, NASM, 1/10/89.
[44]   Mueller to Atwood, 10/27/65, SCP-48-7; Atwood to Mueller, 11/2/65, and Phillips to Atwood, 11/16/65, SCP-50-5.

projects got behind schedule they took short cuts, causing them to miss schedules still further. It became a downward spiral, leading to more short cuts, and excessive overtime. Quality deteriorated, and problem reports grew. As Mueller explained, they needed to stop and regroup, and ask "what is our real problem and how are we going to solve it? ... And usually that fall back and regrouping takes some external force to cause it to happen, because ... you're so close to the problem you can't really see what it is that could be changed or should be changed." But Storms believed he "had chosen THE way" and charged down that road, not stopping to think there might be a better way. Mueller later said "that was one of the problems at NAA. They did not have a very good sense of who was a good technical manager. And if they thought of Stormy as being a good technical manager it's just beyond my real comprehension, once you get to know him. He's a cowboy alright and very loquacious but not very good technically. Not willing to go down to the detail to be good technically." Unlike the way NAA usually operated, NASA took the system integrator role on Apollo, and did not buy a finished product; it bought services to produce the product which was defined as they went along. Because of this approach, Mueller said "unless it is a joint effort, it's not going to succeed very well." And "it was essential for the NASA people to be at least as cognizant of the design, design criteria, the limitations, as it was for ... North American, because NASA was the integrating contractor, and it had to put all those pieces together and make them fly."[45]

In the fourth quarter of 1965 the press and the public remained focused on Gemini, but Mueller had his hands full with all three human spaceflight programs. Gemini flew successfully, but in addition to known problems he suspected some "unknown, unknowns" could affect any one of the remaining missions. At the same time, Apollo was experiencing significant development issues and its two principal contractors – NAA and Grumman – remained behind schedule and experienced cost growth. He was also fighting for Apollo Extension Systems funding, not only with Congress but with the budget bureau and NASA management; consequently, much remained to be done before year-end.

---

[45] Mueller Interviews: NASM, 1/10/89 and Slotkin, 2/24/10.

5

Learning, developing and planning

Turning "potential failures into final successes."
Mueller, January 26, 1966

In a newspaper interview published in October 1965, just before the scheduled launch of Gemini VI, Mueller went out on a limb and said, "I don't think there's anything we haven't thought of." He called the recent simulation of the mission a full dress rehearsal that had gone perfectly, and claimed that the agency had thought of everything, including what to do if Gemini VI could not undock with the Agena target vehicle – it would return with Agena attached to the nose of the spacecraft. He predicted docking would be easy, but proving it had still to be demonstrated. He then described the mission as "a major step forward in space exploration in understanding man's ability to work and live in space," explaining that each flight provided new knowledge. In conclusion, he said that the coming year would be busy with about ten launches to complete Gemini and start Apollo.[1]

The target vehicle sat on an Atlas launch pad, as Schirra and Stafford settled into Gemini VI on a nearby pad. Agena-6 would take off first, followed by GT-6 about ninety minutes later as the target completed its first orbit. The Atlas lifted off as planned and released the Agena at high altitude – which inexplicably exploded as it ignited its engine, forcing NASA to scrub GT-6. Before this flight, the Agena stage had successfully delivered 140 payloads into orbit. Mueller immediately called Webb to fill him in, telling him it would take ten days to complete the investigation, but the agency could move another flight forward in the interim. Before the launch attempt, Mueller told the *St. Louis Post-Dispatch*, "Even if we should fail this time, it would not be an over-all setback. We have planned our program in enough depth to surmount any single failure." He now had to show the public how NASA would respond. Webb called the next day to ask about the reliability of the Agena, and

---

[1]   "Hopes High for Gemini 6," *The Christian Science Monitor*, 10/22/65, GEM-83-4.

Mueller told him that neither NASA, the air force, nor the Lockheed Missiles and Space Company, its builder, had reservation about its performance.[2]

Almost immediately following the loss of the Agena, engineers began to consider rendezvousing two Gemini spacecraft; although after initially discussing it, neither Mathews nor Mueller thought it likely. The idea had come up before, but it required launching two Gemini missions within a fourteen day window, and they did not think this could be done. However, Martin Marietta and the air force worked out a way to prepare a second Titan II for launch in one week, and had the crews available to do it. The Martin engineers briefed the other contractors, and John F. Yardley, McDonnell's Gemini project manager became an advocate for this new mission. However, as Mathews put it, he and Mueller "looked on the proposal with a jaundiced eye," and Gilruth agreed with them. Nevertheless, the question of launching GT-7 before GT-6 became the subject of wide discussion. Could Gemini VII replace Gemini VI on the Titan II that was already on the launch pad? This could be a problem because the longer duration Gemini VII mission made it heavier than Gemini VI. While Mueller still did not support the proposal, Mathews got NASA and Martin engineers together, to take a quick look at what would be involved in doing this. After concluding that it was feasible, they held open the possibility.[3]

Back in Washington, Mueller met with Webb and Seamans to discuss putting spacecraft seven on booster six, but not about the possibility of having spacecraft six rendezvous with spacecraft seven, mainly focusing on schedule issues. However, the contractors did not give up on the rendezvous prospect. Meanwhile, back in Florida, engineers determined that spacecraft seven could not be launched on booster six. After further investigating the possibility of launching two spacecraft within a two week window, they decided that it would be possible. KSC then let MSC and headquarters know that this would not be a launch constraint. Next, Houston figured out how to control two spacecraft at the same time, and Gilruth called Mueller who, although he had warmed to the idea, still "wanted to sleep on it," he said.[4]

Within hours these discussions leaked to the press, and Low warned Mueller who called Seamans that night, and met with him and Webb the next day. They discussed what went wrong on GT-6 and what the agency could do about it. Asked why he did not test the Agena without the launch of a Gemini-Titan, Mueller reminded Webb of the all-up testing philosophy to test the maximum configuration possible at any given time. He discussed the schedule constraints, and the proposal for spacecraft six to rendezvous with spacecraft seven. Webb liked the idea, and asked Mueller if he thought it would work. Mueller wanted to double check and called Matthews, telling

[2] "Gemini flight put off after Agena target is lost by trackers," *St. Louis Post-Dispatch*, 10/25/65, GEM-83-4; Logsdon, *Exploring the Unknown, VII*, 345-346.

[3] Hacker & Grimwood, *On the Shoulders of Titans*, 12-3; Mathews Interview, Sherrod, 2/17/70; Mueller Interview, Putnam, 10/4/67.

[4] Hacker & Grimwood, *On the Shoulders of Titans*, 12-3.

him that "Webb liked the idea and thought it important enough for the President to announce." He warned Matthews that there could be no hedging, because once Johnson made the announcement, this committed the nation. Did he still think it possible? Matthews wanted to check with Low and Gilruth, and Mueller gave him fifteen minutes. Gilruth "counted noses," and Mathews reported it unanimous in Houston, so Mueller told Webb "he had a deal." NASA renamed the mission GT-6A to differentiate it from the canceled GT-6. Webb contacted the president, and on October 28, three days after Agena-6 failed, the White House announced the rendezvous, which the press immediately named Mission-76. The president's press secretary said the mission would be in January, but NASA was hoping to undertake it in December. It was important to push ahead with what was available because, as Mathews remembered, if they had waited for another Agena, "We would have lost five months."[5]

Engineers at KSC stored GT-6, got GT-7 ready for launch and scheduled the steps to prepare GT-6A, starting as soon as GT-7 cleared the pad. Astronaut training did not present a problem because both crews had already been well trained, and NASA could set up the communications network to handle two simultaneous missions. The astronauts wanted to add an EVA, and discussed the possibility of transferring an astronaut from one capsule to the other. That required disconnecting life support umbilicals, which Mueller vetoed as too risky. Mission-76 anticipated the first set of astronauts, Borman and Lovell, spending nine days in space before the Schirra and Stafford rendezvoused with them.[6]

On November 10, as flight preparations continued, Mueller addressed an education conference in Columbus, Ohio. The former Ohio State University electrical engineering professor spoke of the changes that had taken place in science and technology since his departure from academia in 1956. Rockets could send nuclear weapons across continents, and with the launch of Sputnik the space age dawned "as a severe blow to our national pride," he said. The USSR developed increasingly large and sophisticated space vehicles, while the US made significant progress in putting space to practical use. As in past speeches, he said that the investment in civilian spaceflight supported national security. NASA closely coordinated its efforts with the military, and during the past decade had developed the equipment, operational skills, and experience essential to achieving the goal of "attaining and maintaining world leadership in space." While disappointed by the delayed launch of Gemini VI, he announced that the agency would launch Gemini VII on December 4 and nine days later Gemini VI-A would attempt to rendezvous with it. Calling the experience and proficiency that was being developed in the Gemini Program a major step toward achieving the goal of human exploration of the Moon within the decade, he said this capability benefited future spaceflight

---

[5] Ibid.; Mueller Interview, Slotkin, 9/9/09 (second session); Discussion Items for Meeting with Mr. Webb & Seamans, 10/26/65, Notes & Transcriptions, GEM-80-4; Mathews Interview, Sherrod, 2/17/70.

[6] Hacker & Grimwood, *On the Shoulders of Titans*, 12-3 and 12-4.

programs. Using another theme that he would frequently employ, he compared the settling of the nation's western frontier with the new space frontier; and called it a "striking parallel" that provided "an unprecedented peacetime stimulus to science." He praised the academic community for its role in meeting those challenges. But to continue space exploration required a supply of well-trained scientist and engineers. And to increase the supply, the agency was instituting programs to educate students in science and technology, and strengthen research capabilities at universities, which "underlies the future advancement of the nation's welfare, economic growth, and security." He told them the space program "is a powerful creator of knowledge in very tangible ways," resulting in scientists learning more about the universe. To achieve the nation's goals and maintain power, he explained that "trained people are vital elements and ... our nation's schools are a basic resource which must be encouraged, nourished, strengthened, and involved on a broad basis." Then, quoting the president, he said, "No national sovereignty rules in outer space. Those who venture there go as envoys of the entire human race. Their quest, therefore, must be for all mankind – and what they find should belong to all mankind."[7]

On December 4, Borman and Lovell achieved Earth orbit aboard Gemini VII after a near perfect launch. They had not much to do other than conduct experiments and wait for Gemini VI-A. For this mission the astronauts wore special "soft" spacesuits, and requested to remove them for greater comfort in the cramped spacecraft. For safety reasons Mueller ordered that one crew member must remain suited at all times, although he need not wear gloves or seal his helmet, and during critical maneuvers – launch, rendezvous and re-entry – they must both wear them. But wearing their spacesuits became uncomfortable during the long flight, and the crew appealed to Houston for reconsideration. Gilruth agreed with them and asked Mueller to reverse his decision. After checking with Berry, MSC's medical director, Mueller learned that the unsuited astronaut had a more normal pulse and blood pressure than the other one, so in the interest of their health he reluctantly agreed they could fly most of the time without wearing spacesuits. However, this incident resulted in an apocryphal story circulating in Houston claiming that Mueller lacked "compassion" for the plight of the astronauts on the long flight, although his main considerations involved their health and safety.[8]

Back at KSC, the launch team had nine days to get GT-6A ready for liftoff, a task which normally took two months. This would be feasible because the spacecraft and its booster had already been through an integrated checkout. In one day, technicians re-erected the Titan II and mated it with the spacecraft. Making good progress, they had GT-6A ready on December 12 – one day earlier than promised. But when the time came to launch Schirra and Stafford into orbit, the malfunction detection system shut down the Titan's engines a little more than one second after ignition.

---

[7]   Mueller, Joint Conference on School Management, Columbus, OH, 11/10/65. The use of the frontier thesis during the 1960s was not all that unusual. Slotkin, *Gunfighter Nation*, 498-623.

[8]   Hacker & Grimwood, *On the Shoulders of Titan*, 12-5.

Because of the possibility of explosion the astronauts ought to have ejected, but not feeling any motion, Schirra judged that the rocket had never left the ground and did not pull the ejection D-ring. After determining that the booster was safe, technicians got the crew out of the capsule. Mueller later called Schirra's decision, "smart." Adding, "If you aren't moving don't move," because the "ejection system was not exactly the safest thing in all the world." Schirra's decision saved the space vehicle. After the president called Webb to express his disappointment at the apparent failure of Mission-76, the administrator spoke to Seamans and the general manager pointed out that with six days left in the Gemini VII mission the agency might be able to find and rectify the problem and still launch GT-6A in time for a rendezvous.[9]

As Martin Marietta and the air force began preparations for a third launch attempt, they found that a tail plug that was used to supply electricity to the vehicle while it was on the ground had fallen out prematurely, and corrected that problem. While the media clamored for more details, Schirra's action, the fact the malfunction detection system worked as planned, and the seemingly simple problem satisfied them for the moment. Further inspection of the launch vehicle found a more serious problem in the rocket engine – a dust cap had been left in one of the feed lines, and its presence would have caused the engine to shut down even if the tail plug had not fallen out prematurely. The launch team got GT-6A ready once again just three days after the second launch attempt. On the morning of December 15, NASA finally launched GT-6A. The flight plan called for four orbits to catch up to Gemini VII, and Schirra maneuvered to initiate the rendezvous. A little more than five hours into the mission, he saw Gemini VII. Maneuvering slowly, spacecraft six approached spacecraft seven, and at about 200 yards Schirra said that Gemini VII "resembled a carbon arc light," because with the sun reflecting off it the spacecraft was almost too bright to look at. Then spacecraft six floated to within forty yards of seven. Having used less than half their fuel in the rendezvous, there was plenty for other purposes, and for three orbits spacecraft six maneuvered to inspect seven, and at one point took up position a few inches from it. It was evident that if spacecraft six had been alongside an Agena then a docking would not have presented a problem. Gemini VI-A returned the next day after about twenty-five hours in space. The banner headline in the *St. Louis Globe-Democrat* said, "Americans Eye-To-Eye in Space," and quoted Mueller saying "the launch team preformed 'a perfectly professional job.'"[10]

And an editorial in *Aviation Week & Space Technology* said,

> Dr. George Mueller ... whose slight stature and horn-rimmed glasses often obscure to the uninitiated the steely courage that lies inside him, deserves commendation for the swift and effective manner in which he rallied his forces from the sudden disaster of the Agena target explosion in the original Gemini 6 rendezvous mission on October 25. A man of lesser courage and lesser

---

[9]   Ibid., 12-6.; Mueller Interview, Slotkin, 2/24/10; Seamans, *Aiming at Targets,* 62.

[10]   Hacker & Grimwood, *On the Shoulders of Titans* section, 12-6 and 12-7; "Americans Eye-To-Eye in Space," *St. Louis Globe-Democrat*, 12/16/65, GEM-83-5.

confidence in his people and his technology could have easily slid into a cautious delay instead of reorganizing the mission to press on to successful achievement only 45 days after the Agena vanished over the southwest Atlantic. Dr. Mueller represents the very best type of technical leadership that this country requires to realize the full potential of its technology as his actions during the Gemini crisis amply prove.[11]

II

Phillips shared the results of the NAA tiger team review with Mueller in mid-December, describing the situation as critical. He wrote that the review "confirmed beyond any doubt all previous indications we have had that the Apollo Program is severely jeopardized by this contractor." Continuing in this vein, he said the company failed to meet commitments, and he questioned whether NAA had "sincere intent and determination ... to do this job properly." The general considered management of the space division to be weak, ineffective, and "past the point of no return." And he doubted their ability to meet NASA's requirements. Since mid-1964, the agency had expressed concerns about the costs and schedules of the command and service modules and S-II projects; and whereas the spacecraft showed some improvement, serious problems persisted resulting in additional schedule slippage and further cost escalation. Optimistic contractor progress reports masked the true situation, with performance on the spacecraft unsatisfactory and the S-II completely out of control. Phillips found these projects months behind NASA's most pessimistic outlook, putting the whole flight test schedule in danger. The general called the space division overstaffed, and in need of streamlining. Their engineering was inadequate, quality was below standard, and the quality assurance organization was bloated with over three thousand people. What is more, they did not properly use existing management systems, the program management organization was dysfunctional, and the corporate staff played a passive role in managing the division. Phillips recommended that parts of the space division that were not focused on Apollo be transferred to other parts of the company, and Storms be removed as president and replaced by someone who would be "able to quickly provide effective and unquestionable leadership for the organization to bring the Division out of trouble and into [a] position where program commitments will be met." But Atwood protected Storms because, as Mueller saw it, "Stormy worked for Lee for many, many years ... and Lee had a great deal of faith in him because he built airplanes that flew. That did not mean he could build spacecraft that could fly." Nonetheless, Mueller believed it inappropriate to direct Atwood to remove Storms, a technicality that he could have circumvented, though a roadblock to be resolved. So without Atwood taking action on his own, Storms remained in place. Phillips met with

---

[11]   "First Rendezvous," *Aviation Week & Space Technology*, 12/20/65, GEM-83-5.

Storms on December 19, and laid his cards on the table. As expected, the NAA executive saw things differently and later said the general had it all wrong. Storms complained that the review team did not know all the facts because they only examined "the paper trail and paid little attention to the hardware itself." However, the company missed deadlines and had significant cost overruns, though Storms claimed the agency had "no one on the team with a clear understanding of the structural design problems of the [S-II], or the chaos in engineering that NASA's own people created."[12]

Phillips urged Mueller to write Atwood to convey the seriousness of the situation, and drafted a letter for his signature which Mueller sent on the same day as Phillips' own letter to Atwood – December 19. (The Phillips letter, with the review team's unedited notes attached, reappeared as the "Phillips report" during the congressional investigation into the Apollo 204 fire; see chapter 6). They sent these letters, Mueller said, because "We wanted to get his attention," and "we were frustrated." Both letters said that the results of the tiger team review convinced them action had to be taken to correct the situation. Mueller wrote, "I am not sure that you see the performance of [the space division] in the same light that I do." He described some of the problems with the command and service modules – missed schedules and delays, and wrote, "I could go on; there are other things that we've had to accommodate such as cost growth, but I believe that this list gives you some insight into my evaluation of performance in the spacecraft project." Turning to the S-II, he spelled out the cost and schedule issues, and wrote, "It is hard for me to understand how a company with the background and demonstrated competence of NAA could have spent 4½ years and more than half a billion dollars on the S-II Project and not yet have fired a stage with flight systems in operations." He said "results are a function of management and technical competence. I submit that the record of these two programs makes it clear that a good job has not been done. Based on what I see going on currently, I have absolutely no confidence that future commitments will be met." After summarizing a set of recommendations, he wrote in conclusion, "I consider the present situation to be intolerable and can only conclude that drastic action is in the best national interest."[13]

Atwood approached his dealings with NASA cautiously. At the time, he was engaged in discussions to sell NAA to Rockwell Standard Corporation, and with the agency as his largest customer he had to do what NASA wanted. He never argued, though he later called Phillips' allegations that NAA failed to plan and control their projects wrong. "I don't believe that," Atwood said, calling the company's planning and control systems adequate. He blamed NASA's inability to stop making changes and its failure to freeze requirements for many of the things Phillips and Mueller

---

[12] Phillips to Mueller, 12/18/65, CSM and S-II Review, GEM-89-3; "Inputs on NAA People," Undated, GEM-84-4; Mueller Interview, Slotkin, 9/9/09; Mueller to Atwood, 12/19/65, GEM-89-3; Gray, *Angle of Attack,* 201-202.

[13] Logsdon, *Exploring the Unknown,* 629-639; Mueller Interview, Slotkin, 2/24/10; Mueller to Atwood, 12/19/65, GEM-89-3.

complained about. He pointed out that the company was using technologies well within the state of the art on the command and service modules. "We didn't have to invent any real techniques to build it," he claimed, "[b]ut the arrangements inside and the instruments and wiring and all that were in a state of flux for a long, long time" because the agency kept changing things. NASA did not provide "a coordinated set of instructions to the manufacturers. It was a continual sort of rumble of desires of various elements of NASA's technical staff." He pointed out that MSFC had a strong configuration management program which they used on the S-II stage, although MSC placed little emphasis on configuration management. And Storms told Atwood that "Houston wanted to avoid a detailed specification as long as they could so that they would have free license to make changes by waving their arms ... If documentation didn't exist, changes could be made by informal verbal communication." MSC wanted to be in charge and they were, he insisted. NAA found it very hard to accept the way Houston managed the spacecraft project, because on aircraft development projects they worked to specifications and designed to performance, and this added to the friction between MSC and the company.[14]

The S-II stage got to the point where Mueller visited the company and told Storms "to get some real talent into this thing," because the problem impacted the overall Apollo schedule. They had some interesting discussions until Atwood assigned some top people to take over the management of the S-II. However, it took time to work through the problems because of their challenging nature. The CEO assigned Ralph Ruud, a corporate vice president, an expert in tool design, and a peer of Storms, to oversee the S-II development. Atwood also hired retired air force General Robert E. Greer from Martin Marietta to get the S-II back on track, and these management changes began to make a difference. Ruud resolved the S-II production problems with what Mueller called a "combination of technical, management, and inspirational leadership." And by getting the "right person with the right background in place," the company finally got the job done. Ruud implemented new manufacturing technology, and overcame the production problems which were endangering the lunar landing schedule. Mueller got personally involved with the S-II because the schedule was so tight, he said, and "there wasn't a lot of time for reiteration ... for solving the problem several times."[15]

NASA originally considered the Saturn V the pacing item in the Apollo Program. However, Shea pointed out, "the launch vehicle turned out not to be the pacing item, because ... the spacecraft program really got off on the wrong foot." He wanted to "stop making each spacecraft a little bit different and gradually approach the lunar configuration." Nonetheless, until the Apollo 204 fire, when the spacecraft had to be redesigned, MSC would not do that. Shea, like Mueller, attributed the incremental approach at MSC to the way that NACA had developed aircraft, which was similar to how MSFC originally built launch vehicles. When Shea arrived in Houston in the

14   Atwood Interviews: NASM, 1/12/90 and 6/25/90.
15   Mueller Interviews: NASM, 5/2/88 and Slotkin, 9/10/09; Johnson, *The Secret of Apollo*, 145.

fall of 1963, a year after the lunar orbit rendezvous decision, the Apollo spacecraft could not accommodate the lunar excursion module; although, he said, "They were still thinking about … it somewhat." MSC had made a series of design studies but had not settled on the final configuration. Because of inadequacies in its original design, Shea divided the Apollo spacecraft project into two parts: Block I, mainly for Earth orbital test flights not requiring docking; and Block II for flights to test the lunar excursion module and for use in the lunar landings. This permitted NAA to finish the Block I spacecraft while designing the improved Block II, which also incorporated the experience gained from Gemini. Mueller recalled, "I'll never forget the first meeting I had in Houston … about the Block II spacecraft. My first question was why do we need a Block II spacecraft?" This occurred in early 1964, and "we had been working on the lunar orbit rendezvous for over a year and the group down there explained … the tunnel wasn't big enough for the astronauts to get through it and [in any case] there weren't any provisions for docking with the lunar [excursion] module." The Block I never met the needs of the program because NASA had designed it before the lunar orbit rendezvous decision, at a time when it was expected that the entire spacecraft would land on the Moon. This was compounded by the chaos in Houston, with the Apollo Spacecraft Project Office in one building, center management in another, the Gemini Project Office in yet another, and people "everywhere – scattered all over Houston," he said. And, nobody in management knew that "this tunnel wasn't any good for anything."[16]

Mueller said fixing the problems with the spacecraft proved to be more difficult than with the S-II, because "Stormy just couldn't imagine any of his people having a problem. It almost took the Apollo 204 fire to convince him that he had to do something better than he was doing." NASA eventually got NAA to bring more management into the space division, "but it took a long while," as Mueller recalled. In addition to Ruud and Greer, Atwood assigned George W. Jeffs, corporate director of engineering, and a rising star in the company, to serve as Myers' deputy and chief engineer. Jeffs remembered, "We had lots of areas to clean up … lots of areas to get down to the meat and potatoes, and get rid of some of the ice cream," and he reduced the number of engineers working on the spacecraft project by about half, cutting out around 2,000 of them.[17]

As 1965 came to an end, Mueller gave a wide ranging interview to *US News & World Report* magazine. He described the lunar landing mission in detail, though the interviewer repeatedly asked about risks, before asking directly, "Just what are the biggest risks for the astronauts in this Moon landing?" Mueller responded, "There are risks inherent in each of the phases of the program. I would say that there is one area where risk is perhaps somewhat larger than it is elsewhere. That is in the launch phase itself – from Earth. We recognize that this is [very] complex equipment that's involved. The next period of risk is the landing on the lunar surface itself." Then the

---

[16]   Shea Interview, Neff, 2/12/69; Mueller, Notes & Transcriptions, 3/13/69, GEM-79-15; Orloff and Harland, *Apollo*, 36.

[17]   Mueller Interviews: Slotkin, 9/9/09 and 2/24/10; Murray & Cox, *Apollo*, 183.

interviewer asked, "Should Americans brace themselves for possible disaster of our own astronauts?" After giving a routine answer about taking all steps to prevent this, Mueller said, "people must brace themselves and be willing to accept the fact that someday, some way – which we can't describe at the moment – we may lose one of our astronauts, or even a crew of astronauts."[18]

By the close of 1965 NASA had flown five successful piloted Gemini missions, including a rendezvous, and Apollo-Saturn was moving closer to its first flight test. Some people argued for curtailing Gemini but Mueller supported completing the program, telling Mathews it would continue in 1966. Problems with the Agena remained, and the agency needed it for the docking tests. After working round the clock for four months, engineers still had not found what caused the failure of Agena-6. NASA did not know what it would take to fix the problem, or when the vehicle would be ready for a mission. Mueller appointed a Gemini-Agena Target Vehicle Review Board co-chaired by Gilruth and Schriever's deputy, General Ritland, to oversee the effort while various committees and task forces worked on the technical problems. As Mueller kept to his busy schedule, Mathews stayed on top of the day-to-day Gemini issues, keeping his boss updated through frequent calls and face-to-face meetings. Desperate to prove the feasibility of docking Gemini in space, McDonnell's Yardley proposed "a poor man's target" as he called it, by developing an augmented target docking adapter that would be delivered directly by the Atlas, eliminating the need for the Agena. Mueller approved the jury-rigged backup, which became available in early February 1966. Meanwhile, pressure increased to complete the analysis, testing and modifications to the Agena. Problems remained but Mathews wanted to use the Agena, so Mueller gave him additional time to complete the work and obtain approval of the Gilruth-Ritland review board before agreeing to use the adapter. With help from Schriever, Mathews got priority test time for the Agena at the Arnold Engineering Development Center. On March 4, after twenty-two successful tests, the review board approved the Agena for flight. NASA then scheduled the Gemini VIII/Agena-8 docking test for launch on March 16.[19]

In early 1966, NASA's investment in human spaceflight remained substantial. Employment peaked at about 300,000, with 95 percent of these workers in industry or at universities. Already beyond the mid-point of the Apollo Program, decisions needed to be made about how to take advantage of its capabilities in the post-Apollo time frame. The time to order long-lead components for the Apollo Extension Systems was rapidly approaching. Consequently, Mueller said, "We must use it or see its value erode ... If we do not use what we have created, continued expansion of the Soviet program will very likely lead to future Soviet missions that will have the impact of Sputnik." Retrenchment would mean that the US could not develop the scientific and technological knowledge needed. Yet at a small fraction of the original cost, it could continue, and expand human space operations for decades. This had

---

[18] "How US Plans to Conquer the Moon," *US News & World Report*, 12/27/65, GEM-83-5.
[19] Hacker & Grimwood, *On the Shoulders of Titan,* 13-1 to 13-3; Telephone Calls 1/21/66, Call Sheets, GEM-71-1.

significant implications for national defense and the US position in the world. And Mueller noted, continued investment in human spaceflight would be "the cheapest insurance policy the nation can buy" to keep the US secure, because it represented "the most compelling and dramatic demonstration of this country's scientific and productive excellence." Because the administration still supported the MOL project, to avoid conflict with the air force the administrator proposed a more modest Apollo follow-on program than previously planned. And when he testified before Congress in February 1966 in support of the FY 1967 budget, Webb said the post-Apollo plan "reflects the President's determination to hold open for another year the major decisions on future programs – decisions on whether to make use of the space operational systems, space know-how, and facilities we have worked so hard to build up, or to begin their liquidation." John D. Hodge, a member of the original Space Task Group and chief of the flight control division at MSC, called the Apollo Extension Systems "Mueller's baby," and said he rammed it "down everybody's throat." Mueller thought it necessary to keep human spaceflight going, and it "did serve that purpose," said Hodge. And as the number of people working on the space program fell, the original cadre from NACA and the army remained intact, while many of the new people hired from industry left the agency. Yet Mueller knew that to retain good talent he needed interesting programs for them to work on.[20]

III

The agency announced the crew members for Gemini VIII. Armstrong, who had backed up Gemini V, would be the commander, and David R. Scott, a rookie, would be the co-pilot. The main objective was to rendezvous and dock with the Agena target, but the agency planned to test new equipment, including an extravehicular life support system and a backpack with a tank of gas for an extended test of the zip gun used by White. Scott's spacewalk would be the first one since Gemini IV, and as he trained, NASA remained unaware of how difficult it would be to work in zero-gravity. The flight plan called for Scott to egress and make his way to the rear of the spacecraft to don the backpack, which would be no easy task as it turned out, but he spent a lot of time training and gaining confidence in the equipment. But Mueller focused on the rendezvous and docking, which remained critical program objectives. The next five Gemini missions would be more advanced than the first seven, and the new equipment would increase training time. Meanwhile, issues with the fuel cells restricted flight duration to two or three days. Gemini VIII's launch date potentially overlapped the first Apollo-Saturn IB launch (AS-201), but Gemini VIII included rendezvous and docking tests and AS-201 would be an automated mission to begin the process of rating the launcher for human flight, so Mueller gave GT-8 priority.

---

[20]   Mueller, National Dairy Council, Washington, DC, 1/24/66; Compton, W*here No Man Has Gone Before*, 66; Lambright, *Powering Apollo,* 139-140; Hodge Interview, Lambright, 10/5/91.

In the end, no delays occurred, although as the launch date drew near the mission was plagued by equipment problems.[21]

About a month before the launch of AS-201, Mueller addressed a reliability symposium, reporting "excellent progress" in the Apollo Program. Apollo remained on schedule, although he said NASA "must learn much more about what man can do in a spacesuit outside his spacecraft; we must reduce rendezvous and docking to a simple, routine maneuver; and we must develop the capability to control the spacecraft's flight path during re-entry to a degree of accuracy that will permit landing on land, in an area the size of a large airport." Yet despite earlier successes, he said spaceflight remained risky, causing the presence of humans to be essential to make decisions and adjustments to the hardware, which "turned potential failures into final successes," pointing to Schirra's "calm courage and cool judgment" during the GT-6 launch abort.[22]

Citing statistics which resonated with his audience of reliability engineers, Mueller said that the Atlas booster used with Mercury had over a hundred launches before the agency flew a piloted mission, and the Titan had more than thirty launches before being rated for use with Gemini. Saturn on the other hand, would have only a limited number of flight tests prior to the first human mission, and this required more extensive ground testing, with detailed tracking of every part during manufacture, assembly and testing. And by "understanding and correcting each test failure or performance discrepancy, we believe we can achieve the required high reliability" to allow NASA to conduct major missions after a few Saturn V flights rather than the thirtieth or hundredth as in Gemini and Mercury. He explained that the Apollo test program replaced "statistical confidence" with "inherent confidence" by qualifying flight hardware on the ground. AS-501, the first Apollo-Saturn V flight test, would be a major milestone, testing the actual version of the space vehicle to be used in a lunar landing mission, and would fly with each of the three stages live. Test firings of these stages began in 1965, and final testing and checkout of the flight hardware would continue right up to the time of launch.[23]

As the reality of tightening federal budgets became more apparent, Mueller began to downplay the Apollo Extension Systems in his public remarks. In February he said, "under NASA's current budget plans we will hold open the option to employ the Apollo hardware and capabilities in a program extending through 1971." As he had frequently said previously, the post-Apollo program would include flights in Earth orbit, lunar orbit and to the Moon's surface. However, he temporarily omitted talk of the Apollo lunar landing by 1969 being an intermediate goal, and of missions to Mars and Venus by the 1980s, although he still spoke about long duration missions in Earth and lunar orbit. He emphasized the more practical applications of space technology, pointing out that "the first task to be performed, before such

[21]   Hacker & Grimwood, *On the Shoulders of Titan,* 13-4 and 13-5; Seamans, *Apollo,* 64.

[22]   Mueller, Symposium on Reliability, San Francisco, CA, 1/26/66.

[23]   Ibid.; Mueller, Astronomers of UCLA, Los Angeles, CA, 4/29/66; Mueller, Aviation/ Space Writers Association, NYC, 5/26/66.

benefits can be provided, is to learn to operate in space," with the remaining Gemini and the early Apollo flights serving that purpose. Nevertheless, with his staff and advisors, he spoke expansively about post-Apollo missions. He sent a letter to STAC members before their February meeting outlining ideas for a new national goal to land astronauts on Mars by 1980. The new program included piloted Venus orbital missions, which required program planning to begin by 1968. This program would exploit the investment in Apollo and focus on achieving as many intermediate goals as possible. To make the Mars mission possible, NASA needed nuclear propulsion, which could be tested in low Earth orbit; and the agency could test the Mars lander on the Moon. OMSF conducted conceptual studies, and Mueller proposed a series of steps to lead to the Martian mission, beginning with a space station in low Earth orbit. From this orbiting outpost, astronauts could establish a lunar base, perform near-Earth and lunar exploration, and mount Mars and Venus flyby missions. He described the steps necessary to go from the Apollo-Saturn to new space vehicles over the subsequent fifteen to twenty years, but it would require a Kennedy-like commitment to achieve the objective of landing humans on the near planets and return them safely to the Earth. STAC did not take up this proposal at the February meeting, choosing to focus instead on the selection of lunar landing sites, a more modest program to replace the Apollo Extension Systems, and the human spaceflight experiments program. However, they agreed to discuss the broader proposal at their meeting in May.[24]

Early studies of the orbital workshop centered on a spent Saturn S-II stage, and later switched to the S-IVB stage which became the basis for discussion of whether it should be a used stage, the so called "wet" workshop concept; or an unused S-IVB stage, a "dry" workshop. In either case, the orbital workshop would become NASA's first space station, "something to test what you needed to do in a real space station," as Mueller described it. He remained unsure about the feasibility of the wet workshop because of concern about "once you got that thing up there could you actually do all of the things that need to be done to make it into a working workshop." With little experience of working in zero-gravity, the ability to assemble the wet workshop in space remained unknown.[25]

Speaking before a group of prominent business leaders the week before launching AS-201, Mueller warned that manpower dedicated to Apollo had peaked and started to decline. He argued "this team must be kept together for the solution of problems that have not yet developed during the Apollo missions." As spending declined, NASA would effectively "go out of business" before flying its first operational vehicle – the first flight of the Saturn IB had not yet taken place and human flight on this vehicle remained a year away. This led to questions of what to do with the organization and resources established for the lunar landing. He said the FY 1967 budget permitted NASA to "hold open the option for a program to procure

---

[24] Mueller, AASA, Atlantic City, NJ, 2/12/66; Eighth Meeting of STAC, 2/18-19, KSC, FL, attachment: Mueller to Townes, 1/14/66.
[25] Mueller Interview, Slotkin, 2/24/10.

additional flight vehicles beyond those now programmed;" using them through 1971. But if the administration failed to exercise this option during FY 1968, the agency would have to phase down the space program and mothball facilities. Yet as he later pointed out, "Our space capabilities cannot be mothballed. We must use them or see their value erode." The time to decide about the future of human spaceflight was approaching, and decisions had to be made. And tying the space program to competition with the USSR, he told these business leaders that the Soviet Union had just "reminded the world of the competitive vigor of their effort in space by landing [an automated probe] softly on the Moon." But that proved helpful to NASA by demonstrating the possibility of a lunar landing, and showing that the Moon's surface could support the weight of at least a small probe; because some people had expressed concern that a thick blanket of dust would swallow any lander. Fearing the development of a technological gap with Europe, he pointed out that Germany and France built their own satellites, and the Soviets had almost doubled their pace in the past year. However, the Saturn IB that was due for launch in a week's time would place the largest payload of any launch vehicle in low Earth orbit, exceeding the lifting capacity of the largest Soviet booster, the recently introduced Proton.[26]

Mueller said "man will have to live and work in the space environment for some time before [he] can begin to fully exploit this new resource that is becoming available." The Apollo-Saturn would permit the US to gain years of flight experience and develop techniques for future space programs. Thus, he called it time to exercise the nation's option on the future human exploration of space, and at "a small fraction of the initial cost we can continue and we can expand our operations in space for the next ten years or more." He pointed out that continuing human spaceflight had implications beyond the immediate goal, because it impacted national security, world leadership, and scientific advancement. And while the Wright brothers had opened the possibilities of flight at Kitty Hawk, it took years of barnstorming to give way to commercial aviation. Then he cautioned, like other American inventions, we should profit from history and "fully use and exploit these machines we have developed at such a heavy investment of resources," and in that way realize our full capabilities in space.[27]

Plans for AS-201 involved a suborbital flight of about 5,500 miles, reaching an altitude of 300 miles. Delayed by weather, and then by some last minute technical glitches, AS-201 took off from KSC on February 26, with both stages of the Saturn IB live; testing the S-IVB for the first time. The Block I Apollo capsule controlled by an electromechanical sequencer landed in the South Atlantic, testing the separation of the spacecraft from the launcher, while the heat shield withstood the heat of re-entry. AS-201 experienced few problems, and according to space historian Roger E.

---

[26]   Mueller, Business Council, Washington, DC, 2/17/66; Mueller, Treasurers' Club, Columbus, OH, 2/23/66; Mueller, Aviation/Space Writers, NYC, 5/26/66; Luna 9, NSSDC, NASA.

[27]   Mueller, Treasurers' Club, Columbus, OH, 2/23/66; Mueller, Aviation/Space Writers, NYC, 5/26/66.

Bilstein, "All things considered, the two-stage Saturn IB vehicle achieved a notable inaugural flight."[28]

The *St. Louis Post-Dispatch* quoted Mueller as announcing, "It was a completely successful flight and a major step toward our manned lunar landing goal." The paper called the mission "flawless," although Mueller noted that "the only deviation in the flight was a performance slightly less than what was planned for the spacecraft motor." As he later told a congressional committee, with AS-201 "we introduced a new concept in our flight testing procedures – the 'all up' concept." This included the first flight tests of the spacecraft's service propulsion system and the Saturn J-2 hydrogen-burning engine. Using all-up testing, NASA accepted an increase in risk in order to obtain more results on a single flight. The agency scheduled three additional Apollo-Saturn IB flight tests during 1966 – AS-202, 203 and 204, with the first piloted flight scheduled for early in 1967 leading to a rehearsal in low Earth orbit of deployment and docking of the lunar module with the command module. (Around this time, Mueller renamed the *lunar excursion module*, calling it the *lunar module* or LM, because he considered the word "excursion" too frivolous.)[29]

Then when he testified before the House space committee in March, Mueller spoke about a Soviet space probe landing on the Moon on February 3 and another on Venus on March 1. He said the Japanese planned to launch their first satellite and that the Europeans were concerned about American technological leadership. He spoke of a developing technological gap with them, noting that the launch of the second French satellite on February 17 illustrated European determination to narrow the gap. The USSR's quest for leadership in space led them to double the amount spent on their program during the previous year, and they thought technological leadership would lead to ideological leadership in the Cold War. Bringing costs and schedules under control was another success, which Mueller attributed to the conversion of the major Gemini contracts to cost plus incentive fee. Thus, he said, "Through management actions and new contracts in which the profits of the Gemini contractors is tied to their total performance, schedules have been accelerated and costs are under control." He reminded the committee that the time had come to make decisions about the capabilities that could be used following the initial lunar landings, and rhetorically asked, "What comes next?" He went on to say that the agency could accomplish a wide variety of flights in Earth orbit, in lunar orbit and on the Moon's surface, and described a series of missions that would be made possible by the lift capability of the Saturn rockets. But before these missions could fly, he said, we can "operate effectively in space only by doing – by spending time in space." And following Apollo, an important task would involve gaining experience and developing the advanced operational techniques needed for future

---

[28] Press Kit, Apollo Saturn 201, 2/17/66, GEM-64-1; Bilstein, *Orders of Magnitude*, 338-339; Mueller, Astronomers of UCLA, Los Angeles, CA, 4/29/66.

[29] "First Flight of Apollo is Successful..." *St. Louis Post-Dispatch*, 2/27/66, GEM-83-5; Mueller, House of Representatives, 3/11/66; Mueller, Astronomers of UCLA, Los Angeles, CA, 4/29/66; Kraft, *Flight*, 276.

missions. The FY 1967 budget contained options to start Apollo applications in 1968, but if not approved, it would necessitate cutting back and mothballing space facilities.[30]

The Apollo Extension Systems adopted the name Apollo Applications Program – frequently called the AAP – in 1966, and Mueller began formulating plans for its first mission scheduled for the spring of 1968 aboard a Saturn V. As then constituted, the AAP had a number of specific objectives involving orbital assembly, operations and resupply, demonstration of personnel transfer in orbit, long duration spaceflight of up to three months, and extended lunar exploration and associated experiments. Because of its estimated cost, Mueller said the program requirements would be "honed down during the intervening years to a small-scale prototype space station costing less than $2 billion," and in order to reduce the post-Apollo budget still further the agency would assign the lunar orbit and surface experiments originally envisaged as a part of the Apollo Extension Systems to the core Apollo Program itself, eliminate additional hardware, and drop plans for synchronous Earth orbit missions. Webb supported the Apollo Applications Program, but not Mueller's still expansive vision of it. The administrator thought it desirable to take advantage of the knowledge and experience learned from Apollo, and warned that the nation faced a crisis in space planning. Decisions about the follow-on to Apollo had to be made that year, and Webb told Congress, "It is extremely important not to think you can postpone the decision again ... it is imperative to have a through-going national debate on whether we want to go past the point of no return." Workers had already been laid off and could not be easily replaced. Yet within months, tens of thousands of additional people would be released by industry.[31]

Checkout of AS-202 was underway at KSC in preparation for its June flight to further test the J-2 engine and test the re-entry of the Apollo capsule at a velocity equivalent to that when returning from the Moon. AS-203, which would not carry a spacecraft, was to evaluate the behavior of propellants in the S-IVB while coasting in orbit. AS-204, intended as a long duration orbital mission, would occur during the second half of the year. NASA selected the crew for the first piloted Apollo-Saturn IB flight: Gemini veterans Grissom and White would join Roger B. Chaffee. This was scheduled for the first quarter of 1967, although if all went well with the early tests, it was expected that the mission would be advanced to AS-204. However, its main goal remained to verify the compatibility of the space vehicle, crew and ground support, and the detailed plans remained fluid. But to be successful in landing astronauts on the Moon before the end of 1969, Mueller said, NASA had to achieve "100 percent success in every one of the large number of extremely difficult ground and flight tests which are now beginning." He warned that schedules remained tight, with no margin for error. Yet because of what he called the "austere level of

---

[30]   Mueller, House Space Committee, Washington, DC, 3/11/66.
[31]   Schneider to Mueller, History of AAP, 12/5/69, GEM-50-1; Lambright, *Powering Apollo,* 140.

funding," any major setback could derail the schedule necessary to achieve the lunar landing before the end of the decade.[32]

IV

On March 16, NASA successfully launched Agena-8, followed by Armstrong and Scott aboard GT-8. As the astronauts chased the Agena, minor problems plagued their maneuvers, but after three hours they made radar contact with the target vehicle orbiting about 215 miles away. Gemini VIII closed to within 150 feet and remained on station for a little more than a half hour, coming within three feet of the target prior to finally moving in to dock. Although it had never been done before, docking came easy and the spacecraft experienced no unusual vibrations. However, as the astronauts began a series of tests to determine how the docked combination reacted to different maneuvers, it began to roll. Unable to stop this unwanted motion, and believing it to be caused by the Agena, Scott and Armstrong undocked. When the anomalous motion continued, they concluded that it had something to do with their own spacecraft. Nothing they did could stop it. Indeed, the rotation spread to other axes. The onset of the problem occurred while they were out of communication with the network of ground stations. And by the time that they flew within range of the next station, they were tumbling at one revolution per second and communications were intermittent. Scott and Armstrong were becoming dizzy, had tunnel vision, and were approaching their physiological limits of endurance. Realizing that one of the thrusters was misfiring, but unable to identify which one, Armstrong switched off the primary system and activated the re-entry control system in order to stabilize the spacecraft. After reactivating the primary thrusters one by one in order to identify the one that was faulty, he switched the system off permanently. A mission rule was that once the re-entry control system had been activated, a return to Earth must promptly follow; this curtailed the flight.[33]

As always, Mueller went to the Cape for the launch. Aboard a NASA airplane returning to Washington to attend the National Space Club's annual Robert H. Goddard Memorial Dinner, he learned of the problem and immediately returned to KSC. He was met at planeside by G. Merritt Preston, MSC's director of operations at the Cape and later Debus' deputy, who briefed him as they drove to the old Mercury mission control center. They arrived as Gemini VIII began its re-entry. Seamans was attending the Goddard dinner, at which Vice President Hubert H. Humphrey was to be the guest speaker. On receiving a call informing him of the problem Seamans told the audience that the flight would be aborted, and then while

---

[32]   Mueller, Aviation/Space Writers Association, NYC, 5/26/66.
[33]   Hacker & Grimwood, *On the Shoulders of Titan*, 13-5; *Exploring the Unknown*, 47; Harland, *How NASA Learned to Fly in Space*, 154-157.

Humphrey was delivering his speech, Seamans signaled him that all was well, and the vice president announced the crew had landed safely.[34]

After splashing down in the Pacific, the astronauts returned to a hero's welcome in Houston and the spacecraft went to the McDonnell plant in St. Louis to determine what caused the thruster problem. The company never figured out exactly what went wrong, but engineers deduced that an electrical short circuit had caused a thruster valve to remain open. To prevent this from happening again, they installed a change to insure that power could not go to a thruster with the switch in the off position. Yet despite the near disaster, the mission had achieved the first docking of two vehicles in space, which was a milestone in preparation for Apollo. Mueller called Gemini VIII the "closest scrape we had," and estimated the agency came within seconds of losing the crew. The spacecraft almost ran out of propellant, although the astronauts found the problem in time. And he later said, "I must say I sweated that one out; Armstrong was lucky. If it had happened on the first Gemini manned mission the program would have come to a screaming halt."[35]

Mueller continued promoting the Apollo Applications Program in late April, and speaking to a group of astronomers at the University of California at Los Angeles he said "it is becoming possible to place the astronomer in the space environment, near his instruments." He spoke about establishing a human space observatory and using the Moon for astronomical observations, and told them that the Woods Hole summer study groups sponsored by the Space Science Board had "recommended that studies begin as soon as possible to explore the lunar capabilities for astronomy," which could include Earth-based engineering studies, lunar environmental studies, and testing small telescopes on the Moon. But whether these ideas led to actual programs remained up to the scientific community. The first phase of the Apollo Applications Program would use excess Apollo equipment, although of course the lunar landing program had first call on all flight hardware. Additional missions remained under consideration, but moving to a second and more advanced phase of the post-Apollo program would require additional equipment, which would require an increase in funding. He suggested that some of the experiments outlined at Woods Hole could take place during a second phase of the AAP, extending flights through at least 1971. Nonetheless, the examples of possible missions that he cited were modest compared to some that he spoke about in private. It was not that he had given up on flying these missions; he just did not speak of them publicly anymore.[36]

As a follow-up to Phillips' tiger team review, NAA presented a corrective action plan at the end of January 1966. The general said that the company's response was "objective," and gave them credit for taking steps to improve the problems

---

[34]   Hacker & Grimwood, *On the Shoulders of Titan*, 13-5.

[35]   Ibid., 13-7; Seamans, *Apollo*, 64; Mueller Interviews: Sherrod, 11/19/69 and Bubb, 5/22/69.

[36]   Mueller, Astronomers of UCLA, Los Angeles, CA, 4/29/66.

identified. NAA responded to each point, and Phillips wrote that he and Mueller found their initial actions satisfactory, although the proof would be in their performance during 1966. As NAA implemented their corrective action plan, in early spring Atwood asked Phillips if the situation had improved, and the general replied "things are doing much better, much better." Because sufficient progress had been made, Mueller and Phillips said NAA could return to a more "normal" management condition.[37]

In early May, Mueller spoke to PSAC about goals and missions for future human spaceflight, noting that the FY 1967 OMSF budget contained $300 million less than in FY 1966, resulting in the loss of trained manpower in government and industry. And this lower level of funding for the Apollo Applications Program restricted it to the preliminary definition phase only. However, he told the advisory committee "there is a need for a major commitment to the AAP if we are to keep our pipeline for spacecraft and launch vehicles from drying up." Because of the long lead times required for new development, he said, "We cannot delay unduly the initiation of the major new program without suffering a total net reduction in effort in space and loss of the momentum that we have built up." Reflecting on the debate concerning the AAP years later, Mueller said Webb "supported it ... I guess everybody did. But he didn't have the ability to force it through. No one did. Tiger [Teague] supported it and [Clinton] Anderson supported it and everybody wanted to do something but in the face of the Vietnam War and the riots in the streets it wasn't very popular and no one wanted to spend any time on it or spend any money on it." Adding, "I was trying to get some payback for all of the work going into Apollo and bring it into the future. The astonishing thing to me was that no one was that interested in the future."[38]

Mueller continued calling for a new national space goal to take advantage of the infrastructure built for the Apollo Program. And in a speech in May, he cited the National Aeronautics and Space Act of 1958 as requiring continued operations in space, and listed a number of possible goals. He cautioned that it "should represent a major step forward from a technical and competitive standpoint," since the USSR maintained its program at the same rate as the US. Returning to the more expansive goals of past speeches, he proposed landing astronauts on Mars in the 1980s, called the Earth-orbital space station a logical sub-goal, and advocated an expansion of lunar exploration. And he said it would only take modest increases in the budget to achieve a Mars landing, which meant keeping pace with economic growth. To do otherwise, he argued, would lead to the withering away of the investment already made. Although the president still supported the space program, the administration

---

[37] Status Report of the 12/65 NASA Review of NAA and Response Thereto, 9/14/66, SCP-51-1; Atwood Interview, NASM, 8/25/89; Memorandum for Mr. Webb, 4/16/67, NASA Archives, Webb Papers.

[38] "Identification of Goals and Missions for Future Manned Space Flight," Mueller, 5/5/66, GEM-46-2; Mueller Interview, Slotkin, 2/24/10.

5-1   George E. Mueller and Wernher von Braun, Saturn V rollout, May 26, 1966. (NASA photo)

struggled to pay for its other priorities. Then after PSAC gathered the information needed to establish post-Apollo goals, they concluded that after the first few lunar landings, NASA should explore the Moon during the AAP, with missions lasting from seven to fourteen days, once or twice a year.[39]

The agency planned to conduct three rendezvous operations on GT-9, and Mueller made them the first priority for the mission, with EVA work second. When the Atlas carrying the Agena target vehicle failed on May 17, the agency had a

---

[39] "Identification of Goals and Missions . . ." Mueller, 5/5/66, GEM-46-2; Compton, *Where No Man Has Gone Before*, 97.

solution in the form of the augmented target docking adapter developed after the Agena-6 failure. NASA postponed the mission until May 31 and renamed it GT-9A The backup docking target would enable Gemini IX-A to conduct all of the rendezvous and docking tests planned for the original Gemini IX flight, and those not carried out on Gemini VIII. Mueller said even if the adapter did not work, the mission would still perform the scheduled EVAs and test a sophisticated astronaut maneuvering unit developed by the air force.[40]

On May 19-20, when STAC met again, the main topic of discussion concerned the Space Science Board's recommendations for lunar and planetary exploration. Mueller briefed the committee members about his long range plans as a follow-up to the letter that he had sent in February, but the advisory committee again put off discussing it until their next meeting. But as it turned out, the Space Science Board intervened when they issued a report on the space goals for 1968-1975, one which severely disappointed Mueller. While supporting increased expenditures for space science, they failed to say much about the content of the science program, focusing instead on details of program implementation. The report recommended a flat NASA budget overall, called for decreases in human spaceflight and increased space science funding. Mueller wrote to Townes, telling him "the report fails to fulfill its function inasmuch as its fundamental assumption precludes its offering the kind of advice that I am sure the Agency needs." By focusing on funding levels, he said, the board "restricts itself to a course of action which can only result ... in [a] less than adequate space program in the years ahead." The science board report said, "in view of other demands from other segments within our society, and particularly in view of the drain that Vietnam represents, we assumed a static NASA budget for the next few years." Although they called for an expanded scientific spaceflight program, they said that the Apollo Applications Program was "questionable" and asked "whether a similar but even more ambitious program by, say, 1975 would be equally desirable. Manned Martian missions ... may encounter serious problems of man's performance and even [exceed his ability to] endure the space environment for the necessary long periods of time." The report claimed that time did not permit proper experiments to be conducted before 1975. "Continued lunar exploration is undoubtedly desirable in view of the investment already made in the Apollo Program but establishment of a permanent manned station on the Moon by 1975 would require a major additional effort and allocation of major additional resources." And the board also opposed launching a space station into low Earth orbit. Consequently, if plans for a robust human space program after Apollo were ever to have a chance, the opposition of the Space Science Board showed that Mueller still had to gain additional support in the scientific community.[41]

---

[40] Mueller, Aviation/Space Writers, NYC, 5/26/66; Hacker & Grimwood, *On the Shoulders of Titan*, 14-4 and 14-5.

[41] Minutes for the Ninth STAC, Goddard Institute for Space Studies, NYC, 5/19/66, with attachments, Mueller to Townes, 8/3/66, and Hess to Webb, 7/29/66.

V

By the end of May technicians had GT-9A ready for launch, and the Atlas successfully placed the augmented target docking adapter into orbit on June 1. However, telemetry indicated that the shroud covering the docking system had not opened completely, and was still attached. A problem on Gemini IX-A delayed its launch for two more days, but finally, on June 3, with the status of the shroud still unresolved, GT-9A performed a nominal launch. Within minutes Stafford and Eugene A. Cernan began pursuit of the target and after about three hours got their first glimpse of it about sixty miles away. As they closed in they saw the half-opened shroud. Stafford told mission control, "It looks like an angry alligator out here rotating around." When he suggested using his spacecraft to knock the shroud off, Houston advised against it. Gemini IX-A remained about thirty feet away as the crew looked it over and took photographs. A later investigation determined that minor errors on the ground had caused the shroud to remain attached; it was an improvised configuration, two companies were involved, and a supervisor was absent when the work was done. With a docking out of the question, Stafford withdrew and performed the second planned rendezvous using a different technique. After a full day's work, the Gemini spacecraft moved away. The next day the crew made a third rendezvous, performed other experiments and rested.[42]

On June 5, the astronauts got ready for an EVA, opened the hatch, and Cernan stood up. He found the handholds and Velcro installed to help steady him to be inadequate, and had limited mobility in the spacesuit. Mueller recalled, "Space suits turned out to be a real challenge." Cernan "couldn't really do anything mechanical. It was too hard to move his arms and it was difficult for him to get back in the Gemini capsule." Making his way to the rear of the spacecraft and strapping himself into the astronaut maneuvering unit sapped his strength and he found it difficult to maintain his body position. Every move took longer than expected, and occurred in slow motion. His exertions overwhelmed the environmental system of the suit, and his visor fogged over. With Houston's blessing, the astronaut maneuvering unit test was canceled and Cernan returned to the hatch and, after considerable difficulty, managed to ingress. The EVA took much more strength and energy than anticipated and ended after a bit more than two hours, more than a half-hour shorter than planned. However, Cernan's experience finally made everyone aware of the difficulty of working in a spacesuit in zero-gravity. On June 6 Gemini IX-A landed within a mile of the recovery ship and was hoisted onto its flight deck with the astronauts onboard. Seamans called liftoff and re-entry "flawless," although, he added, "much experience was needed before Apollo." It was a successful flight from Mueller's perspective because he considered rendezvousing to be as important as docking, but the media dwelled on the failure to dock with the target vehicle, the fogging of Cernan's faceplate, and his inability to test the astronaut maneuvering unit.

---

[42]   Hacker & Grimwood, *On the Shoulders of Titan,* 14-15 and 14-6; Kraft, *Flight,* 258.

5-2   "The Angry Alligator," Gemini IX-A, June 3, 1966. (NASA photo)

Nonetheless, the ever enthusiastic Mueller later called the mission, "a great success actually," because the astronauts performed three different types of rendezvous and everybody returned safely.[43]

The *Orlando Sentinel* published an article saying "officials are now admitting behind the scenes and cautiously, that problems of docking Earth-orbiting spacecraft are beginning to cause some serious concern." It went on to say, "NASA has a low batting average in docking Gemini spacecraft with target vehicles, and behind their verbal screen defending the Gemini Program, only 27 minutes of docking had taken

---

[43]   Ibid.; Mueller Interviews: Slotkin, 9/9/09 and 2/24/10; Hacker & Grimwood, *On the Shoulders of Titan,* 14-16; Seamans, *Apollo,* 66; Telephone Calls 7/6/66, Call Sheets, GEM-71-2.

place thus far." The *St. Louis Post-Dispatch* wrote, the agency "has found a surefire way to eliminate failures from its programs. It simply quit using the word." Instead of calling it a failure, Mueller used the word "unsuccess" in describing what happened. But what people did not understand was that he did not consider these tests failures for the reason that NASA had learned something from each one. Nonetheless, Seamans, not as sanguine as Mueller, wanted a mission review board to look into the flight, and Mueller appointed one of his deputies James C. Elms to chair it. With Gemini X planned for launch on July 18, the board had little time to make changes regardless of what they found, unless they delayed the flight, which Mueller refused to do. He later called appointing the review board "window dressing."[44]

The second Apollo-Saturn IB flight test in July turned out to be AS-203 since the modified Block I spacecraft for AS-202 was not ready in time for the flight. AS-203 would test the launch vehicle's orbital characteristics, in particular using television cameras to observe how the propellants behaved in their tanks while weightless. A lunar Apollo flight would require the J-2 engine to be restarted in space. NASA gave consideration to restarting the J-2 on AS-203, but MSFC opposed complicating the mission, so Phillips and Mueller agreed not to try it. AS-203 took off on July 7 with an aerodynamic nose cap instead of an Apollo spacecraft, and after the tests had been completed the pressure was allowed to build up in the hydrogen tank to determine when the common bulkhead would fail. NASA hoped to have AS-202 ready for launch in August.[45]

Gemini X would be a complex mission with several objectives. The flight plan called for a rendezvous with two Agenas, first docking with Agena-10, then using it to reach Agena-8, which had previously been maneuvered into a high parking orbit. NASA selected John Young and Michael Collins as the crew, and they expressed concerns about rendezvousing with Agena-8 because it had long since exhausted its power supply and, unable to operate its radar transponder, would require them to perform a visual rendezvous. The flight plan also included an EVA for Collins in which he would connect an umbilical to a valve on the exterior of his spacecraft to draw gas for his zip gun, then cross to the inert Agena-8 and retrieve the meteoroid experiment that Scott was to have collected. In addition, a spray had been developed to prevent fogging of his helmet visor. NASA scheduled the rendezvous with Agena-10 on the fourth orbit, devoted the second day to experiments, and assigned the Agena-8 rendezvous to the third day. With the help of the air force, which tracked the passive Agena-8 using powerful ground radars, NASA calculated the maneuvers that Gemini X needed to perform during the second rendezvous and input this information into the onboard computer.[46]

---

[44]   "Attack Threat By Sea Worries Pentagon," *Orlando Sentinel*, 6/16/66 and "NASA Perfects A Sure-Fire Way to End Failures" *St. Louis Post-Dispatch*, GEM-83-5; Hacker & Grimwood, *On the Shoulders of Titan*, 14-7; Mueller Interview, Slotkin, 2/24/10.

[45]   Bilstein, *Orders of Magnitude*, 340.

[46]   Hacker & Grimwood, *On the Shoulders of Titan*, 14-8.

The Atlas took off on time on July 18, and so did GT-10. As the astronauts made their first maneuvers in the Agena-10 rendezvous, Mueller opened the post-launch press conference by saying, "This certainly is an auspicious beginning ... But we have a long way to go and [a] very ambitious flight plan." Everything worked out well and almost six hours into the mission the astronauts docked with Agena-10. But because they used too much fuel, mission control omitted a test in which they would practice undocking and redocking. Then the crew got ready to light the main engine of Agena-10 to initiate a rendezvous with Agena-8. This required them to boost into a phasing orbit with its apogee at 476 miles, which was "the highest man had yet ventured in space," Mueller later said.[47]

The following day the crew performed another engine burn with Agena-10 to get into the proper orbit to initiate the visual rendezvous with Agena-8 and undocked from Agena-10. Once the range was thirteen miles, Gemini X completed the rendezvous with Agena-8, and took up station about ten feet from it. Collins opened the hatch, ventured outside, plugged in his gas umbilical, and held on to Gemini X as Young moved to within six feet of Agena-8. Collins moved over, maneuvering using his zip gun, and like Cernan had difficulty in working in zero-gravity; though he managed to retrieve the experiment and hand it to Young. The plan to install a new experiment on Agena-8 was canceled. When mission control told Collins to return to the spacecraft, he had difficulty with the life-support umbilical and needed help from Young to get inside. After withdrawing from Agena-8 they performed some experiments and then got ready to return to Earth. Following three days in space, they landed four miles from the recovery ship. Both Gemini IX-A and Gemini X contributed to the Apollo Program by proving the ability to maneuver and dock with target vehicles. However, EVA remained a problem. At the post-flight press conference, Mueller called this mission "extremely gratifying," and said the rendezvous with Agena-8 was a major achievement. The "techniques that have been demonstrated here for rendezvousing with passive objects ... are just the first and minor beginnings to what will become activities in future space programs," he said.[48]

Speaking to the National Space Club once again on August 16, Mueller reported the Gemini Program had already achieved all six of its original objectives. Through Gemini X, the program demonstrated long duration spaceflight, tested techniques for rendezvous and docking, maneuvered from one orbit to another, practiced guided re-entry, performed EVAs, and conducted scientific experiments. During eight piloted flights, astronauts accumulated more than 1,600 hours of flight experience; and both flight and ground crews gained experience in rendezvous, docking and post-docking maneuvers. "Perhaps most important," he pointed out, "we are learning in Gemini what is easy and what is difficult in space operations."

---

[47] Ibid.; Lecture II, Mueller, University of Sydney, Australia, 1/10/67; "Perfect' Launch Hits Fuel Snag," *Today*, 7/19/66, GEM-83-5.

[48] Hacker & Grimwood, *On the Shoulders of Titan*, 14-8; "Young, John W, Collins Back at Cape, Praise Crews," *St. Louis Post-Dispatch*, 7/22/66, GEM-83-5.

Rendezvousing proved easier than anticipated, but EVA "appears a little more difficult," requiring development of techniques to perform work in space. Nonetheless, he said, "Vital tasks remain to be done on the remaining two Gemini missions." The first two Gemini flights flew six months behind schedule, but after the first piloted flight the program had begun to meet its milestones and cost estimates. "From then on," he said, except for an initial delay in docking, "we have accomplished each milestone ahead of or on schedule and we do expect to complete the Gemini Program ahead of schedule in 1966." NASA assigned each mission primary and secondary objectives and, except when they endangered crew safety, Mueller pushed to achieve as many objectives as possible on each flight. Although not all Gemini missions were as successful as Gemini X, "our program planning has been such as to permit us to capitalize on our successes and to learn from our reverses," he said. And according to Mathews, Mueller always told the reluctant individual, "Now go ahead and do it." He pushed the schedule to get Gemini flying as often as possible, which led to completing the program sooner than planned, despite earlier schedule slips.[49]

---

[49]   Mueller, NSC, Washington, DC, 8/16/66; Mathews Interview, Sherrod, 2/17/70.

6

Peaks and valleys

"Mankind is in space to stay."

Mueller, October 25, 1966

Mueller presented a detailed status of the Apollo Program in another speech to the National Space Club in August 1966, and during the question and answer period said a small possibility existed to land on the Moon before 1969, but "it would be idle to forecast when the first landing will take place." This comment fed speculation of an early lunar landing because he also said that Gemini, which started behind schedule, would end ahead of schedule. Then in answer to a reporter's question, he said the US might never develop a system for rescuing astronauts stranded in space because the "Cost, complexities and uncertainties involved indicated that the money and effort would be better spent to make space flights safer in the first place." The press jumped on this remark, and the *Orlando Sentinel* wrote, Mueller "is not given to excessive frankness in public; in fact his tendency in the opposite direction is well known in the space business. Accordingly, Mueller's forthrightness the other day on a particularly touchy topic came as a distinct surprise." William M. Hines of the *Washington Star* penned, "Mueller dismissed, almost with a wave of his hand, hopes for a space rescue capability at any time in the reasonable near future." While NASA had the technology for space rescue, the agency lacked the necessary budget. Because of the adverse publicity his comments generated, Teague sent Mueller a letter with copies of several critical editorials, expressing concern over misinterpretation of his answer, because, he wrote, he knew how much "NASA has labored long and diligently to cope with emergencies in orbit." Mueller replied that he could not do justice to such a complex question in the limited time available at the meeting, and reminded Teague that a "great deal of effort has been expended by NASA in its efforts to assure the safety of astronauts in space flight." Work on space rescue predated his arrival at the agency, and in 1964 Bellcomm had examined the feasibility of using Apollo as a backup rescue system. Nonetheless Mueller later argued that it did not make sense to build a space rescue capability, and in fact NASA never did develop the capability for Apollo.[1]

After several delays, AS-202 finally took off on its suborbital flight on August 25. Following a near perfect launch, NASA brought the spacecraft back to Earth in a way designed to impose a heat load on the command module equivalent to that which would occur during a return from the Moon, and the scorched capsule landed safely in the Pacific. After the recovery, Mueller told a press conference, "The flight was normal and it did reach all mission objectives. All systems performed normally. The results of the flight are exceedingly gratifying." The agency then rated the Saturn IB for human spaceflight, and began to prepare AS-204 for launch on February 21, 1967, with a Block I spacecraft crewed by Grissom, White and Chaffee on a mission which the agency planned to call Apollo 1.[2]

By the summer of 1966, the agency had forty-nine astronauts, including five scientists. Gilruth and Slayton felt NASA had a sufficient number and claimed that astronaut-pilots could be trained to do the science that would be performed on the Moon, though Mueller disagreed. He had sold the Apollo Applications Program on the premise that it would contain more science than the basic Apollo Program, and he pushed MSC to recruit more scientists as astronauts. All the while Gilruth opposed him. But Mueller required the support of scientists to gain congressional support for his post-Apollo program, and his position prevailed. NASA would announce plans to accept applications from scientist in September 1966, to be screened jointly with the National Academy of Sciences. And the agency planned to announce the names of the second group of scientist-astronauts in August 1967. (Though in the end, none in this new group of scientist ever set foot on the Moon.)[3]

The Apollo Telescope Mount experiment became a key part of the Apollo Applications Program. Mueller wanted Marshall to build the telescope because it had to be integrated with the orbital workshop, which MSFC was building. Newell wanted it to be built at Goddard Space Flight Center, and he also wanted a different design. To resolve this disagreement, the two associate administrators went to see Seamans, and after going through their arguments, the general manager, who had become deputy administrator upon the death of Dryden in December, said, "Well, I think we ought to do it down at Marshall." Mueller always did his homework before such meetings, while Newell "tended to be more

---

[1]   Mueller, NSC, Washington, DC, 8/16/66.; "Rescue of Astronauts Viewed as Still Just a Hope," *NYT*, 8/17/66, and "Safety Concerns Mueller," *Orlando Sentinel*, 8/25/66, GEM-83-5; Mueller to Teague, 9/21/66, GEM-44-1; Mueller Interviews: Slotkin, 2/24/10 and Bubb, 5/22/69.

[2]   Bilstein, *Orders of Magnitude,* 340; "Data Indicate Apollo Flight Was a Success," *St. Louis Post-Dispatch*, 8/26/66, GEM-83-5.

[3]   Compton, *Where No Man Has Gone Before,* 68-69; "Where are they Now?" David J. Shayler, *Spaceflight*, 3/1979, GEM-194-7.

emotional and less logical," according to Mueller. Consequently Seamans usually sided with him, as in this case.[4]

Mueller set the final deadline for completion of the Gemini Program as January 1967. If the last two flights had not taken place by then, he planned to scrub them. NASA scheduled Gemini XI for early September and Gemini XII in late October, less than two months apart. Long debated ideas about a rendezvous on the first orbit of Gemini XI to simulate an Apollo lunar module lifting off from the Moon remained under consideration. Meanwhile some engineers proposed tethering Gemini to an Agena to assist in station-keeping and to study the dynamics of tethered objects in space, a potentially efficient method of propulsion which translates rotational energy into velocity. The air force also wanted the astronaut maneuvering unit to be tested on the last Gemini flight. NASA assigned Gemini XI to Conrad and Richard F. Gordon, Jr., and Gemini XII to Lovell and Aldrin.[5]

Ever the scientist, Mueller had an interest in studying the ability of astronauts to do work in weightlessness, and in late 1965 called a small group of NASA engineers together to determine whether there was anything the agency could do to offset its effects. He thought about using an artificial gravitational field, going so far as to patent some of his ideas, and Langley undertook several studies of the concept. Mueller said he wanted to simulate gravity in order "to preserve the physiological wellbeing of man in the exploration of space." He closely followed the research at Langley, and proposed a number of experiments to test his theories. Questions about artificial gravity increased in importance because of plans for the orbital workshop. One leading proposal involved spinning the space station, but research results from Langley convinced Mueller to change his mind about this. "There was quite an argument in the early stages, do you have artificial gravity or not," he recalled. Phillips, Gilruth and von Braun favored spinning the orbital workshop, so to convince them that this would not work they all participated in an experiment at the Naval Aerospace Medical Institute in Pensacola, Florida. Sitting in a slowly spinning room did not bother them at first, but after about a half an hour, Mueller remembered, "the great desire for artificial gravity dissipated." This would not be the only time that he participated in experiments and training exercises. He went up in an aircraft that flew parabolic arcs to briefly simulate weightlessness, and later learned to scuba dive to take part in training sessions in which astronauts exploited underwater neutral buoyancy to simulate working in zero-gravity. He wanted to "understand how hard it really was to do something in that kind of an environment," he said, though he found swimming underwater more to his liking than flying parabolic arcs or sitting in a spinning room.[6]

---

[4]   Mueller Interview, NASM, 5/1/89; Seamans, *Aiming at Targets*, 131-133. Just before Christmas 1965, NASA Deputy Administrator Hugh L. Dryden died, and Webb appointed Seamans as his new deputy.

[5]   Hacker and Grimwood, 15-1; Mueller Interview, Slotkin, 2/24/10.

[6]   Weightlessness, 1965 to 1968, GEM-109-10-11; Hitt, et al., *Homesteading Space*, 17-18; Mueller Interviews: JSC, 1/20/99 and Slotkin, 2/24/10.

Mueller recognized it would be difficult to obtain funding for an extensive human spaceflight program beyond Apollo, yet OMSF continued to sponsor studies of planetary exploration, and proposed Mars surface exploration. Given that the initial Apollo Applications Program was expected to end in 1971, one study laid out a second AAP consisting of integrated planetary exploration which would include a Mars landing as early as 1983, a mission to Venus, and extensive lunar exploration. Although he usually refrained from talking about these internal studies in public, he occasionally gave hints to his thinking in speeches and interviews. And in August, he told *Christian Science Monitor*, "I am sure the time will come when we want a manned expedition to the planets, particularly Mars, just as we now have the Moon program ... There has to be not only scientific justification for it, but there are technological, operational, over-all sociopolitical aspects to such a decision."[7]

As the Gemini flights and Apollo development continued, Aldrin, now assigned to Gemini XII, made significant engineering contributions to the programs. In addition to his work on orbital rendezvous, he helped to develop techniques to train for working in weightlessness by wearing a pressurized spacesuit underwater. Gilruth had "mixed emotions" about training underwater, saying "some of our people don't think the neutral buoyancy work was any good." However, Cernan tried it out in a Houston swimming pool and found it similar to his movements in space. As Mueller remembered, "Buzz Aldrin was busily learning how to operate a spacesuit under zero-G conditions." He learned not to fight the suit, but to let it help to solve the problem. Although MSC did not use this training method in preparation for Gemini XI, they made other changes to improve astronaut maneuverability in zero-gravity.[8]

On September 12 an Atlas took off with Agena-11, followed shortly by GT-11. Once in orbit, Conrad and Gordon turned on the rendezvous radar and found they were less than 60 miles from the target. An hour later, they were station-keeping alongside the Agena, having made the technically demanding first-orbit rendezvous. After practicing docking and undocking several times, they called docking in space easier than in the Houston trainer. Then they briefly fired the target vehicle's engine for calibration purposes, directing the thrust out of plane so as not to disturb their circular orbit. Gordon's EVA came next, and he attached a ninety foot tether to the Agena, in the process experiencing some of the same problems that frustrated his predecessors in trying to work in zero-gravity. Because of this, he returned to the spacecraft after about 33 minutes, cutting the EVA short by more than an hour. Consequently, the agency still did not know if astronauts could perform useful work in space. Gordon later said, "I knew it was going to be harder, but I had no idea of the magnitude."[9]

---

[7]   "Planetary Exploration Utilizing a Manned Flight System," 10/3/66, GEM-73-2; "After the moon – where?" *The Christian Science Monitor*, 8/4/66, GEM-83-5.

[8]   Mueller Interviews: NASM, 6/22/88 and JSC, 1/20/99; Hacker and Grimwood, *On the Shoulders of Titans,* 15-3; Logsdon, *Exploring the Unknown*, 45.

[9]   Hacker and Grimwood, *On the Shoulders of Titans,* 15-3.

6-1   Edwin E. "Buzz" Aldrin training underwater for Gemini XII. (NASA photo)

The next day they used Agena-11 to increase their apogee to 850 miles; their operational ceiling was a thousand miles to stay below the Van Allen radiation belt. This was done to take color photographs for comparison with the black-and-white television from meteorological satellites at that altitude. After they had recircularized the low orbit, Gordon performed a second EVA in which he stood on his seat to do experiments. Once he was back inside, they began the tether experiment. With some effort, Conrad finally got the two spacecraft rotating around their center of gravity The system rotated at a rate of fifty-five degrees per minute, or about one revolution every six and a half minutes, creating a slight, though detectable artificial gravity, and proving that it was possible for two spacecraft to station-keep economically. After running the experiment for three hours they jettisoned the bar that connected Gemini XI to the tether. Because they had plenty of fuel remaining, they conducted another maneuvering test, orbiting behind the Agena to practice long distance station-keeping. After returning to the Agena simply for practice, they set the computer for the program's first automated re-entry to conclude a mission of almost three days. Following touchdown, Mueller expressed pleasure with the results when explaining its accomplishments to the press. But he was concerned by the problems that Gordon had experienced in performing the EVA.[10]

Later that month, Mueller laid out his views about the broader implications of space exploration at the dedication of a science building at a small Pennsylvania college. In the international arena, he said, the space program became a measure of the nation's "technological eminence worthy of emulation." He called the USSR a formidable rival which considered space a factor in its competition with the US, and called technological progress "a basic source of national power" which had become "an important instrument in international relations." Thus, falling behind the Soviet Union in space technology could not be allowed because it would jeopardize national interests. And rather than calling this competition a space race, he said NASA geared its programs to scientific needs consisting of logical solutions to complex scientific and technological problems. Nonetheless the agency kept the Soviet Union in mind when studying the next steps to take. Domestically, the space program focused attention on the nation's education system, resulting in new standards for teaching mathematics and science at all levels. Spinoffs from space R&D had long range implications; and space exploration would answer questions about the solar system and the whole cosmos. However, in addition to its immediate benefit to national security, and as a stimulus for economic growth, he called human spaceflight "a catalyst to the achievement of the goals of our society."[11]

II

Unlike previous missions, Gemini XII lacked concrete objectives except to fly the astronaut maneuvering unit for the air force. However, before trying the next EVA, Mueller had to consider the difficulties previously encountered, and this led to the cancellation of the astronaut maneuvering unit test because of its potential danger. Like a winning coach at the end of an important game, he did not want to endanger the program's reputation by attempting this risky EVA, and chose to run-out the clock. As he wrote at the time, "I feel that we must devote the last EVA period in the Gemini Program to a basic investigation of EVA fundamentals ... through repetitive performance of basic, easily-monitored and calibrated tasks." Removing other potential risks, the final flight plan would not be ready until October 20, three weeks before liftoff. However, Aldrin's underwater training paid off, as he became familiar with the effort involved in working in zero-gravity. In addition to this training,

---

[10]   Ibid.; Seamans, *Apollo*, 67.
[11]   Mueller, Cedar Crest College, Allentown, PA, 10/19/66; Lecture IV, Mueller at the University of Sydney, Australia, 1/11/67.

NASA installed additional restraints, handrails and a waist tether to enable him to move efficiently to a work site and then tie himself in position so that he would have both hands free.[12]

Speaking at Caltech several weeks before the last Gemini mission, Mueller said the first piloted Apollo-Saturn flight would take place in a few months. He claimed that as the US continued to invest in human spaceflight, the USSR made greater investments, "two to three times as great a share" of their domestic product, and he argued that the development of space technology had a positive impact on the US economy. The lesser developed parts of the country, especially the South, received the greatest share of NASA infrastructure and jobs. He then claimed "the space program is not in conflict with efforts to end poverty and improve human welfare," called the "challenges and opportunities" presented by the space program "virtually limitless," and said the "presence of man will be essential in all of these missions of earthly benefit and scientific investigations." Then in remarks made a week before the flight, Mueller said, "Mankind is in space to stay." With 420 US satellites and probes launched since 1957, more than twice the Soviet Union, "Space exploration is a continuing and vital technological pursuit" affecting all Americans regardless of where they live. Since starting Gemini in late 1961, he said NASA had learned what was difficult and what was easy to do in space, and while EVA required "intense concentration," astronauts had conducted useful work during a total of six hours of such activity. Aldrin would perform three additional EVAs on Gemini XII that would almost double this, and Mueller concluded somewhat cautiously, "We hope data collected from this exercise will provide a sound basis for extravehicular activity in Apollo."[13]

On the afternoon of November 11, an Atlas rocket boosted Agena-12 into low Earth orbit, and approximately ninety minutes later GT-12 lifted off. Once in orbit, the radar on Gemini XII locked on to Agena-12, but shortly thereafter it failed. This did not present a problem, however, because it was Aldrin who had developed the backup manual procedures. And after successfully completing the rendezvous, the astronauts practiced docking and undocking several times. Because of a fluctuation in the pressure of the combustion chamber of Agena-12's engine during its insertion into orbit, Houston decided to forgo using it to boost Gemini XII into a higher orbit, a maneuver that had been demonstrated on previous flights. Then Aldrin performed a standup EVA, installed a telescoping pole across to the Agena, and conducted several experiments. The next day he used the pole to move to the Agena to install a tether and perform a number of demonstration tasks, and did not have the same difficulties as the other astronauts because of the physical, procedural and training changes made for this flight. Leaving the Agena, he moved to a "busy box" at the rear of the Gemini spacecraft, stood in a foot restraint, and used tools to turn bolts and cut metal to show that this could be done in zero-gravity. Commenting on his

---

[12] Hacker and Grimwood, *On the Shoulders of Titans,* 15-4; Seamans, *Apollo,* 67.
[13] Mueller, 75th Anniversary Conference, Caltech, Pasadena, CA, 10/25/66; Mueller, Wisconsin Chamber of Commerce, Milwaukee, WI, 11/2/66.

EVA, Aldrin later said, "With this system, I could ignore the motion of my body and devote my full effort to the task at hand." Returning to the capsule after two hours, Aldrin settled back into his seat. Lovell practiced maneuvers, including an experiment to test whether the gravity gradient could stabilize tethered vehicles. The next day, Aldrin conducted a second standup EVA to complete his experiments. Then they prepared to return to Earth. The failure of the rendezvous radar had a positive side benefit of allowing NASA to test backup methods. However, this mission would be remembered for Aldrin's successful EVAs.[14]

At a White House ceremony, the president presented the crew of Gemini XII with NASA's exceptional service award, while Mueller and Mathews each received a distinguished service award for their contributions to the Gemini Program. As Seamans later wrote, "There were many deserving of recognition, but none more so than George Mueller, who had revitalized the [program] soon after his arrival, and Chuck Mathews, the day-to-day manager." About 1,800 days separated the approval of Gemini in December 1961, and its completion in November 1966. Because Gemini and Apollo development took place concurrently, it did not contribute to the Block I spacecraft, but it helped the development of Apollo's fuel cells and the Block II. Many procedural lessons were applied to Apollo, and since most NASA personnel working on Gemini transferred to Apollo, they brought operational experience with them. Moreover, NAA and Grumman used some of the manufacturing and test procedures developed by McDonnell Aircraft. Mueller's method of handling experiments continued, and exchanges of information took place between program offices, particularly in the area of flight operations. Mueller said that converting the Gemini contracts to cost plus incentive fee saved millions of dollars. Although the final accounting showed that Gemini cost $1.15 billion – more than twice the original budget – this was less than the projected run out cost of $1.35 billion when he arrived at NASA in 1963. Thus, under his management, the Gemini Program saved at least $200 million, and probably more, given that without his intervention the costs would probably have escalated beyond $1.35 billion. Gemini helped to develop technology and operational procedures for human spaceflight, in particular rendezvous using a number of methods. Of the fifty-two experiments conducted, twenty-seven involved technology or tested new equipment, seventeen were scientific, and eight involved medical procedures. Astronauts learned how to live and work in space, and the agency acquired ways to deal with unplanned situations, while cutting months off of the flight schedule. Asked the main lesson learned from Gemini, Mueller later said, "Perhaps the most important thing was the demonstration that you could put a program together ... [and] carry it out on schedule ... But neither of those would have been possible without the structure of program offices that we set up."[15]

---

[14]   Hacker and Grimwood, *On the Shoulders of Titans,* 15-5; Seamans, *Apollo,* 67; STAC, MSC, Houston, TX, 12/9-10/66.

[15]   Distinguished Service Awards, GEM-66-5; Seamans, *Apollo,* 67; Hacker and Grimwood, *On the Shoulders of Titans,* "Summing Up;" Apollo Oral History, 7/21/89, NASA Archives, RN 18924; Mueller Interview, JSC, 1/20/99.

6-2   Completing the Gemini Program, James A. Lovell, Jr. and Edwin E. Edwin E. Aldrin, Jr., Gemini XII. (NASA photo)

Vice President Humphrey wrote to Mueller on December 8, congratulating him for Gemini, calling it "a most impressive success from the first flight on April 8, 1964, through the twelfth which made November 15, 1966, such a memorable and historic day." While Mueller said he maintained friendly terms with Humphrey, he did not frequently interact with him or the president. Since the vice president chaired the NASC, they met from time to time and Mueller got to know him better than he knew Johnson. While Humphrey did not get involved in technical details, Mueller briefed him from time to time, but said he did not appear "particularly concerned." Mueller said he did not meet as often with the president. "Obviously whenever we had one of these major events I got to say hello to him . . . or he gave me a 'gong' of one sort or another." But they did not have much interaction because Johnson liked to deal with the "principals," mainly the administrator, and sometimes his deputy. As a result of talking with some people who did interact with him, Mueller added, "I was lucky."[16]

During the ten piloted Gemini flights, sixteen astronauts flew nearly 2,000 hours (four flew twice). They learned how to maneuver the spacecraft to rendezvous and dock, and re-enter the atmosphere to land on target. NASA developed techniques for EVA and conducted experiments, which Mueller called "essential to the Apollo lunar mission and to other space operations in the future." Looking back, he remarked "when you think of the steps that were involved in going from Mercury to Apollo, it was a significant step towards the Apollo Program." Gemini answered some of the questions that the agency would face during Apollo, and it contributed technology and operational procedures. In retrospect, he said, "it was a great decision to move forward, although there was a fair amount of discussion at the beginning of the Gemini Program as to whether we ought to do it at all." Gemini enabled NASA to learn how to operate in space. And while the program did not advance scientific knowledge, it showed that astronauts could "survive in the weightless environment, and still do things usefully ... that was Gemini's contribution," he said.[17]

NASA held a symposium in Houston on February 1-2, to wrap up the Gemini Program, focusing on its accomplishments. Mueller gave a short address recapping its original objectives, then said "our successful flight program has provided vivid demonstration of our achievements in each of these objectives." The ability to fly in space for fourteen days on Gemini VII removed doubt of the ability of astronauts and equipment to perform satisfactorily during the time needed to accomplish the lunar landing. And, he pointed out, "One of our more dramatic achievements has been the successful development of a variety of techniques for the in-orbit rendezvous of two manned spacecraft." He reminded his audience that rendezvous, a primary objective of the Gemini Program, was fundamental to the success of Apollo. During Gemini, eleven astronauts performed nine separate rendezvous, using seven different modes or techniques, and performed nine separate dockings with target vehicles. As a result, he said, "Gemini developed a broad base of knowledge and experience in orbital rendezvous and this base, will pay generous dividends in years to come." He spoke of difficulties performing EVAs, but noted that Gemini XII had shown that with proper training and equipment astronauts could perform useful work in zero-gravity. He cited Mission-76 as an example of the ability to respond to the unexpected, and to execute alternate plans, which led to achievement of all flight objectives. Then he pointed out, "The space effort is really a research and development competition – a competition for technological pre-

---

16   Humphrey to Mueller, Humphrey, 1965-1968, GEM-41-7; Mueller Interview, Slotkin, 9/9/09, both sessions.
17   Mueller, ARCS, Los Angeles, CA, 12/5/66; Mueller Interview, Slotkin, 9/10/09.

eminence which both demands and creates the quest for excellence." In a call to continue the effort, he concluded, "We must now, using this as a foundation, continue to advance. The American public will not permit otherwise – or better yet, history with not permit otherwise."[18]

In an address that he made in December, Mueller recapped his expectations for the Apollo Program over the next three years. He referred to the Apollo spacecraft as a workhorse with the ability to travel the 250,000 miles between Earth and the Moon and return. By 1970 the US would have flown more than 500 "man-days" in space; and astronauts would have circled the Moon, landed, collected samples and returned to Earth. While the Apollo Applications Program remained in its definition phase, he expected to complete the preliminary design during FY 1968. And with the approval of the administration and Congress, Apollo follow-on flights could begin as early as 1968 and run through 1971. But additional funding would be required to continue beyond that point to allow an Apollo to dock with and pressurize an empty Saturn S-IVB stage so that the crew could convert it into an orbital workshop. He discussed building the Apollo Telescope Mount to make astronomical observations, and a lunar mapping system to orbit and investigate the Moon using remote sensing techniques. These experiments could be combined in various ways, and after initial use, the workshop would remain in orbit while the astronauts returned to Earth. Then months later another crew could reuse it. He envisioned two mission categories after the initial Apollo Applications Program – long duration flights and space experiments, and he spoke of extended lunar exploration focused on space sciences and Earth applications. Planned experiments would appeal to the scientific community, while longer duration missions should satisfy engineers. After this first phase of the post-Apollo program, a new generation of missions, including space stations, extended lunar surface visits, and planetary exploration could take place using modified Apollo-Saturn space vehicles. And then, looking beyond AAP, Mueller spoke of a series of missions starting in the early 1970s that would ultimately take astronauts to the planets, although the first step required a space station in low Earth orbit. Steps beyond Apollo and the space station would address other longer term objectives, extending operations in space and supporting Earth observations and research. The state of weightlessness could be used for scientific and engineering purposes, and the vacuum of space would enhance scientific observation and experimentation. All of this would help to expand knowledge and provide other benefits. The space station would be the first step toward long term exploration, becoming a test bed for space operations. After proving the ability to support humans in low Earth orbit for several years, the space station would help to prove concepts to support human exploration of the planets, and be used as a platform for launching long duration missions.[19]

---

[18]   Hacker and Grimwood, *On the Shoulders of Titans,* "Summing Up;" Mueller, Gemini Conference, Houston TX, 2/1/67; Logsdon, *Exploring the Unknown*, 377-385.

[19]   "Manned Orbital Flight in the Seventies," Mueller, 12/5/66.

Mueller spoke at KSC's year-end staff meeting in December 1966, telling them NASA had a lot to be proud of, because "We are doing our job and doing it well." Apollo remained on schedule, costs continued under control, and society was already benefitting from spinoffs. He spoke of Gemini's success, and reminded everyone the astronaut remained "the principal consideration." Adding, "We do not hurry ... nor do we establish flight mission goals that would imperil the astronauts." Then as he always did when speaking publicly, he discussed the Apollo Applications Program, calling the space station an early requirement for whatever post-Apollo program was finally selected, and said these missions would directly contribute to the continuing workload at KSC.[20]

As the year ended, Mueller had the success of Gemini behind him, but problems with the Apollo spacecraft and lunar module remained. Nonetheless, as 1967 began the future looked bright with the first human flight on the Saturn IB scheduled to take place in less than two months. But the development of the lunar module remained particularly vexing because it was experiencing problems with its main engine, landing gear, and radar systems. Grumman found it a difficult vehicle to build because of its sensitivity to weight. Schedule slippage put it on the critical path, as the vendor worked to meet NASA's exacting specifications. The agency assigned engineers to work directly with Grumman, and Mueller said the lunar module's thin skin caused him the greatest concern, because "if we lost the pressure the thing would collapse." He considered the lunar module development to be "a tour de force," more difficult to build than the command and service modules. It would be "the first time we've ever built something to land on another planet," he said. And the agency could not be sure it would work because the first landing test would take place on the lunar surface; something he called, "a new and a wonderful experience, particularly on planet that was three or four days or seven days from where we were."[21]

In early January, Mueller visited Apollo tracking stations in Hawaii and Australia. He made several stops in Australia, and while in the neighborhood flew to Colombo, Ceylon (now Sri Lanka) about 3,500 miles from Western Australia to visit Arthur Clarke. He lectured at the University of Sydney, laying out possible objectives for human spaceflight after Apollo, calling for a balanced Apollo Applications Program, and emphasizing economic benefits and the value of space exploration. He also mentioned studies underway to determine exactly what kind of space station would best meet future needs, and how much it would cost. Then quoting American author William C. Faulkner, he said, "I believe that man will not merely endure; he will prevail. He is immortal ... because he has a soul, a spirit capable of compassion and sacrifice and endurance." Just days before tragedy struck the Apollo Program he added, "In the exploration of space, I believe with Faulkner that man will prevail."[22]

---

[20] Mueller, Annual Staff Meeting, KSC, FL, 12/12 1966, GEM-46-3.

[21] Compton, *Where No Man Has Gone Before*, 88; Mueller Interview, Slotkin, 6/10/10.

[22] Key Personnel Visited, Overseas Trip, 1/19/68, Correspondence, GEM-43-12; Lectures I & IV, Mueller, University of Sydney, Australia, 1/10-11/67; Phone Calls, 3/25/66, Call Sheets, GEM-71-1.

The *New York Times* published a front page article on January 22 written by science writer John N. Wilford saying, "A growing confidence that the United States will be able to meet its Apollo project goal of landing astronauts on the Moon by 1970 has plunged space planners into an urgent debate over where to aim next." He raised questions about the Apollo Applications Program, writing, "Some wags in Washington ask not 'After Apollo, what?' but, 'After Apollo, why?'" Published just before the president submitted his budget for FY 1968, the newspaper pointed out that NASA asked for increased funding for the post-Apollo program, while some in Congress wanted to cut space spending to pay for the war and new social programs. The *Times* said, "The debate is expected to intensify after President Johnson submits his budget for the fiscal year 1968 to Congress," and after listing the different options, Wilford quoted Mueller saying, "we'll do anything the nation wants us to do," but the "logical next step" is an orbiting space station. Then Wilford turned to opponents of spaceflight, including one college professor who said, "I just don't think the Moon is going to be an adequate substitute for the fact that we haven't addressed ourselves to cleaning up the slums." Nevertheless, public support for the Apollo Program remained strong.[23]

The following week, NASA briefed the press regarding their budget request for FY 1968. Mueller said that the Apollo Applications Program would "maintain the momentum" of the team built to support Apollo. And the *Baltimore Sun* wrote, "One of the more impressive aspects of the entire project was Mueller's explanation this afternoon of how it was going to happen ... In public discussions he chooses his words carefully and is not noted for casual forecasts of future space operations. However, today he spoke with confidence and [a] degree of enthusiasm that led at least one observer to conclude NASA has its manned space flight program well in hand, including the lunar landing."[24]

Throughout 1966, NASA struggled to win support for the Apollo Applications Program. While the president supported space exploration, he had a host of other issues to deal with. Johnson assigned the budget bureau the task of reducing non-Vietnam spending to pay for these other priorities, and the agency could not persuade budget director Charles Schultze to increase spending to keep the Apollo industry team in place and avoid layoffs. Although by December, Schultze agreed that a decision had to be made about post-Apollo funding, he remained opposed to the amount that Webb wanted. Finally, the president acknowledged that unless he made a decision, human spaceflight would end with Apollo. Then in late January, Johnson agreed to include $454.7 million in the budget for the Apollo Applications Program in FY 1968. These funds would pay for additional launch vehicles and spacecraft, and for the development of the orbital workshop. Mueller

---

23 "Where to After the Moon? Post Apollo Plans Vague," John Wilford, *NYT*, 1/22/67, 7 GEM-83-6.

24 "NASA Plans to Send Workshop Into Orbit With Shuttling Crew, *Washington Post*, and "3-Year Orbiting Lab Planned," *Baltimore Sun*, 1/27/67, GEM-83-6; Compton and Benson, *Living and Working in Space*, 33.

had what he wanted and announced this decision just a few days before the Apollo 204 fire.[25]

III

"Instead of the glorious day that was anticipated, 27 January 1967 was a tragic day for NASA," wrote Seamans. The president invited space industry executives, NASA officials, the vice president and congressional leaders, and members of the diplomatic corps to the White House to witness the signing of a treaty on the peaceful uses of outer space. The signing would be followed by a dinner arranged by Mueller to mark the end of Gemini.[26]

Meanwhile, NASA was conducting a simulated launch test at KSC in preparation for AS-204. The spacecraft was atop the launch vehicle and all of the systems were operational, but for safety reasons the Saturn IB did not have fuel in its tanks. The test began at one o'clock in the afternoon There were some problems with the environmental control and communications systems. These were frustrating, and Grissom, the mission commander, became angry with intermittent communications problems. But nothing indicated any danger. Other minor issues plagued the test as it continued until 6:31 p.m., when the simulated countdown reached ten minutes to liftoff. Then a fire broke out in the sealed capsule and an astronaut yelled, "We've got a fire in the cockpit." A few seconds later someone yelled, "Hey, we're burning up in here!" Fourteen seconds after the first sign of fire, communications went dead. The increased internal pressure ruptured the shell and flame and smoke erupted from the capsule. Attempts to rescue the crew failed because the heat and smoke prevented opening the hatch for about five minutes.[27]

Just before NASA's dinner in Washington began, while everyone stood around with drinks in hand, Debus received a call from Petrone at KSC seconds after the fire. Webb, who had arrived late, was introducing members of Congress to the industry executives, and when he grabbed Atwood to introduce him to Senator Smith, the CEO, who had just got off the telephone, informed him: "They have got a bad fire at the Cape." Quickly caucusing, the NASA officials told their guests what they knew, and departed for headquarters to plan the next steps. They naturally considered the agency to be in the best position to investigate what had happened, but acknowledged that it would be the president's decision. In the interim, all agreed that Phillips should go to the Cape to take charge of the fire scene, impound all materials, and institute a press blackout until more was known about what had happened. Mueller recalled, "we then immediately moved to set up a review committee, an internal one, and began the process of damage control in the press." In fact, KSC officials could not provide much information to the media because at

---

[25] Lambright, *Powering Apollo*, 141-142.
[26] Seamans, *Apollo*, 74-75; Lambright, *Powering Apollo*, 142-143.
[27] Ibid.; Bizony, *The Man Who Ran the Moon*, 140; Bilstein, *Orders of Magnitude*, Ch. 5.5.

6-3   Scorched capsule after the Apollo 204 fire. (NASA photo)

that point they did not actually know much. Nonetheless, journalists clamored for information and wanted access to the capsule to take photographs.[28]

Following the Apollo 204 fire, Webb did not ban Mueller or the center directors from talking in public, but on February 2, Mueller and Scheer signed a directive instructing all NASA employees not to comment about the fire or the investigation, directing that "the speaker must stay away from those areas about the accident which speculate on causes or probable causes." Nor did the agency restrict employees from presenting technical papers, although they again cautioned staff to stay clear of commenting about the fire. However, as it turned out, the AIAA had a flight test conference at Cocoa Beach scheduled for February 6-8. Per the usual practice, the presenters had submitted technical papers for advanced publication, and two papers prepared by NASA engineers from MSFC raised eyebrows after the fire. Daniel H. Driscoll, Jr. wrote about ground testing of the Saturn, saying, "Because meeting schedule is of prime importance, there is a natural tendency for test engineers and others who influence his task to choose the most superficial answer (quickest solution) to a problem." Clearly, this was not what the agency wanted published by the press after the fire, although as it turned out it would be.[29]

Webb needed the president's agreement about NASA conducting the investigation and he convinced Johnson that the best people to handle it worked for the agency. He told the president that an independent investigation could damage him because of the role his former assistant, Robert G. "Bobby" Baker, had played in awarding Apollo contracts to NAA. He said NASA would do a thorough investigation, find out what happened and determine the best way to proceed. Johnson agreed and the agency had its investigation, though the president expressed concern that the mounting "sense of national excitement" about the space program would now be replaced with "national despondency." He told the White House press corps that the Apollo Program must continue. And the administrator then met with key members of Congress, who agreed not to start hearings until NASA had completed the investigation. Mueller later called permitting the agency to conduct the investigation "clearly the right way." While this was not universally appreciated, he pointed out that someone would have criticized whatever approach was taken.[30]

Seamans, Mueller and the center directors agreed that Floyd L. Thompson, the Langley Research Center director, should chair the review board. An "elder statesman" of the agency, Thompson had little direct connection with the Apollo Program, despite about a quarter of Langley's funding coming from OMSF. A well-

[28]   Murray & Cox, *Apollo, the Race to the Moon*, 201-202; Freitag Interview, Sherrod, 6/11/69; Webb Interview, Sherrod, 6/16/69; Mueller Interview, NASM, 1/10/89.

[29]   Mueller to Center Directors, 2/2/67, GEM-60-1; "Ground Testing of Saturn Launch Vehicle Stages," Daniel H. Driscoll, Jr., MSFC, Huntsville, AL, AIAA Paper 67-234, GEM-60-2.

[30]   Lambright, *Powering Apollo*, 145-148; Mueller Interview, Slotkin, 2/24/10.

regarded technical manager, he had everyone's confidence. The initial list of review board members included six NASA specialists, a representative of the air force, and an explosives expert from the US Bureau of Mines, who was also a member of PSAC. Seamans knew the board needed an astronaut, and after considering Schirra, selected Borman, who was considered to be a good engineer with a reputation for being bright and articulate. Thompson later said he was thankful that the agency put Borman on the board, because he "has a good connection between his mouth and his brain." Thompson also wanted some "Apollo 204 firemen," and eventually got fire experts from the Bureau of Mines, the Federal Aviation Administration, the Civil Aeronautics Board and the navy to help in the investigation. He also made sure the review board focused on the technical causes of the fire, and stayed out of management issues. When members of Congress later asked about this, Thompson told an interviewer, "I wasn't going to air the internal workings of NASA."[31]

After a few hours' sleep, Seamans flew to KSC, stopping at Langley to pick up Thompson. Arriving at the Cape they met with Phillips, Gilruth, Shea and others to discuss what they knew. Mueller remained at headquarters, conducted the planned Apollo executives group meeting, and then spoke to people in Washington "trying to do damage control," he said. He thought it essential to get the administration and Congress to understand NASA's actions in the wake of the fire. Almost immediately, the press started looking for someone to blame, asking "who caused this to happen." During the week following the fire, Mueller told the center directors to "hang together" and avoid blaming anyone or anything. He said "it's my problem, I'm responsible and you guys are going to support me. And that was an essential step to avoid having divide and conquer by the media." Webb wanted to shield people at NASA from the press, especially the members of the review board. All the while, journalists clamored for more information. The agency refused, and issued a press release which said that "a board had been established to review the circumstances surrounding the fire, to establish probable cause, and to review the corrective action and recommendations being developed." Seamans then instructed the review board to document their "findings, determinations, and recommendation and submit a final report to the Administrator [that was] not to be released without his approval."[32]

The media began looking into the use of oxygen in the capsule. This prompted NASA to investigate the original decision to use a pure oxygen atmosphere and to prepare background papers for use in the official inquiry. Mueller later said, "There were a number of communications I discovered about oxygen after the fire

[31]   Lambright, Powering Apollo, 144-145; Seamans, *Aiming at Targets,* 138; Seamans to Apollo 204 Review Board, 1/28/67, GEM-60-1; Myers Interview, Sherrod, 3/31/70; Mueller Interview, JSC, 8/27/98; Langley budget information, Langley, VA., 1967, GEM-77-10; Thompson Interview, Sherrod, 9/10/69.

[32]   Seamans, *Apollo*, 74-75; Skaggs to Slotkin, e-mail, 6/5/11; Lambright, *Powering Apollo*, 145; Mueller Interview, Slotkin, 2/24/10

but none of them that really raised the question to a point where you had to do something about it. They were observations rather than opinions." However, he admitted, "we were complacent about oxygen and not really [cognizant of] the potential disaster and it's just lucky that we didn't have a problem in space." The capsule needed oxygen for the astronauts to breathe, and it had to be at a partial pressure similar to oxygen at sea level, at least three pounds per square inch. The agency used oxygen at five to six psi in space, inflating the capsule to atmospheric pressure on the ground to balance the internal and external pressures. And for the cabin verification check during the countdown test the internal pressure had been raised above ambient. Mueller later called the system a basic design error, and said it was wrong to use pure oxygen at sea level. "It simply is something that the designer did not understand".[33]

As the investigation got underway, the three astronauts, all service members, were buried with full military honors. When Mueller, who received his master's degree in electrical engineering at Purdue University visited the campus in May 1968 to dedicate Grissom and Chaffee Halls, he said, "Every advance, every improvement in man's long struggle against darkness, has exacted a terrible cost – the lives of the noblest men of each era. But all mankind moves forward through the sacrifice of a few ... These astronauts knew and understood the risks of their profession, and they accepted these risks as part of their job. They did not expect to meet death, but they were prepared to meet it if such a confrontation should occur. They were deeply concerned that if an accident would happen, there should be no pause or slackening in our effort to reach the Moon and beyond. As Gus Grissom said, shortly before he died: 'The conquest of space is worth the risk of life.' "[34]

Thompson met and spoke with Mueller and Phillips during the investigation, keeping them informed, and used the resources of OMSF to support the review board. Mueller considered Thompson very smart and an excellent center director, characterizing their interaction during the investigation as "constructive." However, Webb later claimed that Mueller and Shea collaborated to control the investigation, while Webb insisted that *he* worked to protect it from manipulation. There is no evidence Mueller interfered, and when later asked about it he denied this allegation, saying, "I didn't ... I think [Webb] had a great imagination of all sorts of subversive activities going on behind his back. He was under a considerable amount of pressure and his personality was one that said somebody else had to be the problem, not him." The administrator must have made this accusation, Mueller speculated, because he did not understand technical details. "That was Jim Webb. He had to find a scapegoat." The review board operated independently of OMSF, and except

[33]   Diary, 1/31 and 2/1/67, SCP-65-5; Mueller Interviews: Slotkin, 2/24/10 and Bubb, 5/22/69.
[34]   Appointment Book, 1967, GEM-70-3; Mueller Interview, Slotkin, 2/24/10; Mueller, Dedication of Grissom and Chaffee Halls, Purdue University, Lafayette, IN, 5/2/68. Grissom and Chafee had received their bachelor's degrees from Purdue following World War II; and Mueller earned his master's in electrical engineering at Purdue in 1940.

for receiving reports from Thompson, Mueller personally had little direct contact with the investigation. Webb concentrated on containing the political fallout, while Seamans interfaced with Thompson, and Mueller kept the Apollo Program going and helped people overcome the trauma. Some felt guilty, and he tried to help them get over those guilt feelings. And he said, "Everybody felt we should have been able to prevent this." He tried to restore morale to keep the program going and make the necessary fixes, but it was a very stressful time and the loss of momentum delayed the lunar landing by a year or more. [35]

Mueller asked Leroy E. Day, Apollo's director for test operations, to check what tests the agency conducted on spacecraft materials flammability. Day told Mueller that a materials selection committee issued guidelines for spacecraft cabin materials, requiring that fire must not support combustion in a pure oxygen atmosphere. NASA tested and approved over four hundred materials, with most tests taking place at 5 psi, although some occurred at 15 psi, slightly higher than atmospheric pressure of 14.7 psi. After the fire the agency conducted additional tests at sea level pressure, and Mueller said he was "shocked" and "amazed" by how quickly the fire spread. He admitted "if there was a failure, it was a failure to test all of the equipment under the conditions it was going to be used." Under space conditions, at 5 psi, there would not have been a problem; though testing at atmospheric pressure with pure oxygen, created a different set of physical circumstances. He later said, "ideally you test every part under the conditions it's going to be used, and at extremes on either side of that condition ... [but] the one thing that we did not test was the fire resistance of a capsule" at sea level with a pure oxygen atmosphere. The use of 100 percent oxygen originated in Mercury, and continued in Gemini and then Apollo. Mueller said "it's one of those things where you don't really question it till you've got a problem ... because it was part of the lore." NASA depended on prior experience, never thinking there could be a problem, and he said "if you don't imagine you're going to have a fire, well, then you don't test for it." He called this a problem when defining any test program, making sure to stand back far enough and objectively look at the possible problems to test for. And you have to do it "with a great deal of reasoning ... because you can very easily use much more money testing for things that aren't really important, and delay the program almost indefinitely."[36]

IV

The administrator decided the review board should not issue interim reports, only a full report at the end of the investigation. However, because the president, Congress, and the press would not wait for answers, Webb told Seamans to meet with the

[35] Diary, 2/1-28/67, SCP-65 folder 5; Mueller Interviews: Slotkin, 2/24-25/10 and 2/22/11, and NASM, 1/10/89; Webb Interview, Sherrod, 4/28/71; Shea Interview, Sherrod, 5/16/71.
[36] Notes re flammability, 1/28/67, Apollo 204 Fire, GEM-60-1; Mueller Interview, NASM, 1/10/89.

board weekly, draw his own conclusions and prepare a written report. When Seamans made his first report, Webb told him to remove everything of an emotional nature, striking out whole sections and reducing it to a basic statement of facts. Once he approved it, the administrator personally took it to the president, sent copies to the chairs of the congressional space committees, and later distributed copies to the press. Prepared on February 3, the first report said the agency had recorded a significant amount of telemetry data during the test, and it contained a timeline showing that fourteen seconds after Chaffee detected the fire, flight controllers lost contact with the crew. The cabin pressure rose and shortly afterward the capsule ruptured. "The cause of death was asphyxiation due to smoke inhalation," Seamans explained.[37]

Meanwhile, Mueller issued a press release on February 3, announcing the three Apollo-Saturn missions scheduled for 1967 – AS-206 (later redesignated AS-204R because the automated test would use the Saturn IB that had been assigned to the first piloted mission), AS-501 and AS-502 – would proceed. He told NASA contractors to assume that the first manned flight would be a Block II spacecraft on a Saturn IB. Phillips and Shea met with contractors in Houston to tell them about plans to continue the flight test schedule, and Shea said "Our job is straight forward. Keep the program moving, find out what happened, fix it, convince the world that the fix is right and press on."[38]

Although the congressional leadership had agreed to await the outcome of the NASA investigation, some members pushed for an independent review. Teague silenced members in the House, telling them to give the agency time. However on January 31, Senator Anderson wrote to Webb asking for information, and asked for Seamans, Mueller and Berry to testify in executive session on February 7. Webb agreed. In his opening statement Seamans told the senators, "In parallel with the Board investigation and review, the NASA Manned Space Flight Program Office . . . is continuing its assigned responsibilities in the Apollo Program . . . [It] retains a major responsibility in the collection and analysis of accident data for use in its own assessment of the accident and the possible corrective actions that it may be required to take." He assured the committee no major decisions would take place before completing the investigation, and NASA would inform Congress prior to any public announcement. This session satisfied Anderson, and he told the press that he had full confidence in NASA's investigation. The headline in the *Baltimore Sun* on February 8 said "Senate Committee Says Fire May Not Delay Space Race." But the *New York Times* highlighted "NASA Engineers Criticize Test Schedule Pace," and quoted Driscoll's AIAA paper. After the Senate space committee released NASA's testimony, the *Times* published an article which quoted Mueller denying accusations that the agency had decided to use pure oxygen for the capsule in order to speed up

---

[37]   Seamans, *Apollo*, 75.

[38]   Press Release 67-21, 2/3/67, GEM-60-2; Holcomb to distribution, Meeting of Apollo Program Managers, NASA and Industry, 2/3/67, MSC, Houston, TX, SCP-65-4.

its development, calling the charge "serious" and "unfounded." To the contrary, he said the program "proceeded at a deliberate pace so that it would, in fact, reasonably, economically, but certainly safely, arrive at a set of equipment capable of carrying out the mission."[39]

People hesitated to bring things to Webb because of his verbosity, which Sherrod called "amazing," naming him "the most prolific memo writer in history." Seamans, Mueller and others faced a dilemma in dealing with him: he wanted to be kept informed of significant issues, especially anything having to do with the press, while not wanting to get into the technical details; though after the fire he tried to dive into technical matters. Yet he still could not get too involved because he did not have a technical background. Before the Apollo 204 fire, Seamans had found a way to work with Webb, although after this the administrator did not act like the man he had known for the past six years. "He was gruff ... I hadn't seen him that way since the last days of Brainerd Holmes ... Everyone at NASA was feeling the strain," Seamans explained.[40]

Following the fire, Gilruth claimed that Shea did not tell him what went on in the Apollo Spacecraft Project Office, complaining that he reported directly to Mueller in Washington. However, Mueller insisted, "in all honesty that wasn't the case. Bob likes to think that he didn't have anything to do with it but in fact he was informed and did participate." Gilruth signed off on all of the key documents that certified equipment being ready for flight. And according to Shea, he kept the center director informed, despite having difficulty getting him "to sit still to listen long enough". Shea claimed that he became "an intermediary, without ever even appearing so, between Mueller and Gilruth. I could talk to Gilruth, at least until the accident, he was happy and trusted me, and I was his guy, even though I was also Mueller's guy." Shea said Gilruth and Mueller were totally different. Gilruth was inarticulate and used emotional arguments, whereas Mueller wanted everything laid out in a logical manner. This led to difficulty between the two, and they could not communicate effectively with each other. Gilruth had no answers for Mueller, Shea said, leaving "no basis" for useful communication. Mueller said some people make decisions "by feeling rather than reasoning," people like Gilruth, Newell, and Storms. And he could not understand what caused them to make decisions that way. They made emotional rather than "reasoned" arguments, while he remained logical, believing that logic always overcame emotion; "if the logic is good," he added.[41]

Seamans' second interim report, which was published on February 14, described

---

[39] Lambright, *Powering Apollo*, 149; "Statement of Seamans," 2/7/67, Senate Space Committee, GEM-60-2; "NASA Current News," 2/9/67, and "Haste in Space," *Wall Street Journal*, 2/20/67, GEM-60-3; "NASA Rejects Race To Moon As Factor In Astronaut Deaths," *NYT*, 2/12/67.

[40] Vogel Interview, Sherrod, 10/23/69; Mueller Interview, Slotkin, 2/24/10; Seamans, *Aiming at Targets*, 140.

[41] Mueller Interviews: Sherrod, 4/21/71 and Slotkin, 2/22/11; Shea Interview, Sherrod, 5/16/71.

the composition of the review board, noting that about 5,000 scientists, engineers and technicians participated, and that Thompson established twenty-one expert panels to conduct the inquiry. Seamans wrote about fire propagation, and described the steps involved in the technical review of the fire. And the report said, "All three spacesuits were burned through ... [although] the cause of the fire has not been determined." Grissom had the most severe injuries, with burns over sixty percent of his body, though his death was caused by inhalation of toxic smoke.[42]

In the middle of this crisis, on February 20, Mueller gave a speech at an IEEE meeting saying, "We are making a maximum effort to learn the cause of the fire, so that we can make whatever future precautions are indicated as the result of the findings of the investigation." And then he reiterated NASA's position that work on "the unmanned flights" in the Apollo Program would proceed. Human spaceflight would await the results of the investigation, and all testing that involved pure oxygen at high pressure would remain restricted. And he said that until the review board released its report, he could not comment about the fire. Apollo-Saturn testing would continue, following a series of critical decision points. Emphasizing safety, he said, "Before and after each of these critical events a complete review of vital systems, expendables and other aspects of the mission status is made." And after each review, NASA would make a decision about proceeding to the next step. Much had been accomplished, but there was a lot to learn about human capabilities and limitations in space. Nonetheless, the agency would continue planning the post-Apollo program using capabilities of the Apollo-Saturn system. Despite uncertainty in the near term, he said NASA was planning a program of space exploration which would result in human exploration of the solar system and, "ultimately, the search for extraterrestrial life itself." Mueller made this speech to change the subject, because in the aftermath of the fire he could only talk about the future. "You know it was not the kind of thing that was easy to explain and certainly not to the public that any spark would have caused [the fire] and we had designed it so [the crew] couldn't get out – that was very hard to explain ... And we really never planned on the fire."[43]

Mueller did not lose confidence. However, he recognized the risks, calling human spaceflight "serious business." Although as a nation the US usually succeeded in the things that it put its minds to, he wanted people to recognize the risks involved in Apollo. But he never failed to assure people "we will land a man on the Moon by the end of the decade." He did not waver, at least publicly, because he argued, "I didn't have any reason to ... I really did understand the program fairly well at that time and knew what contingencies we had. Now you can't be 100 percent sure that you're going to be able to land people on the Moon and bring them back in one piece but as sure as you can be, I was." In his own way of carefully choosing his words, he said,

[42]   Seamans to Webb, "Further report on Apollo 204 Review Board Activities," 2/14/67, GEM-60-3; Thompson Interview, Sherrod, 9/10/69; Seamans, *Apollo*, 75.

[43]   Mueller, IEEE, Washington, DC, 2/20/67; Mueller Interview, Slotkin, 2/24/10.

"It was just reasoned confidence. It wasn't blind confidence. It was reasoned," and "we did" land the astronauts on the Moon and bring them back safely. Apollo involved far more than his personal pursuit of a goal, it involved a national commitment. While dissenters existed, the administration, an overwhelming majority in Congress and the public supported the goal. So when it came to the point of having to make a decision, the "program won," he said. The nation accepted the goal and despite losing the AS-204 crew, support for Apollo remained strong. And he added, "It would have been a disaster if we had given up the goal at the time of the fire because that would have destroyed much of everything we've done in terms of space leadership and put us behind the Russians for sure."[44]

Seamans issued his third and final report on February 25. Based on prior testing and because of using a pure oxygen atmosphere in the past, it had appeared to all that there existed little chance of a fire. "Neither the crew nor the development and test personnel considered the risk of fire to be high," he wrote. Webb made no public statements after the first two reports, but following this one he issued one of his own. He said he had met with three members of the review board – Thompson, Borman, and the fire expert from the US Bureau of Mines, Robert van Dolah. Based on these interviews, and Seamans' reports, he had concluded that the degree of risk of fire could not have been foreseen, and that NASA had never realized how difficult it would be for the astronauts to escape from the pressurized capsule on the ground. Velcro, used in much abundance, caused an unseen danger. Melting of aluminum joints in the oxygen distribution system had spread the fire, which burned in a way that prevented the astronauts from reaching the hatch. Webb concluded that many items in the design and performance of the environmental control system required reexamination and redesign. Borman told him that he would have entered the capsule on January 27 without concern, although from what he learned since "there were hazards present beyond the understanding of either NASA's engineers or astronauts." In response to these reports, Anderson scheduled a hearing of the Senate space committee for February 27. The House held off until the review board issued its final report, at which time Teague's subcommittee planned to hold hearings on behalf of the full House space committee.[45]

As the Senate hearing opened, Seamans, Webb, Mueller and Phillips faced the committee. Mueller spoke first, reading a long statement, describing what happened and what NASA intended to do based on the review board's initial findings. He explained that the simulated launch test on January 27 used conditions and procedures in accordance with seven years of operational experience, which led them to consider the potential for a fire in the spacecraft cabin remote. However, he called NASA's approach to preventing fires inadequate, and told the committee that the agency had already taken actions to reduce the probability of ignition in the

---

[44] Mueller Interviews: Slotkin, 2/24-25/10.
[45] Seamans, *Apollo*, 76; Mueller Interview, Slotkin, 2/24/10; Lambright, *Powering Apollo*, 150-151; Statement by Webb, 2/25/67, GEM-60-3.

6-4   Senate Hearings, Apollo 204 Fire: Robert C. Seamans, Jr., James E. Webb, George E. Mueller, and Samuel C. Phillips, February 27, 1967. (NASA photo)

capsule, though they could not absolutely eliminate the risk. NASA explored ways to reduce the ability of fire to spread, and studied how to minimize its consequences should one occur again. The investigation received the agency's full support and resources, with analyses and studies underway to either change designs or to develop procedures to minimize fire danger. Once NASA understood what had actually happened, the agency planned to simulate different phases of the fire in order to increase their understanding, and planned to review the spacecraft design in view of what they had learned before making any changes. Then they would evaluate the redesign and new procedures to insure their soundness.[46]

Mueller discussed various methods to extinguish fires aboard the spacecraft. It would not be possible to completely eliminate accidental ignition in the capsule, but NASA would select new materials and arrange them to minimize burning and thereby prevent the spread of fire. The agency had conducted a complete review of

---

[46]   Lambright, *Powering Apollo*, 151; Statement by Mueller, Senate Space Committee, 2/27/67.

the spacecraft design from the perspective of emergency egress, and a redesign of the hatch was already underway. A combination of unusual events during Mercury and Gemini had led to the initial Apollo hatch design. The hatch on Grissom's Mercury spacecraft blew off after landing, causing the capsule to sink. Mueller said Grissom "opened a hatch when he shouldn't have, so we fixed it so that it was harder to open ... and it was sealed better." That turned out to be the wrong thing to do because internal pressure sealed the capsule and impaired emergency egress. Another reason for the hatch design came from the Gemini EVAs. Astronauts had difficulty sealing the hatch, and testing showed that gases escaping from materials newly exposed to the vacuum of space created pressure that kept it open. Before the fire the spacecraft had two hatches, one opening inward and the other outward. The most promising new design, and the one eventually adopted, involved a unified hatch that could be opened outward in just a few seconds, not the ninety seconds required by the previous one. And reexamination of the environmental control system resulted in replacing soldered joints that had melted during the fire with mechanical joints, and replacing aluminum tubing with steel.[47]

Mueller concluded his testimony by saying that the changes would be introduced in the first flight version of the command and service modules, the first one equipped with all of the features required for a lunar landing, the Block II Apollo spacecraft. After receiving the review board's final report and completing additional studies and tradeoffs, NASA would make decisions about which specific changes to make. After the modifications and several months of checkout, the spacecraft would be shipped to KSC for four months of preparation, testing, and further checkout in preparation for launch. In other words, human spaceflight would be delayed until at least the first quarter of 1968. In the interim, the three "unmanned" flight tests planned for 1967 would proceed. These would test the changes affecting flight safety, insuring that all spacecraft changes would be prequalified in advance of the resumption of piloted flights.[48]

In the questioning following Webb's testimony, Senator Walter F. Mondale asked about a report prepared by Phillips concerning the performance of NAA. The administrator said he did not know the answer, and referred it to Mueller who called the first Apollo spacecraft of higher quality than the first Gemini spacecraft, and said disagreements between the government and contractors occurred all the time, while NAA performed no better nor worse than any other contractor. Mondale continued to press, "Is it your testimony that there was no such unusual Phillips report? Is the rumor unfounded?" Mueller replied, "I don't know of any such unusual report." The senator persisted, asking questions about the existence of a report by Phillips, and Mueller could not recall having seen such a report. As he later explained, the point that he tried to make in his response to Mondale was that critical contractor

[47]   Ibid.; Mueller Interview, Slotkin, 2/24/10.
[48]   Statement by Mueller, Senate Space Committee, 2/27/67.

reports were routine in the space program, and "none of them were any more significant than the other."[49]

Seamans later wrote, "I knew George meant that there hadn't been a formal bound report. However, I suspected that Mondale may have had in his possession a version of the results of an informal 'tiger team study,' led by General Sam Phillips." What Mondale called the Phillips report had never formally been issued, so according to Seamans, Mueller gave a technically correct answer. Webb later accused Mueller of protecting himself when asked about the Phillips report. "He wanted to make sure he didn't get the blame." Rees, one of the primary authors of the document, called it a collection of notes; but various sections carried the label "report." Mueller later said "I didn't think it existed. I really didn't. I was pretty emphatic with Sam that he had to get rid of it. [But y]ou've got to realize that you can never get rid of anything." Adding, "It was later that I really discovered that it still existed ... So then I had to go tell Seamans and Webb that it was a report that Sam had prepared ... [and so] we were faced with battle damage." Seamans told Mondale that from time to time NASA conducted reviews of contractor progress and the information he referenced might be in that category. Webb thought he went too far in his testimony when he told the senator, "We do look at the performance of each one of our contractors, and I can remember discussing the possibility of a second source on parts of Apollo." Mondale kept returning to the Phillips report, and Webb tried to change the subject, first telling him the preliminary findings of the Thompson review board did not point to the cause of the fire being the responsibility of any contractor. Then he suggested that the senator might have confused the report with something else. No, Mondale insisted, he referred to the Phillips report issued in 1965.[50]

Although Webb insisted that he did not know about the report, Mueller had personally discussed the results of the NAA tiger team review with him, "not the whole report but the essence of it and what we had recommended," he said. Nonetheless he insisted that neither of them knew such a written report existed. On reading a draft, he had told the general "I don't think that's something we want to pass on to anybody, and I'm certain it's not something we want to have anywhere in the files." Mueller thought that other than a few personal copies, notes of the tiger team review did not exist. But as he later learned, quite a few copies did exist. NASA had intended the tiger team reviews to be constructive, because if they were not constructive then they were not worthwhile, according to Mueller. You do not want to tell someone they "screwed up," he said, you want to tell them what they had to do to improve, with enough specificity to enable them to know what problems you were trying to solve. However he claimed, "in all honesty I wasn't sure from the

---

49   Bizony, *The Man Who Ran the Moon,* 148-149.
50   Seamans *Aiming at Targets,* 141; Seamans, *Apollo,* 76; Webb Interview, Sherrod, 6/8/69; Rees Interview, Sherrod, 2/6/71; Mueller Interview, Slotkin, 2/24/10; Bizony, *The Man Who Ran the Moon,* 149.

Mondale question, which is where this whole thing started, just what in the world he was talking about."[51]

While internally Webb expressed unhappiness about being blindsided, publicly he supported NASA against accusations of incompetence. Let down and bitter, the two main targets of his wrath became Seamans and Mueller. From that point on, Mueller said Webb began a "search for the guilty party" in an attempt to get rid of the person or persons who had let him down, and he and Seamans became the primary targets. According to Sherrod, "Webb's antipathy for Mueller had deep roots," but he never tried to get rid of him. Nonetheless, after Webb resigned from NASA in October 1968 and people were promoting Mueller as his replacement, Webb said he would testify against Mueller – although as he said later, "I knew he would never get the appointment from President Johnson anyway." Meanwhile, he spoke about Mueller's and Seamans' shortcomings behind their backs and, bypassing them, gave directions to their subordinates. Seamans, always the engineer, said that Webb, a nontechnical person, "believed that the technical staff let him down," and as "the bridge to the technical people ... [I] had failed and needed circumvention."[52]

After finally seeing the Phillips report, the administrator called Seamans, Mueller and Phillips to his office and reproached them. The media focused on the agency's shortcomings, and blamed NASA for risking the lives of the astronauts in their effort to win the space race. As Webb stewed about the Phillips report, he had the agency's general counsel collect everything in Phillips' files about NAA. Mueller recalled "we had a mess on our hands really trying to keep the people in North American motivated and going in the right direction to get this thing rebuilt and built properly. And obviously there was a great deal of concern." However, he aimed to protect the program, not NAA, and if he could have changed contractors he would have. But, he said, that "would have been a disaster in terms of getting the program done because shifting contractors in the middle of the thing is very, very expensive both in time and money." He called the episode of the Phillips report, painful "because it surfaced at the wrong time and made it a very much more difficult set of hearings than it otherwise would have been." However, he pointed out, because NASA had friends and enemies in Congress, the agency never knew which side of an issue any member would take, and "you never know what quicksand you're stepping into."[53]

Mueller called Webb masterful in how he worked with the administration, so "the President didn't become so overtly alarmed that he wanted to do something about it," although "there was the problem with the Congress." NASA "had really staunch friends [in Congress], but after the first week or so, rumors about the Phillips Report began to circulate." The agency had a credibility problem, because "officially" there was no Phillips report. Mueller insisted that while the report "had almost nothing to

[51] Mueller Interview, Sherrod, 4/21/71 and JSC, 8/27/98.
[52] Seamans, *Apollo*, 76; Bizony, *The Man Who Ran the Moon*, 150; Mueller Interview, Slotkin, 2/24/10; Webb Interview, Sherrod, 6/8/69; Seamans, *Aiming at Targets*, 145.
[53] Lambright, *Powering Apollo*, 151-157; Mueller Interview, Slotkin, 2/25/10.

do with the fire itself … it did have a lot to do with the competence of the management of the company." And because the administrator had backed Mueller when he had denied the existence of the report, it "caused a credibility gap particularly between Jim Webb and Congress."[54]

---

[54]   Mueller Interview, NASM, 1/10/89.

7

Recovery

"The space program is founded on faith and on belief."

Mueller, following the Apollo 204 fire

Mueller wanted the congressional hearings over quickly in order to get on with the Apollo Program, while Webb thought the agency should let them play out over several months because if they finished too soon and more facts came to light, NASA would be accused of whitewashing the situation. The administrator wanted the hearings to last long enough for all the relevant facts to surface, and because it took quite some time for everything to come out this strategy proved wise. Mueller and Webb also disagreed about the release of the Phillips report. Mueller wanted to keep it from public disclosure in order to protect the relationship with North American Rockwell,[1] suggesting instead that the agency summarize it for Congress, while the administrator did not want to volunteer any information. Nonetheless, the report leaked, the press published parts of it, and eventually NASA released the whole report to Congress. Webb found it difficult to maintain his objectivity after the fire, wanting "quick answers and straightforward and simple answers to questions, and problems" as Mueller recalled. It was a stressful time because Mondale was implying that NASA had covered up and colluded with NAR, and that he wanted to get to the bottom of the scandal. Even Teague, usually a "tower of strength," began to worry, according to Mueller.[2]

However, reports like the one Phillips prepared were not all that unusual. William E. Lilly, OMSF director for program control and an "old hand" in Washington, said that NASA produced many such reports. The agency conducted plenty of contract oversight, and when something went wrong it sent in a tiger team to review the situation, helping where they could, but very critical when necessary. Mueller and Phillips conducted regular contractor audits. Sometimes these were typed, while at

---

[1] NAA merged with the Rockwell Standard Corporation in March, adopting the name North American Rockwell – NAR.
[2] Mueller Interviews: Sherrod, 4/21/71 and NASM, 1/10/89.

other times they were not. Phillips laid down the law with NAR, and producing such tiger team reports was normal operating procedure. Nonetheless, Lilly said, "Phillips and Mueller were naïve about how these reports could be misused. And, according to Lilly, Webb knew about the content of the Phillips report without being told directly about there being a "Phillips report." He knew about problems at NAR, and that Mueller and Phillips were working to resolve them, even if Mueller never got into the "nitty-gritty details" with him.[3]

It was rumored at the time that Phillips threatened to resign from NASA unless the agency defended his communications with NAR, believing he had to be totally honest with contractors. While it would have been out of character for him to make this threat, if he felt strongly enough he could have requested a transfer back to the air force. Mueller did not think Phillips would have resigned outright, although he said "that was a time of considerable stress and everybody was willing to resign." As Webb remembered, "the military method of going into a contract was to crack down but not to reveal any of the company's business. Phillips felt honor bound to stick to this position. He probably would have left the program if we had revealed the whole report, I was told, though he never said so ... If I hadn't gone along I would have lost my people." Phillips did not resign, although he became more proactive after the fire. And in late March he told Gilruth to think about sending a team of engineers to NAR "for as long as necessary to insure rapid action," which resulted in placing a large contingent of agency personnel on-site at the space division.[4]

NASA's culture called for a careful engineering review of the facts surrounding the fire to determine what went wrong and to determine what had to be fixed, which is what the Thompson review board did. However, Webb did not want to investigate the engineering, he wanted to know *who* had failed him. He felt personally betrayed, and began meeting privately with Phillips. This did not create problems for Mueller because the general remained conscious of the chain of command, and kept everyone informed of what went on. The administrator acted "paranoid" and "lost trust in me," Mueller said, yet he "hung in there" and did his best under the circumstances. Webb did not confront him, so he had to figure out what he had to do and keep doing it, although, he said, "all of the machinations he did were just a waste of time really."[5]

There were rumors Webb wanted to get rid of Mueller, and after he resigned as administrator Sherrod asked why he had not fired Mueller. Webb replied, "I need[ed] him." Mueller had strong allies in Congress, especially Teague, and both the House and Senate space committees respected him. But around this time, the administrator began relying on Phillips. "I don't know why," Mueller said, "but

3   Lilly Interview, Lambright, 7/18/90; Freitag Interview, Lambright, 6/27/91.
4   Lambright, *Powering Apollo*, 151-157; Mueller Interviews: Slotkin, 2/24-25/10; Webb Interview, Sherrod, 4/28/71; Phillips to Gilruth, 3/30/67, and Phillips to Center Directors, 4/24/67, SCP-64-9.
5   Seamans, *Aiming at Targets*, 144; Mueller Interviews: Slotkin, 2/24-25/10 and Lambright, 9/21/90; Low Interview, Sherrod, 9/7/72.

Sam was of course a very even tempered guy and I guess Jim was probably not willing to take me on. I had enough support outside of the Agency ... although I know he was seriously thinking of dumping me but he never figured out how to do it." Webb thought people kept important information from him, and feared things happened that he did not know about. Then when Thompson did not uncover the exact cause of the fire, things got worse, because Mueller said, he had the "kind of mental state [where] the less you find the more suspicious you become." Their relationship became rocky and swung back and forth, but by the time the administrator resigned from NASA, Mueller claimed it had improved; though it was never the same as before the fire.[6]

After the initial shock of the fire, there were many problems. No longer trusting the people around him, Webb began to pull power from the program offices to tighten his personal control. He let people know about his dissatisfaction with Seamans and Mueller, and the animosity grew. He pulled Lilly from Mueller's organization, promoting him to assistant administrator for administration. But while Mueller thought highly of Lilly and would have preferred to keep him in OMSF, it did not hurt to have a friend in the administrator's office. Webb appointed an independent management task force led by retired air force General Frank A. Bogart, OMSF's deputy associate administrator for management, a man he still trusted, and asked him to review the Apollo Program. He also asked the head of his secretariat, army Colonel Lawrence W. Vogel, to lead another group to go through OMSF's files searching for other damaging reports. These reviews did not turn up any smoking guns, and Webb said "What we found in the files was not venal. It just showed [that NAR] were saying, 'just leave us alone; we'll run our own show.'"[7]

Shea began showing serious strain after the fire and Mueller considered replacing him as head of the Apollo Spacecraft Project Office. He also thought about replacing Gilruth as center director because he took the fire so hard and found it difficult to move forward. However, Low contained the problems with Gilruth, and Mueller did not want to make him a scapegoat, believing that would have led to a feeding frenzy by NASA's opponents and "our enemies around the world," he said. While helping others cope, Mueller fought to control his own emotions, and one evening just after the fire, being a recreational painter, he painted a canvas totally black in oil paint and added a barely visible figure. He kept the painting as a remembrance of those dark times for the rest of his long life, and after showing it to your author, said, "I always have done a little painting from time to time. And this was one evening when things looked rather less rosy than normal blacker than black ... I took out some frustration by painting." Adding "here we were in this dense, completely opaque fog, trying to figure out how to get this program done ... I was painting the blackness of the whole program and then I put a figure in it to try to make it bleed it to light and so it was simply frustration." The figure represented "A leader leading to the light."

---

[6]   Mueller Interviews: Slotkin, 2/25/10 and Sherrod, 4/21/71.
[7]   Lilly Interview, Lambright, 7/18/90; Webb Interview, Sherrod, 6/16/69 and 4/28/71.

Although he did not know who that leader could be, "If there was any leader that could have taken [us] out of the darkness it was me; just a moment of frustration."[8]

To deal with these problems and continue working, Mueller said "you have to compartmentalize it and concentrate on the positives where you can do something about it and not try to dwell on what might have been ... And of course part of the problem is that many people get themselves involved in self-recrimination;" not a very constructive position to be in, he believed. The recovery from the fire meant recovering from the trauma as much as recovering from the physical situation. Although, he explained, "working on the physical problems was a constructive way to do something to avoid having that kind of problem again." Adding, "Of course you always ask yourself well why didn't you think of that before." Nonetheless, he considered that a useless exercise, and "being realistic about it there wasn't anything you could do beyond what we had been doing."[9]

Reluctant to replace Shea, he recognized that his friend's mental state impeded his ability to perform his job, and listed the facts as he saw them in an outline of possible alternatives. He wanted Mathews as Shea's replacement, but could not get him to move back to Houston after having just relocated with his family to Washington in December. So, with everyone's agreement, he appointed Low to manage the Apollo Spacecraft Project Office and moved Shea to Washington as his technical deputy. Making the announcement on April 5, he said, "These changes reflect our efforts to give us more breadth and depth in the management of the manned flight program." When considering replacing Gilruth, he looked at several alternatives such as moving him to Washington in some capacity, and appointing Phillips in his place, or perhaps taking the Houston job himself. But in the end he left Gilruth in place.[10]

When Shea returned to headquarters, Mueller thought him to be on the verge of a nervous breakdown but considered him "too good a person to let him self-destruct." In retrospect he said, "I guess with all of the good things you get some bad as well and in his case he had a problem with his mind and he was so smart that he could manage it most of the time. It's only when he was under a great deal of stress that he lost control." With Shea in Washington, it soon became apparent that it would be dangerous if Congress were to call him to testify, so Webb and Mueller used all of their political capital on Capitol Hill to keep him from testifying. After the agency announced these changes, the *New York Times* published an article that said "Space agency officials conceded that the change was 'accident-related' but denied that it involved any intention to fix the blame for the fire." Unhappy with his removal from the spacecraft project office, Shea remained at NASA for only a short while before

---

[8]   Bizony, *The Man Who Ran the Moon*, 174; Mueller Interview, NASM, 1/10/89 and Slotkin, 2/24-25/10.

[9]   Mueller Interview, Slotkin, 2/24/10.

[10] Handwritten Notes by Mueller, 5/27/67, OMSFO&M, GEM-84-13; Bizony, *The Man Who Ran the Moon*, 153-154; Mathews Interview, Sherrod, 2/17/70; Lambright, *Powering Apollo*, 160-161; Mueller Interview, Lambright, 9/21/90; NASA Press Release 68-83, Manned Flight Personnel Changes, 4/5/67-Newsclippings, SCP-72-2.

returning to industry, and eventually he joined D. Brainerd Holmes at the Raytheon Corporation.[11]

One night, sitting alone in a hotel room at the King's Bay Inn in Seabrook not far from MSC, Mueller thought about the meaning of his work. And in his own hand, wrote:

> [T]o a remarkable extent the space program is founded on faith and on belief. On the faith that there is a future for mankind and on the belief that the future is one that will be good for all the people of the world both as individuals and as nations. On the faith that elsewhere in the universe life and intelligence exists and on the belief that finding and sharing knowledge and experience will be good for each race. On the faith that learning more about the stars and about the solar system will improve life here on Earth and on the belief that as man learns the secrets of space travel he will use these ships to explore and eventually inhabit other planets [and] tour the stars.[12]

The next few months would determine whether it would be possible to overcome the turmoil of the aftermath of the fire and put the Apollo Program back on track.

II

Mueller's relationship with Webb deteriorated after the fire because he operated too independently for the embattled administrator, who accused him of not keeping him informed. While Mueller tried to keep Webb updated, the administrator's lack of technical knowledge contributed to his inability to understand what he was being told. As Mathews observed, "Things were all right as long as OMSF ran smoothly. But when things went wrong ... Webb didn't understand it." And the administrator thought Mueller should accept more of the blame for the fire, while Mueller insisted that he did. "I did ... It was my group that did it." Webb also thought Mueller did not tell him what went on in OMSF, and his reporting system did not function as it should have. Part of the problem had to do with Mueller's personality. "He was a cold fish," Freitag said. "Not a dynamic kind of man, very subtle." He was very firm, but understated. Once Mueller made up his mind, he became relentless in pursuit of his objective. And after the fire, while Webb focused entirely on the lunar landing, Mueller continued looking beyond Apollo, which created another issue to separate them.[13]

The accident caused a great deal of stress in the organization and Mueller had difficulty keeping everyone calm, organized and moving forward. As he recalled,

---

[11] Mueller Interviews: Slotkin, 2/24-25/10; Shea Interviews: Sherrod, 5/6 and 5/16/71; "Top Apollo Aides Shifted by NASA," *NYT*, 4/6/67.

[12] Action Items, 11/2/66-12/18/67, GEM-48-1.

[13] Mathews Interview, Sherrod, 2/17/70; Mueller Interviews: Slotkin, 2/24/10 and NASM, 5/1/89; Freitag Interview, Lambright, 6/27/91.

"Congress was investigating, the newspapers were investigating, the White House was investigating, everybody was investigating." However, he believed the problems uncovered were simple and straightforward – something you would never do it if you thought about it: the use of pure oxygen in an enclosed space was dangerous, and the danger became greater as the pressure increased. The cause of the fire had to do with characteristics of complex systems where you never know the weakest link. And it turned out to be using pure oxygen in the capsule at sea level pressure during ground testing. The agency conducted similar ground tests earlier in the program, and did not encounter any problems, and it would not have been a problem in space because of the lower cabin pressure. It only became a problem on the ground and Mueller pointed out, "in a complex system like that it's so easy to overlook the obvious and so you spend all your time on the hidden problems," the "unknown unknowns."[14]

Speaking publicly for only the second time since the fire, Mueller addressed the Explorers Club in New York City on April 1. Quoting Arthur Clarke, he posited, "What we seek in space is not just knowledge, but wonder, romance, novelty – and above all, adventure." Unable to talk about the present, he looked to the future, telling them the first Mars flyby mission would take place in the late 1970s, launched on a Saturn rocket with a nuclear upper stage; and the early 1980s would witness the first astronauts landing on Mars. He predicted that "twenty years from now, we can have landed men to explore Mars and have at least begun the exploration of Venus" using the infrastructure designed for Apollo with only small modifications. Exploration of the planets would employ space vehicles derived from projects begun in the 1960s, and use Apollo spacecraft modified to carry up to six astronauts. Decreasing the cost of spaceflight by using existing technology remained a key objective of the Apollo Applications Program scheduled to begin in 1968, even before the first lunar landing; but post-Apollo funding awaited congressional approval. He claimed "orbiting space stations and a permanent lunar base are clearly within our grasp ... they are a vital foundation for the next major steps in manned flight." The orbital workshop would go a long way toward understanding the requirements for permanent space stations which someday would orbit the Earth. Its first mission would be a flight of up to four weeks, and as NASA gained experience the workshop would increase the time spent in weightlessness. Experiments to advance science and medicine would lead to a better understanding of the impact of prolonged spaceflight on the human body. And following the initial lunar landings, the Apollo Applications Program would include longer stays on the Moon, for periods of two weeks or more. And before the end of the 1980s the agency could erect a permanent facility on the Moon and shuttle astronauts back and forth from Earth on a regular schedule. "No doubt there will be a Soviet base, located not too far from ours and about the same size," he added. Mueller spoke of real programs, with firm goals, already underway, and concluded by telling the Explorers

---

[14]   Mueller Interviews: JSC, 8/27/98 and Slotkin, 2/24/10.

Club members that the Apollo Applications Program, which NASA had already presented to the Congress with the President's approval, would build on the base established by the Apollo Program.[15]

Competition between the centers over which would manage the orbital workshop concerned the administrator, who said "this rivalry had long been a problem." Gilruth worried that when von Braun completed work on the Saturn, "Huntsville would reach for some of Houston's functions." That is just what Mueller had in mind. The orbital workshop represented a major new role for MSFC, particularly in the spacecraft area. And according to Mathews, it became Mueller's way to get "a nose under the tent." Mueller did not care about the exact configuration of the workshop; he just wanted to get it going and, Mathews recalled, the "practicality of the whole approach was pretty questionable." Yet Marshall wanted to take this questionable, though visionary idea for the orbital workshop, and do something practical with it. Then during the spring and early summer of 1967, Mathews said, "I think we developed something like 57 separate program plans" for the Apollo Applications Program, but unfortunately the budget situation "made this thing run hot and cold."[16]

The Thompson review board delivered its final report on April 5, and the administrator immediately released it to the White House, Congress and the press Webb did not see the actual report before officially receiving it, although Mueller briefed him about its tentative findings and preliminary recommendations as early as February 24. According to the report, because the cabin had combined combustible materials with a pure oxygen atmosphere at 16.7 psi – 2 psi above ambient to conduct a leak check – the "test conditions were extremely hazardous." The board's recommendations ranged from reducing the amount of combustible materials in the capsule through to redesigning the hatch and improving procedures. It recommended revamping various systems and subsystems, suggested specific tests, and then concluded "the Apollo team failed to give adequate attention to certain mundane but equally vital questions of crew safety." The exact point at which the fire began could not be determined, yet based on physical evidence and the damage in the cabin, the report concluded that electrical arcing occurred. People speculated, but no one conclusively identified the cause of the fire. Nonetheless, a spark ignited some flammable material near Grissom's couch, and the pressurized oxygen atmosphere quickly spread the fire. Although burns contributed to the deaths of the astronauts, the autopsies showed that they lost consciousness as a result of asphyxiation soon after the fire began.[17]

At an "eyes only" review of the Thompson report, Mueller agreed to implement

---

[15]   Mueller, Explorers Club, NY, NY, 4/1/67.

[16]   Webb Interview, Sherrod, 6/8/69; Mathews Interview, Compton, 7/15/75.

[17]   Lambright, *Powering Apollo*, 163; "Report of Apollo 204 review board," Findings, 4/5/67, GEM-60-10; "Tentative Findings and Preliminary Recommendations review board," 2/24/67, Mueller to Webb, SCP-68-8; Compton, *Where No Man Has Gone Before*, 93; Seamans, *Apollo*, 76.

all the recommendations that were not already corrected by the Block II spacecraft. NASA treated the fire investigation like a major aircraft accident, so they carefully cataloged everything, and a tremendous effort went into trying to reconstruct what caused the fire. Mueller speculated "either somebody threw a switch that arced or pulled a wire out by mistake" and caused the arc. However, he pointed out, "once an arc occurs in a pure oxygen environment, it's just amazing how fast those things burn." Later adding "none of the reliability analyses really take into account the human part of it. They are all assumed to be perfect. Yet when you look at what really happens it's always some guy that makes a mistake that ends up with the problem, 'human error.'" While Thompson chaired the review board, Borman acted as its spokesman and public face, articulately presenting the information and answering questions pertaining to the investigation. Borman said that if the findings were followed then he would have no problem stepping into the capsule himself; which he later did. The report attributed the fire to faulty engineering judgment, not to management or administrative problems. Seamans admitted "we were all guilty;" a conclusion Mueller agreed with. However, neither of them believed it necessary to make major changes in the management of the Apollo Program.[18]

The severity of the report shocked Webb, who felt let down by industry, and Thompson said Webb would "never forgive them." Nonetheless, NAR's Atwood, highly critical of the way NASA conducted the investigation, became very dissatisfied with the board's findings. "I have never seen an accident investigation like the Apollo fire investigation," he reflected later. "I've never seen skating over the fundamentals and putting the blame [on workmanship which] was only marginally responsible, if at all." He argued that testing the capsule in a pure oxygen atmosphere at sea level *caused* the fire, and NASA had made the same mistake on Mercury and Gemini, though luckily it had no fires. He claimed the agency failed to test the capsule for flammability, despite a short circuit during the Gemini Program, implying NASA covered it up to avoid embarrassment. And while the review board did not blame NAR for the fire, Mueller did, and Atwood reacted to the report as if it did too.[19]

Mueller said, "Stormy took a lot of the whoops for what was not very good workmanship. In fact it was lousy workmanship in many instances ... and it's easy to place blame but hard to really have the right blame placed in the right places. And in the case of the fire you really had to go down to the guys that were actually doing the wiring and were really responsible for much of the problem." Consequently, he said, "we went back and redid all of that and ... we did all of the testing of what happens when you have a fire in one-G and pure oxygen ... In retrospect it was something we should have done earlier but then there's lots of things we should have done earlier and this is just a key one that we missed."[20]

---

[18]   "Draft-Eyes Only," 4/11/67, GEM-62-12; Mueller Interviews: NASM, 1/10/89 and Slotkin, 6/8/10; Seamans, *Apollo*, 76.
[19]   Thompson Interview, Sherrod, 9/10/69; Atwood Interview, NASM, 1/12 and 6/25/90.
[20]   Mueller Interviews: Slotkin, 2/24-25/10.

"NASA's April Fool Report," is what the *Washington Star* called the report by the review board. "Now that the shabby farce of NASA investigating NASA in the Apollo tragedy is coming to a close, Congress is preparing to take over, and perhaps the truth will come out. Perhaps." And the *New York Times* editorialized that "Any adequate investigation must go beyond the disaster that took the lives of the three astronauts. It must penetrate the politics and economics of the nation's vast space-industrial complex." On the other hand, *Aviation Week & Space Technology* wrote, "The harsh, scathing report of the Apollo 204 review board has effectively laid to rest charges that the board was assembled to conduct a whitewash of the January 27 fire, which took the lives of three astronauts." Calling the report "a broad indictment of NASA and North American and the whole program," Teague scheduled hearings for April 10, and the Senate space committee planned another set of hearings during the same time frame. Yet the role of NASA management remained unanswered in Thompson's report, because he avoided looking at the management of the Apollo Program. Mueller agreed that physical not management problems caused the fire, and later rhetorically asked, "[What does] Thompson know about the management?" He attributed speculation about management problems to rumors and "all sorts of wild things" under discussion at the time. He did not think the program had management problems, saying, "It's not a management problem, except in the sense that [poor] workmanship is a management problem." And he argued that those who blamed the fire on management problems did not really understand what had happened.[21]

Webb kicked off NASA's testimony at the House hearings by stating that the agency would correct the errors uncovered by the investigation, and go on to achieve the lunar landing goal; although the committee no longer uncritically accepted his arguments. The House hearings lasted three days, with the members cross examining NASA and NAR executives. And as Mueller recalled, about half of the committee remained dubious of the Apollo Program and considered canceling it. While Teague and Miller kept the committee under control, Mueller later said "we used up all our good will at the time" to keep the program going. The turning point in the House came at a private meeting in Chairman Miller's office one night when Borman, who had a lot of credibility, said, "I have been thinking this thing one end to the other. Knowing what I know today, I'd still get into the spacecraft today. I think the fire was a random failure." Teague asked him to testify to that in open session, and after he did so the mood in the House changed, although it did not have the same impact in the Senate.[22]

---

[21]   "NASA's April Fool Report," William Hines, 3/23/67, *Evening Star*, "Unanswered Questions on the Apollo Tragedy," *NYT*, Harry Schwartz, 4/17/67, and "The Apollo 204 Review," *Aviation Week & Space Technology*, 4/14/67 – News Clippings, SCP-72-2; "Head of House Panel Says Report is an 'Indictment' of NASA and its Contractors," *NYT*, 4/10/67; Lambright, *Powering Apollo*, 165-167; Mueller Interview, Slotkin, 2/25/10.

[22]   Lambright, *Powering Apollo*, 165-167; Webb Interview, Sherrod, 6/8/69; Mueller Interview, Slotkin, 2/25/10; Freitag Interview, Lambright, 6/27/91.

Seamans, Mueller, Phillips and Berry testified once again before the Senate space committee on April 13. The deputy administrator said, "North American had not always shown sufficient dedication to the engineering design or workmanship on the job." Although after NASA pointed out their deficiencies, they made changes which produced major improvements in their work. Mueller reminded the committee that the review board had addressed deficiencies in the Block I spacecraft, yet even before the fire NASA planned to replace it with Block II, incorporating "all we have learned in this program, and which would be so designed as to facilitate the incorporation of any changes required as a result of our test flights and other development work." He called the "basic design and fundamental organization that underlie our planning ... sound." Nonetheless, he told them OMSF undertook a study of the Thompson report and promised to present any resulting changes to the committee. However, some members still called for an independent investigation, and wanted to cut NASA's budget. And meeting in executive session, they debated canceling Apollo, although they chose not to do so.[23]

The period following the fire was tense and busy, requiring Mueller to spend a considerable amount of time convincing the administration and Congress that the lunar landing remained doable. Journalists wrote that landing on the Moon by the end of the decade was no longer possible. So, Mueller said, "we had to make sure that people realized it was possible but it wasn't easy. And we really cashed all of the chips we had to keep that program going." However, Teague's steadfast support was the "saving grace" and opponents in the House could not overcome that. Yet Mueller conceded that after the fire "It was touch and go whether we were going to be able to finish the program. It was life or death."[24]

The congressional committee members asked politically motivated questions and they would have preferred simple sound bites in response, but NASA's witnesses answered each question with long detailed explanations of the underlying reasons behind them. Testifying as the agency's leading technical expert, Mueller tried to answer the question: "What went wrong?" He said the review board answered part of the question, although doubts about the adequacy of NASA policies, practices and procedures remained, and he tried to address them. In addition to the steps that had been outlined in previous testimony, he said the agency planned to use the report to determine what further corrective actions to take. He told the committee much of the success of Mercury and Gemini came from the same principles that formed the basis for the Apollo Program. From the beginning, he explained, "We have done our best to design safety and reliability into the system," and NASA conducted extensive testing to assure the safety of the astronauts. He discussed the different test regimes and the process used to certify results, and said that all anomalies were explained and resolved before moving to the next step. He insisted that despite tight schedules, the

---

[23]   "Contractor Criticized ..." *St. Louis Post-Dispatch*, 4/14/67, GEM-83-6; Mueller Statement, House NASA Oversight Subcommittee, 4/11/67, GEM-62-12; Mueller Interview, Slotkin, 2/25/10.

[24]   Ibid.

agency had a "firm policy" of only flying humans in space when "technically and procedurally ready to carry out a safe and successful mission." He argued before the Senate, just as he did in his public remarks, that Apollo was an orderly, not a crash program. He reviewed the approach used to design the spacecraft, and explained how NASA made the decision to use pure oxygen in the spacecraft cabin. Considerations of weight and reliability, whilst meeting crew physiological requirements, drove the decision. Prime concerns involved preventing nitrogen narcosis, the decompression sickness popularly known as "the bends", and the avoidance of equipment failure. Gemini showed that humans tolerated a single gas atmosphere without problems; and after engineers evaluated the experience of Mercury, and the design of Gemini, they decided to use the same atmosphere on Apollo. He said that since the start of the Mercury Program, engineers had conducted more than 20,000 hours of spacecraft and spacesuit testing with pure oxygen, including 914 hours at sea level or higher pressure, leading to the conclusion that the agency could safely use a pure oxygen atmosphere on the ground. He went through similar discussions about fire hazards and emergency egress, explained the design and development of the hatch, and the justification for having it open inward. The new heavier unified hatch was capable of being opened in a few seconds, had a manual release to allow it to be opened from either side, and had safeguards to prevent it from being opened accidentally.[25]

Mueller told the committee, "It is important to keep in mind that the test program embodies a variety of types of risk. Uncertainty, and therefore risk, cannot be eliminated entirely from projects that seek to advance technology and explore the frontiers of science. All space programs must be planned and developed with less than full knowledge." He explained that risk taking "is inherent in each management decision from the inception of a program to its completion." However, he insisted that quality standards for Apollo were stronger than those on Mercury and Gemini, and discussed the position of quality in the program management organization, from the contractor up to NASA headquarters. Yet with one and a half million parts, the Apollo spacecraft had more complexity than Gemini and Mercury. And while Apollo had about 20,000 parts fail in testing before its first flight, it represented a quality improvement of almost four times compared to Mercury's first flight. Nonetheless, using Thompson's report as a guide, the agency had reexamined its quality control procedures at all levels. He enumerated the steps taken to review and checkout equipment, and described improvements that had been made to the Apollo safety plan. Concluding, he promised that OMSF would follow-up on all of the board's recommendations. Then he noted that, "In every research and development program, the people and organizations taking part progress along a learning curve ... In Apollo we have learned much in the six years since the program's initiation. Astronauts Grissom, White and Chaffee paid with their lives for a great deal more

---

[25] Mueller, House Subcommittee on NASA Investigations, 4/10-12/67, AS-204, NASA Archives, RN 31579; Brooks, Grimwood, and Swenson, *Chariots of Apollo*, Ch. 9-4.

that we have learned as a result of this accident and the resulting review. We are at a stage in Apollo comparable to where we were prior to the first manned flights in Mercury and Gemini. The people and organizations involved are motivated and able to complete the learning process."[26]

III

NASA and NAR executives became targets for critics of human spaceflight who used the Apollo 204 fire to question the lunar landing goal and claimed schedule pressure had contributed to errors which cost the lives of three astronauts. Webb, Seamans and Mueller bore the brunt of the congressional criticism, but successfully defended the Apollo Program. However, the fire permanently damaged NASA's standing with Congress and the agency lost its reputation for being infallible. The congressional hearings showed NASA to be neither negligent nor incompetent, and did not find Apollo's program management system at fault. And the major lesson learned from the fire was to test everything at a subsystem level under the conditions in which they were to be used. However, Mueller also noted that it was necessary to put Apollo into historical context. In the 1960s, people accepted more risk. Planning for human spaceflight began fifteen years after the end of the Second World War by people who had lived through the war years. And he pointed out, "If you're at war and you're one of those guys out on the front line you're taking risk and people were willing to accept that." They took more risk because they knew that to accomplish things beyond their safety zone they had to accept it.[27]

Then amidst the tumult that followed the fire, Mueller received word from the National Academy of Engineering of his election to membership. The announcement concluded, "The Academy thus recognizes and honors your important contributions to engineering and your leadership in the field." With less than two hundred members at the time, Mueller joined his peers at the top rung of the engineering profession. His NAE class included Schriever, von Braun, NASA's first administrator T. Keith Glennan, and the former president of MIT and first presidential science advisor James R. Killian, Jr.[28]

By mid-April, MSC identified the required changes for the Block II spacecraft based on the findings and recommendations of the Thompson review board. Low prepared a list of almost sixty major changes that were scheduled for completion by production of the second Block II spacecraft. In all, the agency made about five thousand changes to the command and service modules to fix many things that could

---

[26]   Mueller, House Subcommittee on NASA Investigations, 4/10-12/67, AS-204, NASA Archives, RN 31579.

[27]   Compton, *Where No Man Has Gone Before*, 93-94; Johnson, *Apollo*, 146-147; Mueller Interview, Slotkin, 2/25/10.

[28]   National Academy of Engineering, 2/19-4/28/67, GEM-80-19.

have interfered with future flights; but as Mueller said, they had "no way of knowing one way or another whether [the Block II] would have succeeded in the form it was or not. And it certainly was a lot . . . safer once we finished the retrofit particularly of the capsule, although you've got to recognize we went through the entire stack while we were doing that to be sure we hadn't overlooked something else somewhere." Thus in this period of "intense introspection," the agency fixed many other problems in addition to the ones identified by the review board.[29]

On April 19 Mueller made his third speech since the fire, addressing a meeting of materials and process engineers. Despite the success of Mercury and Gemini, the fire made it necessary to reexamine crew safety in both flight and ground testing. In his only reference to the accident, he explained that studies, tests and design efforts that were then underway would pinpoint additional changes needed to prevent another fire. Focusing on the post-Apollo period, he expanded on the points made in his two previous post-fire speeches, saying that the initial phase of the Apollo Applications Program would reduce the unit cost to put payloads into orbit, and would bring "the astronomer and his telescope up to where the light is – above the atmosphere." The post-Apollo program would add applications of direct economic benefit, expand the exploration of the Moon, provide new scientific knowledge, enhance the international standing of the US, and "maintain the momentum and capability of the space 'team' that has been so painstakingly assembled over the past decade."[30]

Following the release of Thompson's report, Mueller organized OMSF's response and planned its follow-up actions. He assigned responsibility for each finding and recommendation, specifying both the actions to be taken and their completion dates. Once completed, the results would be documented and reviewed with the chair of the pertinent technical panel to assure that their finding and recommendation had been correctly interpreted. He wanted NASA's responses to meet "the spirit and intent" of the review board, and asked each panel chair to certify completion of each item. He also had the program control organization monitor and track the results, because he did not want it said the agency did not take the appropriate actions in the wake of the fire.[31]

When speaking to a group of electric power engineers later in April, Mueller gave an address similar to his other recent speeches, telling them the space program had a major impact on the national economy, and would continue to do so if Congress provided funding beyond Apollo. The cost of the lunar landing program had peaked in FY 1966, and with the decline in FY 1967 employment had already fallen by about twenty percent. Yet, whereas the USSR devoted about the same proportion of their economy to their space program as the US, their investment continued to increase while ours decreased. He dwelled on the practical applications of spaceflight, but called for more modest goals than in the past. Despite the Vietnam

[29]   Phillips to Mueller, 4/18/67, GEM-62-12; Low to Phillips, Block II Changes, 5/19/67, SCP-64-9; Teague Interview, Sherrod, 4/1/70; Mueller Interview, Slotkin, 2/25/10.

[30]   Mueller, Society of Aerospace Materials and Process Engineers, St. Louis, MO, 4/19/67.

[31]   Mueller to Phillips, Gilruth, Debus, 4/21/67, GEM-62-12.

War and domestic unrest, he said that the US needed to continue its investment in space exploration, if for no other reason than to keep its options open. And unless the agency received additional funding it would lose the benefit of previous investments, calling it a "terrible mistake to close up shop." The *Chicago Sun-Times* reported this speech, although not how Mueller expected. The newspaper quoted him saying "we can achieve the mission objective of manned lunar flight and safe return in this decade." But it then stated, "This statement from the conservative Mueller before one of the most sophisticated groups of engineers in the world was tantamount to a written guarantee." The *Sun-Times* speculated that astronauts might land on the Moon on the fourth Saturn V flight, but ignored Mueller's plea for additional Apollo Applications Program funding.[32]

Webb and some members of Congress could not understand how Mueller and his team did not recognize the fire hazard before the accident. After all, the agency and their contractors described the program management system in glowing terms, yet this system must have failed for the fire to have occurred. Mueller did not have a good answer for them, and Webb reacted by imposing new requirements on OMSF. And because Congress received copies of internal agency reviews, management surveys and documents from contractors, as well as the Thompson review board notes, they had increased visibility into the program. Yet while knowing more of the technical details, Mueller wrote, they failed to recognize the fact that "problems are necessary to progress."[33]

Webb considered himself a management and organizational expert, and after the fire he moved some offices around on the organization chart in an attempt to improve the flow of information. He appointed Harold B. Finger as associate administrator for organization and management with increased responsibility for the budget process, giving him functions that previously reported to Seamans in an attempt to reign in Seamans and Mueller. However, Finger did not equal Mueller's political acumen. Mueller quickly moved him aside, later claiming this reorganization had little impact on OMSF, which represented the bulk of the agency. Mueller called it "an overlay which didn't add any particular new or improved operation." It only represented another attempt by Webb to gain a better understanding of what went on, because he thought he should have known about the potential for the accident; he believed that if only he had had a better reporting system then he would have known about these problems and perhaps could have prevented them. "But other than requiring a new set of reports, it didn't really change anything," Mueller said. Webb "lost confidence in me, and he lost confidence in all of the management team." And around this time the administrator concluded that because his deputy

---

[32] Mueller, American Power Conference, Chicago, IL, 4/26/67; Seamans to Mueller, 5/5/67, GEM-62-12.

[33] "Subjects for Center Director's Meeting," 5/15/67, and "Action Notes from NAA," 6/5/67, GEM-48-1.

had not kept him fully informed; Seamans did not run the agency the way he wanted, so he began to take the brunt of Webb's criticism.[34]

According to Atwood, Mueller had suggested that he ditch Storms several times, although that did not occur until after the accident when Mueller told him Storms had not gotten the job done. Concentrating on "just getting hardware built" after the fire, Mueller said that "getting the right people working on it was a challenge. And with Stormy it was almost impossible to get the right people in the right places." Atwood did not consider Storms a part of the problem. He felt the space division had made significant progress developing the command and service modules, though pressure from the agency eventually made it impossible to keep Storms as the space division president. And when the CEO met with the administrator at the end of April, Webb demanded a change. If Atwood resisted, he would take the Apollo spacecraft contract away. To prove his point, NASA asked five companies to submit bids for the Apollo spacecraft contract. But as Mueller pointed out, it was a bluff because had NAR not backed down it would have been impossible to complete the lunar landing by the end of the decade. Webb wanted Storms fired, and Atwood replaced him on May 1. Then the administrator dropped his threat to seek a substitute Apollo contractor. Still in a weak position, Atwood agreed to make all the changes to the command and service modules that the agency wanted. Although he later said, "if it hadn't been for the fire, [the changes] would hardly have been considered necessary."[35]

Following the accident, Webb became more involved in the day-to-day operations of NASA, deeming it his personal responsibility to run the agency. The administrator began calling Phillips directly. The general began meeting with him more frequently, and participating in activities that he took the initiative on. The five prime contractors invited to bid to replace NAR met with Webb and Phillips, and the administrator asked each what they would do if they were to receive the contract. This process did not take long, and after hearing back, Webb discussed it with Phillips and asked him what he thought the agency should do. By that time, the general had made up his mind to write a sole-source contract with Boeing to provide technical integration and evaluation support. And like Mueller, Phillips thought the agency had to keep NAR in place. He believed it necessary to work more closely with them, but this would require some additional help. Because NASA had received the input of five qualified contractors, Phillips said this satisfied the federal requirement for competition and the agency would be justified in making a sole-source award. Webb said, "Okay, go do it." But prior to advising the administrator, the general had discussed this with Mueller, Seamans and the center directors.[36]

Mueller considered hiring Boeing "a step in the right direction." The company

[34]   Bizony, *The Man Who Ran the Moon,* 205; Mathews Interview, Lambright, 10/4/91; Mueller Interview, NASM, 1/10/89.

[35]   Atwood Interviews: NASM, 8/25/89, 1/12/90 and 6/25/90; Bizony, *The Man Who Ran the Moon,* 170-171; Gray, *Angle of Attack,* 254-256; Seamans, *Apollo,* 77-78.

[36]   Phillips Interview, NASM, 9/28/89; Mueller Interview, Slotkin, 2/25/10.

provided NASA with resources to oversee communications between the different organizational elements and insured that everyone worked together. They served as a system engineering group, reviewing the entire space vehicle to insure it worked well together. Webb needed to show he was doing something different, and hiring Boeing served that purpose. A more direct thing would have been to replace NAR, but the agency could not do that and complete the mission by the end of the decade. Boeing gave Webb the opportunity to claim he did something, while Mueller called hiring them completely "cosmetic." Nevertheless, Phillips trusted Boeing, and the company gave him an independent communications link to NAR. "Whether they had any positive impact is another question ... it was another set of eyes looking, and [Boeing] certainly had some competent people," Mueller said.[37]

On May 9, Webb, Seamans and Mueller again testified before the Senate space committee, and then they met with Teague's House subcommittee the next day. The administrator spoke of awarding the Technical Integration and Evaluation contract to Boeing, a contract which broadened the company's responsibility from integrating the Saturn V to include the entire Apollo-Saturn space vehicle. However, Boeing never became the integration contractor Webb imagined. That would have replaced the role of the centers and the Apollo Program Office. Rather, they became advisors, only doing work not performed by another contractor. Webb told Congress about his plans to change how the agency managed the Apollo Program, and called awarding the Boeing TIE contract a logical decision because they performed the same work on the Saturn V, and had done similar tasks for the Minuteman. In addition, Boeing had staff immediately available and transferred most of the TIE contract staff from other company projects.[38]

Mueller's testimony to the two congressional committees contained lots of technical details. He submitted three written reports, and these became an important part of the official record. His oral testimony summarized the actions that the agency had taken, and highlighted the steps that would need to be taken before the first human flight of the Block II spacecraft. He said the agency fully accepted all the recommendations of the Thompson review board and, after reading many pages of detailed testimony, he described what still needed to be done. If NASA flew the first piloted Apollo mission by February 1968, it represented a twelve month delay, but by conducting "unmanned" flight tests during the delay, the 1969 lunar landing remained possible.[39]

NASA made many changes after the fire, both redefining the test program and redesigning the Apollo spacecraft. NAR rewired the whole capsule, and it "took time and energy and effort. Just simply getting everybody going in the same direction at the same time was a major challenge," Mueller said. He went to great lengths to tell

---

[37]　Ibid., 2/25, and 6/9/10.

[38]　Seamans, *Apollo*, 78; "Statements on Apollo Project Reprogramming," Senate Space Committee, 5/9/67, GEM-62-15; Mueller, AIAA, New Orleans, LA, 5/16/67; Levine, *Managing NASA*, Ch. 4.

[39]　Mueller, House Subcommittee on NASA Oversight, 5/10/67.

Congress how the agency managed projects, what they did and how they tried to prevent fires because some of the members still thought NASA had problems with program management. Remaining the point man for human spaceflight, Mueller repeatedly admitted "it was my fault that this occurred" and promised he would do whatever he could to insure it did not happen again. He took the heat because he wanted to take the pressure off of Debus, Gilruth and von Braun, to allow them to do their jobs. He and Phillips did most of the testifying for OMSF, minimizing the center directors' involvement, keeping them "pretty well shielded from the turmoil at headquarters," he said. He wanted them focused on fixing problems and returning to human spaceflight, and not distracted by these "peripheral matters". Fortunately, he said, the Apollo Program "was coherent enough so that you could change some cosmetics up at headquarters" without impacting the work at the centers.[40]

Nonetheless, while Congress remained concerned, a public survey conducted by the *National Observer* showed that fully seventy-seven percent of those questioned approved of the Apollo Program, and only seventeen percent opposed. A *Trendex* poll, taken annually during the 1960s, showed a majority of Americans supported Apollo. Despite concern about the cost of new social programs, and the domestic violence and unrest resulting from the increasing economic and social impact of the Vietnam War, the general public remained excited about the prospects of landing on the Moon in that decade.[41]

IV

Addressing an AIAA meeting in mid-May, Mueller spoke about how the agency "solved the problems encountered in the tragic accident of January 27." Just four and a half months after the fire, he wanted to move forward, and as he had done in his other speeches following the fire, he discussed the benefits and spinoffs of human spaceflight. Recovery from the fire, though, became the underlining theme of this address, and he said that the review board's findings focused NASA on problems in spacecraft design, quality control, checkout procedures and safety. But, he argued, "Every major research and development program has problems ... [and] the solution of problems is the fundamental task of research and development ... But this in no way lessens the emphasis we must place on assuring that our problems are solved."[42]

On June 15, NASA gave Boeing a letter contract extending their work on the TIE contract for an initial period through December 31, 1968. The agency remained the prime contractor for the Apollo Program while Boeing would help with program management by supporting system engineering, hardware evaluation, integration of flight hardware, and engineering evaluation. The company augmented agency efforts with technical *assistance*, not *direction*, despite what Webb wished everyone to

---

[40] Mueller Interviews: NASM, 1/10/89 and Slotkin, 2/25/10.
[41] Mueller, AIAA, New Orleans, LA, 5/16/67.
[42] Ibid.

believe. Mueller publicly supported the TIE contract, and in early September told the *Houston Chronicle* that the "philosophy of a single contractor bearing the final responsibility for the entire Apollo spacecraft and its launch vehicle 'is working out as well or better than any of us expected.'" The *Chronicle* wrote that Boeing was to look "over the shoulder" of NAR and Grumman, but conceded that the company did not have final responsibility for anything.[43]

Within their scope of work, Boeing "did an excellent job, particularly in the restructuring of the program, where we had a whole myriad of new schedules to mesh and new interfaces to build," Mueller recalled, admitting that NASA needed additional support in that area. He also called the TIE contract "good for external relations," because, even its name showed that Boeing would "TIE" all of the parts together. However, unlike the role of Space Technology Laboratories in the ballistic missile program, Boeing was not the integrating contractor for NASA. They provided liaison and expertise where requested, but did not assume any of the responsibilities of the centers. The company assumed responsibilities for revising the interface specifications originally prepared by Bellcomm, but integrating the various systems continued to be NASA's responsibility. Boeing's work, Mueller said, involved "making sure it was getting done ... they were a communications device, in my view, to be sure that everybody was communicating [and] that the same facts were present throughout the system." Boeing sat in on the project reviews, and served as secretary for many of those reviews, which made them aware of what went on in the project and enabled them to pinpoint discrepancies.[44]

NASA's FY 1968 budget request, which was down by about $500 million from the previous fiscal year, still contained $439 million for the Apollo Applications Program. The president endorsed that funding level, and submitted it to Congress. Designed to return "the maximum in benefits for the minimum cost," this would continue production at the rate of four Saturn launchers per year, and enable flights to begin as early as 1968. Nonetheless, Mueller received direction from Seamans to give the Apollo Program priority for all resources until the first lunar landing had been achieved. All hardware would be configured for the mainline Apollo Program, and the release of equipment would be subject to the deputy administrator's approval. Any additional contractor support could only be used for the Apollo Program, NAR would be relieved of tasks not directly supporting the lunar landing, and new phases of the Apollo flight test schedule would not begin until the successful completion of the previous phase. Mueller's directions also said that "mission and crew safety will not be jeopardized in order to meet predetermined schedules." NASA officially established the target date for the first piloted flight test of the Block II spacecraft aboard a Saturn IB for March 1968, and the effort to qualify the Saturn V for human spaceflight would occur "as soon as prudent," wrote Seamans.

43   Levine, *Managing NASA*, Ch. 4; "Contractor Idea Working For Apollo," *Houston Chronicle*, 9/10/67, NASA Archives, RN 10689; "Paper on the Boeing TIE Contract, 10/1/68, NASA Archives, RN 10689.

44   Levine, *Managing NASA*, Ch. 4; Potate Interview, Ray, 6/6/72.

Table 7-1: Seamans' Schedule for "mainline" Apollo flights (as of May 8, 1967)

	Seamans' schedule			Actual results	
Year	# of flights	Launcher	Planned flight test	Launch date and mission name	Mission type*
1967	2	Saturn V	CSM development (U)	11/9/1967 AS-501/Apollo 4	A
				4/4/1968 AS-502/Apollo 6	A
	1	Saturn IB	LM development (U)	1/22/1968 AS-204R/Apollo 5	B
1968	1	Saturn IB	CSM evaluation (M)	11/11/1968 AS-205/Apollo 7	C
	3	Saturn V	Flight tests (M)	12/21/1968 AS-503/Apollo 8	C'
				3/3/1969 AS-504/Apollo 9	D
				5/18/1969 AS-505/Apollo 10	F**
1969	6	Saturn V	Lunar missions (M)	7/16/1969 AS-506/Apollo 11	G
				11/14/1969 AS-507/Apollo 12	H
				4/11/1970 AS-508/Apollo 13	H
				1/31/1971 AS-509/Apollo 14	H
				7/26/1971 AS-510/Apollo 15	J
				4/16/1972 AS-511/Apollo 16	J
1970	4	Saturn V	Lunar missions (M)	12/7/1972 AS-512/Apollo 17	J
				5/14/1972 AS-513/Skylab 1	–
				Not assigned (AS-514 and AS-515)	

Notes: M = Piloted ("manned") and U = Not piloted ("unmanned");
*type = see chapter 8;
**Mission E, flight test in high Earth orbit was canceled because it was not needed due to mission C'.

He provided Mueller with a flight schedule for the remainder of the Apollo Program "subject to change based upon actual accomplishment" (Table 7-1). And Seamans cautioned that the Apollo Applications Program could only be conducted "without interference with, or deleterious impact upon, the mainline Apollo effort."[45]

Mueller publicly talked optimistically about the resumption of human spaceflight. However, because Webb was concerned about how fast the agency could move in that direction, he decided that a position paper should be developed which cautioned NASA employees not to discuss moving forward too quickly. The Apollo Program Office drafted the paper that broadened into a general policy on "discussions which involve an appraisal of the Manned Space Flight Program schedule attainment of established milestones." It contained guidelines on what could be said in public and cautioned that "NASA personnel should refrain from discussing the Apollo Program in more optimistic terms." Signed by Webb, it included terminology suitable for use when discussing the Apollo schedule. Yet in most cases the dates contained proved extremely pessimistic. For instance, AS-503 had to be talked about as having "a low probability of being the first Saturn V manned mission." In fact, not only did the first Saturn V carry a crew, it sent them to orbit the Moon.[46]

In early June at another AIAA meeting, Mueller spoke extensively about the Apollo Applications Program. "If approved by Congress," he told them, "this would represent a national commitment to meet the challenge" of the post-Apollo period. Piloted Earth orbital flight tests remained scheduled in early 1968, but based on the findings of the Thompson review board, the agency would concentrate on preparing for the lunar landing, while minimizing scientific experiments until the start of the Apollo Applications Program. The Douglas division of McDonnell Douglas (the two companies had merged) received a contract to outfit two S-IVB stages for the orbital workshop, with Huntsville developing the subsystems. Von Braun had a neutral buoyancy tank under construction to train astronauts in how to operate the multiple docking adapter in zero-gravity. Most long lead items were already under contract and, Mueller said, with the Apollo Applications Program "we take a long step toward a low cost, high-return payoff from investments in space." And in conclusion he said that "the greatness of our nation will be measured in large part by our response to the challenge of developing and exploiting space."[47]

The Senate space committee called Mueller to testify in early June to review Grumman's performance. He called their work during the early phases of the lunar module project typical of NASA's experience, because initially progress was slow. MSC conducted a tiger team review of Grumman's work in June 1966 and

---

[45]   Compton and Benson, *Living and Working in Space*, 99; Mueller, AIAA, New Orleans, LA, 5/16/67; Lambright, *Powering Apollo*, 193; For the Record, Seamans, Apollo Program Decisions, 5/17/67, NASA Archives, Webb Papers.

[46]   Ibid.

[47]   Mueller, National Capital Section, AIAA, Washington, DC, 6/6/67; Trip Report, "ATM Experiment...," S. H. Levine, 6/26/67, GEM-65-8.

provided an *oral* report to the company's president. During FY 1967 Grumman overran their negotiated costs, but took corrective action resulting in significant cost improvement. Then in April 1967, NASA conducted a quality audit, uncovering about two thousand deficiencies, although only ten would have affected mission success and none would have impacted flight safety. Not only did Grumman have development problems early on, their subcontractors had them too. And while some problems remained, the company planned to deliver the first flight version of the lunar module equipped for automated operation on July 1, about seven months later than originally scheduled. Nonetheless, he advised the committee, "contractor performance is improving [and] many difficult technical problems have been overcome."[48]

The top management of the two main Apollo contractors exhibited significantly different attitudes, according to Mueller. NAR did not pay enough attention to problems, while Grumman did. Mueller said, "The physical problems were real but the solution for them was quite different," primarily in the approach they took. NAR used changing requirements as an excuse for cost increases and schedule slippage, whereas Grumman looked at changing requirements as a way to solve problems; a difference more in attitude, he said. Adding, "Grumman didn't think it knew all about everything whereas North American thought it was the expert about everything and really proved it wasn't." The key executives of the companies were very different as well. Llewellyn J. Evans, Grumman's president, "was right down there talking to people," Mueller said. And Joseph G. Gavin, Jr., the lunar module program manager, "was on top of it ... and they were all working their butts off making this thing happen." However, NAR's "top management didn't spend much time on the problem until we finally got their attention" after the fire, he said.[49]

During the congressional inquiries, NASA's budget remained on hold, as the cause of the fire was investigated. But by July action became necessary, though the committees no longer gave Webb everything he asked for. Congress cut the agency's budget, and the magic of Kennedy's challenge of landing on the Moon did not carry over to the Apollo Applications Program. Mueller called it "strange but human that once you've done something it's no longer a challenge and it's no longer something that captures the imagination. And we were never able ... to find that historic [goal] that Kennedy found."[50]

Toward the end of July, Seamans sent Mueller a directive saying "the planning processes in the Office of Manned Space Flight are not adequate ... to support the needs of the Office of the Administrator." He wrote, OMSF lacked an "agreed upon base underlying estimates and assumptions," it suffered from "centralized direction to planning that forbids candid review and analysis," and lacks center participation. "Above all, there appears to be a lack of candid communications within OMSF and

[48]   Statement before the Senate Space Committee, Mueller, 6/12/67, GEM-62-2.
[49]   Mueller Interview, Slotkin, 2/24/10; "Reacher for the Moon..." 1/19/68, *NYT*.
[50]   Lambright, *Powering Apollo*, 184; Mueller Interview, Slotkin, 2/24/10.

between OMSF and other organizational elements." The deputy administrator gave detailed directions and asked for a written response.[51]

Gilruth once told an interviewer that he went around Mueller when necessary, candidly admitting "I used to go to Webb and talk about my problems sometimes ... I found that I could not get a feeling of mutual transmission of thoughts back and forth very well ... with George Mueller. We just sometimes weren't on the same wavelengths." Gilruth considered these differences a "clash of personalities" rather than a difference in their technical backgrounds. After the fire, he spoke frequently with Webb, complaining about Mueller's (and Shea's) failure to keep him informed. Since Mueller and Seamans had frequent, candid and open communications, in all likelihood the deputy administrator's criticism of OMSF program planning resulted from the administrator's conversations with Gilruth; and Seamans prepared the memorandum at Webb's direction, something consistent with the administrator's indirect management style. He chose not to discuss these issues directly, and used Seamans to find fault with Mueller.[52]

Mueller talked about this memorandum with his headquarters staff and responded on July 25, saying he would implement the recommendations that required further coordination and communication with the centers. However, he added, "Prior to your letter I have never received any formal complaints that either my staff or the staffs of the centers were not being completely responsive to requests from the Office of the Administrator." He attached documents showing his responses to past requests, and added, "I feel confident that your review of the way in which we conduct our planning will allay any misgivings that you have that we are being parochial in our outlook and not making every attempt to be completely open and responsive to your requirements." He noted that a high degree of cooperation existed at the working levels, though he admitted "we do have a breakdown in communications among the senior level." He carefully worded the response, and when he sent it on August 2 he attached information about the Saturn/Apollo Applications Mission Planning Task Force established on July 13, 1966, which had a broad cross section of headquarters and center participation, together with the minutes of relevant meetings. Because Seamans had accused Mueller of not coordinating, he listed all the memos showing his record of coordination. He essentially said "yes I am coordinating and I am sorry you don't think I have coordinated, but you must be mistaken." He did not argue, but reading between the lines his response is clear. He complied using the bureaucratic skills that he had developed at NASA, and inundated the administrator's office with hard copies of documents supporting his position (in the days before e-mail).[53]

---

[51]   Seamans to Mueller, Program planning process in the OMSF, 7/20/67, and Mueller to Seamans, Program Planning in OMSF, 7/25/67, GEM-85-7.

[52]   Gilruth Interview, NASM, 3/2/87; Mueller Interview, Slotkin, 2/24/10.

[53]   Mueller to Seamans, Program Planning, 8/2/67, GEM-85-8.

V

At an off-site meeting held at Lake Logan, North Carolina in early August, Mueller spoke about "the hard facts of the Agency's budget position." Telling the OMSF senior staff, "At no time has it been more critical for us to come to grips with the rock-bottom requirements to get the job done and to use our capabilities to develop and implement a good follow-on effort." He predicted FY 1969 would be no better than FY 1968 – and "perhaps worse," because Congress planned to authorize $4.87 billion for FY 1968 compared to NASA's request for $5.10 billion, and the final appropriation could be even lower. He said, "The fiscal 1969 situation, in view of the considerable domestic and international priorities and uncertainties, will not hold open any hope for alleviation of the budget constraints," and went on, "The period of austerity has begun and the luxury of desirable, but not mandatory, program content is not open to us. As a result, we must close ranks to survive and conduct a sound agency program within extremely limited funding levels."[54]

On August 21 Congress cut the FY 1968 NASA Authorization Bill by more than $500 million below the administration's request, appropriating $4.59 billion. This sacrificed agency funding in order to pass the president's tax bill. NASA's budget fell about ten percent below FY 1967, the largest reduction since 1964. This shortfall had little impact on the Apollo Program, but slashed post-Apollo funding. In essence, Mueller said, "they elected not to have a follow-on program." He said that Congress did not understand the Apollo Applications Program and would not fund it beyond preliminary studies until the agency defined it better. However, Mueller attributed the budget cut to "horse-trading," and said the president "traded NASA dollars for some of the other programs he wanted." He also believed that these cuts resulted from dissatisfaction with NASA after the fire. Discussing the budget at a management council meeting, Mueller warned, "FY '68 will be an exceedingly tight money year;" though the exact impact remained unknown. The council agreed the Apollo schedule should remain unchanged, and decided that if the first two Saturn V launches were successful then the third flight would be piloted. They hoped that would free up the remaining Saturn IBs for the Apollo Applications Program, but Mueller explained that any money saved would not be available for other programs, which meant they would have to find other sources of funding. The council expected the USSR to be active in 1969, flying some kind of a "spectacular" to offset the anticipated impact of the lunar landing. And Mueller concluded, "National prestige and the future of the space program dictate that we be prepared to move out on the Apollo Applications Program during 1969," although he did not know how the agency would get the money.[55]

---

[54] "Second Manned Space Flight Management Council Conference," 8/3-7/67, Lake Logan, NC, SCP-65-12.

[55] Lambright, *Powering Apollo*, 184-184; Mueller Interviews: Sherrod, 4/21/71 and Lambright, 9/20/90; Management Council, 9/8/67, SCP-65-11.

8

All-up testing

"The only question is whether this nation will prevail in space ... or will we abandon the future to others?"

Mueller, February 12, 1968

Webb held a series of meetings in mid-August 1967 to determine what to do about the budget cuts. He wanted to preserve the agency's capability, and expressed concern about keeping the Saturn V production line open. NASA studied the impact of stretching out schedules and placing spending controls on contractors. Boeing evaluated the anticipated impact on the Saturn Program, and the administrator appointed an internal working group to review the probable effect on the Apollo Applications Program. While funding the post-Apollo program remained tenuous, Mueller still pushed it, advocating putting many of the planned experiments together in a highly integrated fashion. According to Mathews, "he was not too worried about the technical efficacy of the idea, as much as to enforce the idea that this was going to be an integrated program." He wanted to dock the Apollo Telescope Mount with the orbital workshop and studied missions lasting twenty-eight to fifty-six days, calling them the "best of all worlds." He outlined these missions on the basis of little definition. However, knowing Mueller as well as he did, Mathews said that "he was deadly serious." Although questions about the feasibility of these plans remained, to Mueller the real question concerned human survivability during long periods of weightlessness, and he wanted to use the orbital workshop to expose astronauts to zero-gravity in order to test what work they could do in space. He also initiated extensive space shuttle studies, because to get to Mars and the planets he realized that NASA would need to do it more economically.[1]

Newell somehow convinced Webb that he was ready for a bigger job and tried to take over as the agency's general manager, although he did not know how to manage large programs. Newell's lack of experience left him ill-equipped to direct Mueller

---

[1] For the Record, John R. Biggs, Program Alternatives, 8/22/67, SCP-64-9; Mathews Interview, Compton, 7/15/75; Mueller Interview, JSC, 8/27/98.

and the other program directors. While Webb had to pull in the reins on Mueller, he had to push Newell to get him more involved, and consequently, he never functioned as general manager. Nonetheless, the administrator wanted more people "around the table" to balance Mueller's strong views, particularly when it came to the Apollo Applications Program and the role of science in the space program, seeing Newell's promotion in that light. But Webb did not give Newell line responsibility, and the program associate administrators continued reporting directly to the administrator's office, which Newell did not become part of. Mueller later called this promotion, "Characteristic of Webb's lack of understanding," and Newell was "relatively ineffectual because he really never seemed to grasp the total picture."[2]

Mueller spoke to the Texas Society of Washington, DC on October 2, paying tribute to the sons of Texas who walked the halls of Congress, mentioning the amount of money that NASA spent in the Lone Star state, and commenting on the close relationship the state had with the space program. Then stepping through the accomplishments of the Mercury and Gemini programs, he said, to some people "the meaning of this achievement is seen only dimly and superficially. The purpose of space exploration and discovery are no clearer to many men in this age than they were in the days of Galileo." Some saw space as an opportunity for international cooperation, economic benefit, or a source of new knowledge. And, he noted, "today, knowledge, as well as guns and butter, measures the true power of modern states." Such knowledge led to economic benefits, new industries and products. And it also improved the understanding of the human body, advanced medical research, and helped to educate the people. Thus, he claimed, the space program "is not in conflict with efforts to end poverty and improve human welfare." Quite the contrary, "it contributes to the fundamental solution of these problems." Mueller called the Apollo Applications Program a "modest program of new undertakings that will return maximum benefits to each of us here on Earth at minimum additional cost." It would continue human spaceflight after Apollo and send astronauts to explore cislunar space, while the orbital workshop would increase the time spent in space, thus providing additional benefits. But with limited resources, there was not much that he could say about the future of human spaceflight beyond Apollo.[3]

The relationship between Seamans and Webb continued to deteriorate. But while the relationship between Mueller and the administrator also grew worse, it remained tolerable. Seamans finally recognized that "it was clearly time to leave," and so on October 2 he submitted his resignation. Webb immediately informed the president, and Johnson asked for a list of possible successors, people without any association with the fire. The administrator asked Seamans to remain at the agency until the end of the year, and announced the resignation effective January 5, 1968. When he heard about this, Mueller realized it would increase pressure on him because the deputy administrator served as his primary interface with Webb. And, he recalled, "Webb

[2]   Mueller Interviews: Slotkin, 2/25/10 and 2/22/11; Lambright, *Powering Apollo*, 190; Management Council, 12/4/67, GEM-100-11.
[3]   Mueller, Texas Society of Washington, DC, 10/2/67.

had accomplished what he set out to do; he had found the guilty, the scapegoat." Mueller did not expect Seamans to resign when he did, for the reason that while the administrator needed a scapegoat, "he also needed some support. I was surprised that he let Bob go at all because Bob was the flywheel back there at headquarters as far as Jim Webb was concerned." Nonetheless, being a political operator, the administrator needed to make changes so that it would appear that he had done something about the fire. Mueller thought highly of Seamans, and said he "really understood what was being done and was good at evaluating whether or not the things were going the way they ought to go. His main strength was in just thoughtfully deciding who ought to do what and then backing it up." And for as long as Webb remained at the agency, Mueller remained under the kind of pressure that led to the deputy administrator's departure. But with Seamans going, Mueller thought it important for himself to stay. He later said "if both Bob and I resigned the program probably wouldn't succeed. Now that sounds egotistical but I think it was true." He considered it essential "to keep that momentum going." And with Webb in the state of mind that he was in, it became even more important to see the program through. However, the administrator did not understand how much he needed Mueller and never asked him to stay. "He probably wished I had resigned" Mueller speculated. "He didn't ask me to leave either ... [He] was like that with everybody who he could possibly conceive of being responsible for this fire." The administrator and the president wanted someone from outside NASA as the new deputy administrator, someone with sufficient technical skills to advise Webb, and who could also assume the top job if necessary. The administrator worried about his ability to control Mueller, and while respecting his technical abilities, he no longer trusted him and felt the need for someone that he could trust as his deputy. Rumors spread that Mueller wanted the job, and Webb suspected that too. Although Mueller later insisted he did not want to be deputy administrator, a job which would have required him to have close interaction with Webb. He preferred to meet with the volatile administrator as little as possible, and when asked about it, he later said emphatically, "No ... Webb's deputy? No. It never occurred to me."[4]

After a long search, Webb selected Thomas O. Paine, an R&D executive with a Stanford doctorate in physical metallurgy and head of General Electric's Center for Advanced Studies, who was approved by the Senate on February 5 and sworn in on March 25, 1968. Mueller called the choice "excellent," and "as good as you could do." He considered Paine "technically quite capable and politically reasonably good. He wasn't in the same ballpark as Webb [politically] but then he didn't have any of the failings of Webb either." Paine's work at GE gave him the kind of background needed to run the space program, and Mueller quipped that Paine remained deputy just long enough for Webb to retire. Mueller liked the new deputy administrator and they worked very well together. Perhaps that was because Paine "worked alone" and

---

[4]   Seamans, *Aiming at Targets*, 148; Webb Interview, Sherrod, 11/15/68; Mueller Interviews: Slotkin, 6/10/10 and 2/24-25/10; Lambright, *Powering Apollo*, 190-191.

"didn't try to direct us," Mueller said. Nonetheless, Mueller thought his relationship with Webb improved following Seamans' departure because the administrator had found somebody to blame for the fire. In that sense, Mueller said, Seamans became the "sacrificial goat ... That was too bad because of all the people Bob didn't deserve any discredit for that fire."[5]

In October 1967, exercising his new authority as head of the Apollo Spacecraft Project Office, Low implemented configuration management; freezing all changes to the Block II command and service modules and the lunar module beginning with the third vehicle without first obtaining approval of MSC's Apollo Senior Configuration Management Board, which he established and chaired. Low announced that he would only approve mandatory changes, and instituted other limitations. So four years after Mueller first introduced configuration management as part of the Apollo program management plan, Houston finally adopted it.[6]

II

On November 9, NASA planned to conduct the first all-up flight test of the Saturn V. It took four flight tests of the Saturn I first stage before von Braun added a live second stage, possibly delaying that program by two years. Yet the first all-up flight of the Saturn V would test all three stages at one time, along with the instrument unit and a modified Block I command and service modules. Everything would be live on Apollo-Saturn 501 (AS-501), except for the lunar module. This flight would symbolize the recovery from the fire, which occurred nine months earlier. Mueller designated the mission Apollo 4, and though it never flew he retroactively called the doomed spacecraft involved in the fire Apollo 1. He later explained that he had decided to consider AS-201 and AS-202 as Apollo 2 and Apollo 3 respectively, even though they flew prior to the fire. The task of Apollo 4 was to test the space vehicle, evaluating its structural integrity, the compatibility of the stages, and its ability to handle flight loads, with engineers closely monitoring subsystem operations and each stage separation. After completing two orbits, the third stage engine would reignite to place the spacecraft into a higher orbit, and then the service propulsion system would drive the capsule down into the atmosphere to evaluate the heat shield under conditions similar to returning from the Moon. Each stage had been test fired before arriving at KSC, where technicians again checked them out and got them ready for launch.[7]

A bit more than four years passed between Mueller's all-up testing decision and the first flight test of the Saturn V. Describing the scene at launch operations,

---

5   Lambright, *Powering Apollo*, 190-191; Mueller Interviews: Slotkin, 2/24-25/10, 6/8/10 and 6/10/10; Orloff and Harland, *Apollo*, 54.

6   Low to Phillips, 10/14/67, SCP-64-9.

7   "Unusual features of this mission," 10/11/67, Apollo 4, GEM-50-12; "Press and media Kit, Apollo 4," 10/27/67, GEM-50-11; Mueller to Slotkin, e-mail, 3/20/2012.

8-1   Apollo 4 (AS-501), November 9, 1967. (NASA photo)

authors Charles Murray and Catherine Bly Cox wrote, "Up in the V.I.P. viewing area, von Braun yelled, 'Go, baby, go!'" However, always reserved even as he watched one of his greatest decisions prove correct, the authors observed, "Mueller looked pleased." The countdown was interrupted by numerous small anomalies, but once the vehicle was off the ground it performed according to plan. In addition to testing the all-up concept, the flight qualified the launch complex and its support equipment. The space vehicle carried a large amount of telemetry apparatus, allowing NASA to take more than four thousand inflight measurements. The reignition of the third stage J-2 engine worked as planned. Apollo 4 successfully re-entered the atmosphere, and the navy recovered the capsule from the ocean near Hawaii. As Seamans wrote, von Braun's rocket team was "astounded ... [and] George Mueller was vindicated for his bold planning and execution." The president told the press, "The whole world could see the awesome sight of the first launch of what is now the largest rocket ever flown. This launching symbolizes the power this nation is harnessing for the peaceful exploration of space."[8]

At the post-launch press conference, von Braun said, "No single event since the formation of the Marshall Center in 1960 equals today's launch in significance ... I regard this happy day as one of the three or four highlights of my professional life – to be surpassed only by the manned lunar landing." And Seamans told the reporters, "The most important result of the mission ... was the clear indication that we do have the capacity in this country to be pre-eminent not only in space, but in all human endeavor involving science and technology." The spacecraft performed "flawlessly." And the ever reserved Phillips told the press, "Yesterday I would have said that I think we have a reasonably good chance of accomplishing a lunar landing before the end of 1969. Today I think that reasonably good chance is maybe a notch above reasonably good." And the *New York Times'* headline read "Doubts About Moon Landing, Caused by Fire on Apollo in January, Swept Away." Seamans later recalled, "The Huntsville people, in particular, were absolutely aghast ... They had said, 'It'll never work.'" Although right after the launch he saw von Braun "shaking his head, and saying, 'I never, never thought it'd be possible.'" Then in an essay published in 1975, von Braun wrote, it "is clear that without all-up testing the first manned lunar landing could not have taken place in 1969."[9]

Two months before Apollo 11, Mueller said that the most exciting moment of the program occurred when "501 went off ... And that thing worked all the way. That, for me, was the most exciting moment of the whole thing." Then four months after the first lunar landing, he told an interviewer "the one that gave me the most relief was 501 ... It was key. If it hadn't gone we'd have been in deep trouble." Recalling the first all-up flight test more than forty years later, Mueller explained, "if it had

[8]   Bilstein, *Stages to Saturn*, 357; Murray and Cox, *Apollo*, 249; Benson, et al., *Moonport*, 19; Seamans, *Apollo*, 80.

[9]   Bilstein, *Stages to Saturn*, 357-359; "Goals Achieved," 11/11/67 and "Saturn Success Spurs US Hopes," 11/11/67, *NYT*; Seamans Interview, JSC 2, 11/20/98; "Saturn the Giant," von Braun essay in *Apollo Expeditions*.

failed it would have really created a problem," because NASA had "a lot riding on it," and he called having that many parts all work together "a remarkable achievement." Adding, "The success of the all-up test gave us a new confidence. We felt the recovery was truly taking place." His relationship with Webb improved somewhat after AS-501, and NASA's dealings with NAR got better as well. Mueller sent Atwood a letter congratulating the company for the results of Apollo 4, and in response the CEO said "[the] success of the flight was a tremendous boost to the morale of all our people." Mueller wrote similar letters to the other Apollo contractors, emphasizing that Apollo 4 was "a very large step forward. It is in my view, the most significant milestone of the Apollo-Saturn program." After this achievement, attention turned to a lunar module test on a Saturn IB as Apollo 5 and the second Saturn V flight test designated Apollo 6.[10]

Despite the success of the Apollo 4 mission, Webb continued to stew about what he considered to be the lack of teamwork, and on December 4 he announced what he called a "new management concept" for the agency. Writing to the program associate administrators, he said, "Under this mode of operation, the four program directors, reporting directly to the administrator, are expected to be full participants in the total management progress rather than being concerned primarily only in managing and promoting programs and activities for which they bear line responsibility." This meant Mueller and the others should take a "NASA-wide view point" when advising Webb; a collegial approach to the management of the agency. The administrator also tasked Newell to implement an agency-wide program planning process. Yet while Newell's efforts took time, and generated lots of paper, they had no real impact, and Mueller called this a "great waste." Nonetheless, he insisted, he tried to support it but it "was not supportable ... [and] it seemed like it was going to be a disaster." Then following a series of management meetings in early December, Webb sent Mueller a memorandum which said "somehow there was a gap in communications between the Office of Manned Space Flight and my own office, and that in the future I expect the burden to be on you ... to make sure that I understand how you were conducting the two-thirds of NASA's business that is entrusted to you rather than that I would have to audit and take strenuous efforts to seek information as to what is going on." Later Mueller called this another example of Webb's "thrashing around," but at the time he took the memorandum sufficiently seriously to discuss it with Teague and other close advisors.[11]

Planning for the post-Apollo period ran in parallel with the recovery from the fire, and it picked up momentum with the return to flight status after a one year hiatus during which NASA "fought the battle of the fire," as Mueller described it. He thought Newell's planning efforts a failure, and felt the need to get NASA working on an integrated long range plan if they were going to create the requisite political

---

[10] Mueller Interviews: Bubb, 5/22/69, Sherrod, 11/11/69, Slotkin, 6/8/10 and Lambright, 9/20/90; Atwood to Mueller, 12/1/67, GEM-39-11; Bilstein, *Stages to Saturn*, 360.

[11] Management Council, 12/4/67, Teague, Olin E., GEM-100-11; Mueller Interview, Slotkin, 6/9-10/10.

support. He used the advanced programs offices at each center and in OMSF to develop a long range plan for human spaceflight, with Bellcomm providing "the glue" to bring it all together and tie it into the science community. After the fire, Webb opposed long range planning for space exploration, particularly the idea of going to Mars. The administrator said, "First we've got to do the Moon before we begin to put into effect a longer-range plan." Mueller acknowledged this, pointing out that "for every person you got to support it, you'd have ten people finding ways of shooting it down." He could not build a consensus to do anything after the lunar landing because, he said, "no one really believed we could get to the Moon" in the first place. And Webb did not want to plan beyond Apollo, telling him to stick "to our knitting" and not produce "grandiose plans" for the future. Seamans supported long range planning and encouraged Mueller's efforts, believing that planning was necessary in order to understand where they were going; though he would leave the agency at year-end.[12]

Increasing the frequency of his public speaking, Mueller gave a talk to physics students at a colloquium in early December, telling them, "A milestone of the first magnitude in the Apollo Program was accomplished on November 9, with the successful first flight of the Apollo-Saturn V space vehicle." He then expressed confidence that Apollo would succeed by following a carefully planned series of steps which would permit the alteration of plans during a flight. Returning to the rhetoric that he had used before the fire, he again compared space exploration to Columbus' journey to the New World and Lindbergh's flight to Paris. And while acknowledging the other challenges that the nation faced, he said the agency's goals remained the same as in 1958 when Congress established NASA; and the reasons for space exploration were as valid in 1967 as at the dawn of the space age. Finally, he called human spaceflight "an investment in our future ... [which] provides us with the dimensions of a great challenge – to explore space for the benefit of all mankind."[13]

In late December, Webb sent Mueller a copy of an article published in the *Columbia Journalism Review*: "The Apollo Story – What the Watchdogs Missed." Highly critical of the agency, it quoted Hines of the *Washington Star,* who wrote, NASA's initials stood for "Never A Straight Answer." He criticized the agency's "myth of invincibility," created with the collusion of the news media, and then he quoted liberally from NASA's numerous critics in the press. With injured pride, and bitterness, Webb accused Mueller of avoiding responsibility for the fire, and told him to "take the lead in showing how we discharge our responsibilities, the progress we are making in Apollo, and the nature and basis of the decisions we have made in respect to Apollo applications." He told Mueller to explain what had happened and answer questions raised by the fire, including the alleged "inadequacies of NASA's management ... and other actions we have taken with respect to North American." Webb wanted Mueller to spend more time on these activities "as a senior NASA official as well as head of OMSF in answering what we have done and will be doing

---

12   Mueller Interview, NASM, 11/8/88.
13   Mueller, Physics Colloquium, Harvey Mudd College, Claremont, CA, 12/12/67.

to ensure that the inadequacies at North American ... are either overcome or progress is being made." Calling himself the agency's "chief political officer," Webb said that he had attempted to answer the attacks over the past ten months, without satisfying the critics. With Seamans' departure, he would have to rely more on the next level of management, and wanted Mueller to take the lead in this area because "this next year will afford an opportunity for you to fully report on your own stewardship in this program in such a way as to make the record as clear as can be. This will require admitting mistakes where they were made and showing what steps have been taken to best guarantee success in the future." Then meeting with Bogart to discuss OMSF budget and contract issues, Webb said "in making decisions we better make it on the basis that George Mueller is not going to be supporting me but that I am going to be supporting him ... I was out in the front to keep the heat off [the Office of] Manned Space Flight. Now it is important for me to step aside and for George Mueller to put aside all of his importance placed on engineering and get into this type of thing."[14]

Nonetheless, Mueller continued to ignore Newell's new role while focusing on the Apollo Program, dealing with Webb's criticism, and the added bureaucracy. The next mission, Apollo 5, would test the lunar module using the same Saturn IB that the agency had intended for Apollo 1 (AS-204). Despite the fire and sitting on the launch pad for months, the booster remained in flight condition, and after some refurbishing was designated AS-204R. Because the Saturn IB could not lift the complete Apollo system, the lunar module would be housed on top of the S-IVB stage and the conical adapter would be capped by a nose cone. The lunar module was mated in November, and a flight readiness test was performed in December. Set for liftoff on January 22, 1968, Apollo 5 would test the ability of the lunar module's descent stage engine to start, throttle, stop and restart in space, test the jettisoning of the descent stage and the ignition of the ascent stage engine. It would also test the lunar module's structure, instrumentation and control systems, and further qualify the Saturn S-IVB third stage. After a nominal launch, the lunar module tests were initiated, but the decent engine shut down seconds later due to a computer error. Flight controllers followed alternative plans and successfully achieved all the required tests. As MSC's Kraft recalled, "At the end of the day, LM got nothing by high marks. There was a lot to be done before men could fly it. But we knew that the design and the lunar module systems were solid." Quoting Mueller's post-flight press conference, the Associated Press said that Apollo 5 represented "a remarkably good next step on the way to the Moon." The press agency wrote, "Mueller and his colleagues were so pleased that they all but ruled out a proposed duplicate test in late spring." Phillips told reporters, "I'm very bullish. I think the Apollo Program is in good shape and we're going back into a mission sequence. This flight today ... is a very critical one as all of them are, but it does constitute testing the last major piece

---

[14]   Webb to Mueller, *Columbia Journalism Review*, fall 1967, "The Apollo Story – What the Watchdogs Missed." SCP-64-9; Bogart and Webb, 12/27/67, Teague, Olin E., GEM-100-11.

of flight equipment ... we're coming along in good shape at this point." A *New York Times* editorial called Apollo 5 "a significant and important step forward in the continuing American effort to put a man on the Moon." And the paper said all of the equipment needed for the lunar landing had been successfully tested, concluding that "there is little reason to doubt that the capability to land men on the Moon and return them safely to Earth is almost within grasp."[15]

III

In early January 1968, Webb again admonished Mueller to represent all of NASA not just OMSF, calling his office "an organization that has developed much of its own élan from a somewhat independent point of view." Continuing, the administrator wrote, "You will, I am sure, recognize that when you give the President's explanation of his budget before Congress, you are not just speaking for manned space flight but as a senior NASA official justifying and defending the President's recommendations." This missive perhaps proved to be the last straw, leading Mueller to consider whether the time had arrived for him to leave the agency. Writing and calling several friends, including the CEOs of several large aerospace firms, he sought advice on how he might minimize the impact of his departure, writing to one CEO, "At what point in the progress of the program may I or should I leave?" Encouraged to stay, he would remain at the agency through the initial lunar landings. But as Mueller recalled, "that's when Webb was going in circles. So it wasn't ... that I considered leaving as he was considering my leaving." Adding "that was one of those periods of time when with Jim you were never sure what your situation was, and since he was in the process of getting rid of Bob Seamans why I was next in line."[16]

On January 23, Mueller informed Webb that the data from Apollo 5 indicated that all of the primary objectives of the mission had been accomplished. The eight H-1 rocket engines, which were about four years old at time of launch, "verified our judgment that it would be possible to store Saturn IB's for extended periods of time and fly them successfully." The flight data indicated that all of the primary objectives had been "substantially achieved." Phillips followed up with a message to the center directors, telling them the results of Apollo 5 indicated a second LM flight test "may not be required." The final decision about a second lunar module flight test took

---

[15]   Bilstein, *Stages of Saturn*, 341; Apollo 5 Press Kit, 1/11/68, GEM-51-2; Seamans, *Apollo*, 107-108; Kraft, *Flight*, 280; "Lunar Ship Performs Well..." 1/23/68, GEM-83-6; Apollo 5 Post-Launch Press Conference, KSC, FL, 1/22/68, SCP-85-4; "Advance Toward the Moon," *NYT*, 1/26/68; The "R" in AS-204R stood for "redesign." Orloff and Harland, *Apollo*, 139.

[16]   Webb to Mueller, 1/8/68, GEM-44-7; Mueller to Allen, 2/6/68, GEM-39-16, and Mueller to Doolittle, 12/10/67, GEM-41-1; Mueller Interview, Slotkin, 6/8/10.

place at a management council meeting on February 6, and after a design certification review NASA rated the lunar module for human spaceflight.[17]

Mueller gave a major address about the need to support NASA's budget to the Economic Club of Detroit on February 12, pointing out that von Braun's team had launched the first US artificial satellite, Explorer I, aboard the Jupiter C, a variant of the Redstone ballistic missile in 1958, and in the ten years since then the program had come a long way. As a result of successfully flight testing each component of the Apollo-Saturn, NASA had "a reasonable chance to achieve our national goal of landing men on the Moon and returning them safely to Earth before the end of the decade." Yet the US faced the choice of either continuing its investment in human spaceflight or mothballing this "national resource" just when it was approaching the real pay-off. The agency budget request for FY 1969 hit the lowest level since FY 1963, a drop of more than $1.3 billion from its peak in FY 1964. Again raising the specter of Soviet competition, and citing numerous statistics to buttress his points, he told the members of the club, "These severe restrictions have occurred during a time when our chief competitor in space, the Soviet Union, is increasing its investment in space activities at a regular rate." Consequently, the possibility existed for the USSR to pull ahead in the second decade of the space age. While acknowledging "real and urgent problems" at home, he argued, "we also recognize the necessity to maintain our technological leadership on the broadest front possible during this critical period – and our space program is a prime factor in this leadership." The budget then before Congress would barely maintain capabilities at a minimum level, and further reductions "would cripple our space program for years to come."[18]

If adequately funded, however, the Apollo Applications Program would provide the information needed to make the right decisions about the next steps in exploring space. Existing budgetary restrictions would produce a limited program, as compared to Apollo. Furthermore, with the long lead times needed for new equipment, a budget shortfall would "cast an uncertain future" on even that limited objective. The nation required a low cost way to put payloads into orbit, which could be achieved using reusable space vehicles that would yield a significant pay-off in future years. And in the minds of people around the world, he said "space accomplishment has become the prime measure of a nation's stature in all areas of earthly endeavor … Man will prevail in space … The only question is whether this nation will prevail in space … or will we abandon the future to others?"[19]

When testifying in support of the FY 1969 budget before Teague's committee the next week, Mueller said it would be the third straight year of declining funding, a decrease that he attributed to the completion of Gemini and a shift in Apollo from design and ground testing to flight testing and operations; although sharp cuts in the Apollo Applications Program in 1968 and 1969 also had an impact. The president's

---

[17] Mueller to Webb, 1/23/68, and Phillips to distribution, 1/29/68, and LM-3 DCR, 3/6/68, SCP-81-3.

[18] Mueller, Economic Club of Detroit, MI, 2/12/68.

[19] Ibid.

budget for OMSF would continue the space program at a minimum level, but the flights of Apollo 4 and Apollo 5 showed how far the US had come. He called Apollo 4 "the free world's largest and most complex space vehicle," which "worked together on the first try without error." NASA was planning a second Saturn V flight test in March, and by year-end the agency would fly humans aboard an Apollo spacecraft launched by a Saturn IB. Only then would it fly a Saturn V mission with the complete spacecraft. In 1969, NASA would fly five Saturn V's, "and if all goes well," he said, the agency would complete the first lunar landing. Making the case for the post-Apollo program once again, he argued that "if adequately funded, [it] can give us the information we need to make intelligent decisions about the next steps forward." He called the orbital workshop the area of "most promise," noting it was the first step in establishing a space station, and "the next logical, practical, and necessary advance, beyond the Apollo Program, in the development of our national capability for manned space flight." However, "In order to phase our Apollo Program activities economically and effectively into the logical sequence ... the preliminary effort on this ... workshop must begin now."[20]

When testifying before the Senate several days later, Mueller spoke about the management of the Apollo Program. His oral testimony repeated much of what he had just told the House, and he pointed out that all support elements, from mission control to the ships and aircraft necessary to recover Apollo capsules, had been fully tested. NAR made progress in preparing the command and service modules, but he acknowledged that the rate of change was progressing more slowly than hoped. He told the senators that he still had some concerns about the need to reduce the weight that was added to the spacecraft as a result of the changes introduced by fire; though he assured them that the weight margins were adequate. He described problems with the guidance software, calling Apollo's computer systems the area where the agency pressed hardest against the state of the art, warning that this could become the critical path item delaying the lunar landing. On the positive side, the analysis of Apollo 4 showed fewer problems than initially thought. And the AS-502 flight test scheduled in late March as Apollo 6 would serve to verify the launch vehicle design. The first piloted Apollo flight would take place in the last quarter that year, using a Saturn IB. NASA planned to test the lunar module before the end of the year, on the first piloted Saturn V flight. Then, using the cautious guidance dictated by Webb, he said that as the agency gained flight experience with the Apollo-Saturn V, "There is a *reasonable probability* that we will be able to compete the lunar landing with Saturn 509 which is scheduled to be flown at the end of 1969." (However, the first lunar landing would take place on AS-506, the sixth Saturn V flight.)[21]

Mueller told the senators that budget cuts would delay the planned launch of the orbital workshop until 1970 and the Apollo Telescope Mount until 1971, and that funding for the Apollo Applications Program in the FY 1969 budget would only

---

[20]   Mueller, House Space Committee, 2/19/68.
[21]   Mueller, Senate Space Committee, 2/28/68.

"begin to lay the ground work for an eventual space station." He said the orbital workshop would meet the nation's needs in "science, civil and military applications, and serve as a research outpost for a number of years." Then, using surplus Saturn boosters, the agency could conduct follow-on missions and preserve the investment made at minimum cost during the early 1970s. Despite funding cutbacks, the Apollo follow-on program proceeded through the early study and design phases in advance of entering the final phases of design and then production. However, he noted that the budget contained less than half of the amount necessary to fully utilize Apollo's capabilities. And finally he lectured the senators, "If leadership is our intention, then the costs must be accepted."[22]

Mueller again made the case for the Apollo Applications Program in a public address in early March. Acknowledging the requirement to solve the nation's other problems, he said "we also believe in the necessity to maintain our technological leadership on the broadest front possible during this critical period – and our space capability is a prime factor in this leadership." The budget approved by Congress maintained the space program at a minimum level, but did not assure pre-eminence or leadership in space. Yet "further reductions in this budget would cripple our space program for years to come." In reporting the speech, the *Rocky Mountain News* quoted Mueller saying, "It will make little difference scientifically whether we or the Russians are the first to land a man on the Moon, but if we're not first our role as a World leader will suffer a serious setback in the eyes of all other nations." The Denver newspaper also noted that Mueller was "highly critical" of the cutbacks.[23]

MSFC completed work on their neutral buoyancy facility and von Braun used it to convince Mueller that the S-IVB *wet* workshop concept would not work. Von Braun built this facility, which was large enough to submerge an S-IVB stage, to see what it would take to assemble the workshop in space. When shown to Mueller, he wanted to learn firsthand how to assemble the workshop – and learned scuba diving in the process. After being fitted out with diving equipment, he went "down there and see how hard it was to move the valves," and this convinced him it would be too difficult to assemble the wet workshop in orbit because it was too difficult to do underwater, even without a spacesuit. "If you're in a spacesuit, it's even worse," he said. Then as he later explained, "after that, and after we got a little further along in the design, I made the decision to fly [the dry workshop] on a Saturn V instead of [the wet one] on the Saturn-IB."[24]

Webb continued trying to manage the agency by consensus, and while achieving it on Apollo, NASA's leadership disagreed about the Apollo Applications Program. Mueller and von Braun kept pushing to build the orbital workshop, while Gilruth opposed it. Mueller needed it to test ideas about orbiting space stations. However, he

---

22 Ibid.
23 Mueller, Chamber of Commerce, Denver, Colorado, 3/8/68; "...U.S. Space Prestige at Stake in Space," *Rocky Mountain News*, 3/9/68, GEM-83-6.
24 Mueller Interviews: Slotkin, 2/24/10 and Bergin, 8/27/98; Compton and Benson, *Living and Working In Space*, 109.

quickly realized that its "real justification" was as a "node" in a space transportation system. Scientific experiments alone would not justify it. So its main purpose became determining the ability of humans to survive in space for long periods of time, and serving as a node in a space transportation system that would permit more efficient access to space. But this would require the development of an inter-orbital transfer vehicle to move between low Earth orbit, the Moon and eventually Mars.[25]

Newell coordinated studies that looked at piloted versus robotic missions for the Apollo Applications Program. When Mueller disagreed with the conclusions, Webb asked Thompson to review the issues, look at the science planned, and resolve the wet versus dry workshop controversy. The Langley director established an agency-wide committee, which found the Apollo Applications Program objectives to be "in line with post-Apollo needs," but concluded that it attempted too much and so needed a narrower focus. Although the Thompson committee deemed the wet workshop to be marginally adequate, and therefore preferred the dry workshop, it was concerned about the costs. In addition, it observed that Houston and Huntsville were unable to work together because MSC objected to MSFC having primary responsibility for the workshop. Finally, in view of the budget constraints, the committee recommended continuing with the wet workshop.[26]

From the beginning of his time at NASA, Mueller had run planning activities in parallel with the Apollo and Gemini program offices in order to provide the basis on which to decide how best to exploit the Apollo technology for follow-on space activities. Without budget constraints, he said, this effort would have led to "a consistent and continuing program" using the Apollo-Saturn V. However, political pressure led to budget cutbacks and a much reduced Apollo Applications Program. It would end up becoming a choice between continuing production of the Saturn V and developing the next generation of space equipment. Newell's program planning exercise became bogged down in its own bureaucracy, but Mueller's parallel planning continued. Newell later complained, "The problem with manned space flight ... was that they were in the habit of going it alone, they wanted to go it alone, and they intended to go it alone." So competition between Newell's and Mueller's program planning continued.[27]

IV

When Apollo 6 was launched on April 4, the Saturn V suffered a number of serious problems. After about a minute and a half the S-IC first stage developed a violent longitudinal oscillation, experiencing what engineers call the "pogo effect" because it

---

[25]   Lambright, *Powering Apollo*, 194; Mueller Interviews: NASM, 11/8/88 and JSC, 8/27/98.

[26]   Ibid.; Compton and Benson, *Living and Working in Space*, 97-98; Mathews Interview, Lambright, 10/4/91.

[27]   Mueller Interview, NASM, 11/8/88; Lambright, *Powering Apollo*, 194.

made the rocket vibrate up and down like a pogo stick. This lasted for about thirty seconds. During this time huge chunks of the honeycomb structure of the adapter in which the lunar module would be carried on future missions peeled off. The five J-2 engines on the S-II second stage ignited, but after several minutes first one and then another engine unexpectedly shutdown. The instrument unit tried to compensate for the lack of thrust by swiveling the remaining engines on their gimbals and burning them longer, but its programming was designed to recover from only a single engine-out failure and so it veered off course. When the S-IVB third stage took over with its single J-2 engine, the instrument unit managed to steer into an elliptical orbit. The plan was to coast for two orbits and then reignite the J-2 to place the spacecraft in an elliptical path that had its apogee at lunar distance. But when mission control sent commands to reignite the engine, it would not restart. As per the contingency plan, the spacecraft separated from the S-IVB and fired its service propulsion system to put the capsule into position to enter the atmosphere almost as if it had returned from the Moon, as a further test of the heat shield. While the failure of three J-2 engines caused major concerns, von Braun said, "Had the flight been manned, the astronauts would have returned safely ... [Though w]ith three engines out, we just cannot go to the Moon."[28]

Mueller's initial reaction to Apollo 6 was "wonder" as NASA tried to determine what went wrong. It took some time to figure out why the engines failed, because the cause was not obvious. Yet whereas others expressed deep concerns, he remained confident the agency could track down the problems, recalling that "it didn't look like it was going to be a major problem." He said the agency should have known "about the possibility and probably if we had been thinking constructively we would have been prepared for it." In the days before computers had sufficient processing power to analyze the vibrations that cause the pogo effect, engineers used empirical methods to predict the reaction to external forces, and modeled the whole launch vehicle to calculate what would happen. But in the case of the J-2 engines, Mueller said the only way to have found the problem, "unless you thought about it in the first place ... is by flying it."[29]

In response to a reporter's question at the post-flight press conference, Mueller said he "didn't know whether it was a disaster or not," but "we might have learned enough to feel confident about going forward." And he later pointed out, "I had more headlines for that piece of reckless abandon than I had about anything else I said." As a result of this run-in with the press, Alibrando, the OMSF public information officer, advised him "to go down the middle – being neither too optimistic nor pessimistic" about the results of Apollo 6, and advising, "It is important that we do not try to wish away the problems we had on the S-II and S-IVB performance. We don't sound very convincing when we try to make every

---

[28] Bilstein, *Stages to Saturn*, 360-361; Skaggs to Slotkin, e-mail, 6/4/11; Apollo 6, Mission Director's 24 Hour Report, 4/4/68, SCP-85-6.
[29] Mueller Interview, Slotkin, 6/8/10.

function sound like a 'bonus' ... I think we should say in a fashion that is unmistakably clear that it will be some days before the Apollo 6 flight is completely evaluated and the impact on the program assessed."[30]

The United Press International published an article with the headline "Apollo 6 Launching Officially a Failure." Quoting Mueller, it reported that "the mission was officially a failure because of trouble with the Saturn's engines," and speculated that another Saturn V flight test would be needed before it could safely carry astronauts. Not as kind, Susan Butler, a reporter for a local newspaper from Daytona Beach, Florida, criticized Mueller for saying Apollo 6 was "on balance a most successful mission." She called the flight a fiasco wasting millions of dollars that could have been better spent elsewhere, and claimed to have been amazed at Mueller saying he might "still let astronauts ride the very next Saturn V." Adding, "Sorry, George, but I for one, don't believe for a moment you meant what you said." But Mueller insisted Apollo 6 was not a total failure, because it "did almost everything that it was set up to do," and engineers learned what had to be done to insure the launch vehicle would not fail again. As a result of this flight, engineers went through the booster in enough detail to fully understand the launch dynamics because they knew the first Saturn V flight had been flawless. And he added, AS-502 "gave us enough confidence when it failed," because "we understood what was going on." The problems were finite, and NASA had confidence after Apollo 4, knowing it *could* work.[31]

On April 20, despite the ongoing investigations, Boeing recommended "providing [that] the necessary corrective actions and assurance activities can be satisfactorily completed," Apollo 7 should carry a crew. A design certification review took place at MSFC the next day to examine the impact of Apollo 6 (a Saturn V flight test) on Apollo 7 (a Saturn IB mission) – not least because they both used the S-IVB stage. As chairman of the review board, Phillips communicated the decision to continue preparations for Apollo 7, and said that AS-503 should fly as Apollo 8 and also carry astronauts.[32]

The Senate space committee held a hearing in executive session on April 22 to look into the problems experienced on Apollo 6. This was attended by Webb, Mueller, Phillips and George H. Hage, who was the deputy Apollo program director. Following Webb's opening remarks, Phillips and Hage described what happened and discussed the corrective action plan. Pleased with what they heard, at the end of their testimony Senator Anderson and several committee members commended NASA. Then the committee chair asked Mueller if he had anything to add, and he had three

---

[30]   "Apollo: Looking Back," Caltech, Pasadena, CA, 5/18/71, SDC Speeches; Alibrando to Mueller and Phillips, 4/4/68, Apollo 6, GEM-51-5.

[31]   "Apollo 6 Launching Officially a Failure," UPI, 4/5/68, and "Was NASA Man Euphemistic," *Daytona Beach Sunday News-Journal*, 4/7/68, GEM-83-6; Butler, Oral History of KSC, 6/25/02; Mueller Interview, Slotkin, 6/8/10.

[32]   Stoner to Phillips, AS-503 Mission considerations, 4/20/68, SCP-81-3; Phillips to distribution, Saturn V Simi-DCR/AS-502 Evaluation, 10/6/68, GEM-92-8.

observations to make. First, he claimed that at no time during this flight would astronauts have been in danger. Next he said "we fly these vehicles in an unmanned mode in order to be sure that we understand how they work," and repeated that the agency had learned more from Apollo 6 with all its problems than from the successful flight of Apollo 4. "[W]e learned that it is in fact, safe to proceed with two engines out on an S-II stage ... We learned that we could shut down an engine in flight ... successfully, even though it was in a mode where it was failing. So that we did demonstrate that it was safe to shut these engines off when they malfunction in flight." From what the agency knew, the fixes could be made and tested on the ground, so another all-up flight test of the Saturn V would not be necessary. Despite the hearing being closed, Anderson wanted to release the testimony and Webb agreed. Responding to follow-up written questions from the committee staff, NASA officially declared Apollo 6 a failure, because only four of five primary objectives were achieved. The agency then began preparations to fly Apollo 8, retaining the option to conduct another flight test without astronauts pending final resolution of all the Apollo 6 anomalies. However, Mueller never considered Apollo 6 a failure. Despite multiple problems, the second Saturn V flight successfully completed most of the planned tests, and because of the amount of telemetry NASA had the ability to diagnose and fix the problems. The pogo effect was not a new phenomenon; it had occurred in the development of other launch vehicles, including the Titan II. Engineers expected to quickly eliminate the vibrations, but the failure of three J-2 engines looked like it would take more time. Mueller called this "a much more subtle problem and not one that was immediately evident. You can't really test engines in a vacuum ... to duplicate all of the things you find in space. So to some extent your first few flights are flight tests to make sure you understand what the problems are."[33]

NASA assembled a team to look into the engine problems, and soon identified the cause as the failure of an igniter line in the upper part of the engine. The first engine had failed on the S-II because of leaks in the fuel line feeding the augmented spark igniter, and telemetry from the S-IVB showed that its engine had begun to suffer in a similar manner but it had managed to complete its first burn before the line actually snapped once in orbit and made a restart impossible. The second S-II engine failure occurred because of a wiring error – when the instrument unit sent the command to shut down the ailing engine, this was routed to one of the good engines. After engineers had figured out exactly what happened, further testing determined that the igniter fuel lines had ruptured because a bellows in the line designed to accommodate flexure had been disrupted by vibration – it was metal fatigue. Rocketdyne replaced the bellows with an alternative design in which the pipe allowed flexure. The team investigating the pogo effect found that it occurred because vibrations had led to irregular fuel flow, and a simple helium buffer was installed in the liquid oxygen feed line to damp out pressure fluctuations and thus prevent thrust oscillations. A

---

[33] Apollo 6 Mission, Senate Space Hearings, 4/22/68, SCP-85-7; Bilstein, Stages to Saturn, 361-362; Mueller Interview, Slotkin, 6/8/10.

separate problem caused the failure of the honeycomb structure. Moisture had penetrated into the honeycomb, and when boiled by the heat of the aerodynamic airflow the steam had blown off chunks of the material. This was remedied by providing holes in the skin of the honeycomb to enable vapor to escape harmlessly. In August, the agency successfully demonstrated the engine fixes at the Mississippi Test Facility. This cleared the way to proceed with a human flight aboard AS-503 – providing of course that all went well with the piloted test of the Block II command and service modules launched as Apollo 7 aboard a Saturn IB.[34]

Although Gilruth and his management team at MSC sought another Saturn V flight test without a crew, Mueller, von Braun, and most significantly Paine, convinced Webb to approve launching Apollo 7 as planned because the J-2 engine problem looked to have been resolved. Unlike Webb, who lived through the fallout of the fire, Paine agreed with the "ever optimistic" Mueller that if they could resolve all of the problems in time, the agency should proceed with Apollo 7, and continue planning for Apollo 8. Confident of the decision to fly AS-503 with astronauts, von Braun published an article "The Detective Story Behind Our First Manned Saturn V Shot" in the November 1968 issue of *Popular Science*, in which he wrote, "A string of mishaps ... beset the second flight last April. But the diagnosis of these has been so conclusive, and the remedies so successful, that the unmanned trial will not need to be repeated."[35]

Mueller said Apollo 6 was "a great learning experience." The Apollo Program Office and the technical groups which supported them "worked exceedingly well in terms of identifying the problems and curing them." He explained, "We produced the problem on the ground and made the corrections necessary and then we went through the entire history of the two flights, to be sure there weren't other anomalies in them, in as much depth as I've ever seen programs reviewed, and on the basis of that we felt confident enough that this thing was going to fly." However, a minor brouhaha erupted in late May when Mueller publicly called Apollo 6 a "partial success." Based on NASA's binary definition of success, a mission could not be a partial success. Yet, whereas the agency officially classified Apollo 6 a failure, Mueller considered it a successful test flight precisely because it had revealed problems.[36]

V

Mueller gave a background briefing to the Aviation/Space Writers Association at the beginning of May, telling them that prior to Apollo 6 NASA had decided that if it achieved Earth orbit "it would be possible to make a technical evaluation that the Saturn V could be man-rated." After acknowledging "some technical difficulties"

[34]   Bilstein, *Stages to Saturn*, 361-362; Orloff and Harland, *Apollo*, 151-159.
[35]   Lambright, *Powering Apollo*, 197; "The Detective Story..." *Popular Science*, 11/68.
[36]   Mueller Interviews: JSC, 8/27/98 and Sherrod, 3/20/73; Maglinato to Mueller, 6/3/68, SCP-81-4.

during the flight, he said that since the agency knew what happened these would be overcome "with a minimum of time and funds." Then he discussed the seven phases of the lunar landing development program called missions A through G, each to validate a single piece of equipment or procedure which could not be tested on the ground (Table 7-1). These missions did not necessarily represent discrete flights, and NASA would fly as many times as required to finish the full test program. He spoke of an open-ended mission concept, treating each as a separate experiment with alternative procedures which could be changed during the flight. Then he discussed decision points or plateaus built into the initial lunar landing (a mission-G), stepping through each in detail. And he noted, "On almost every plateau opportunity exists for alteration of the profile as a result of consultation between astronauts and the ground." With the completion of the first mission-G, and the crew's safe return, he explained, "we will have met a major challenge of the space age ... But a single goal doesn't make the game. In order to use wisely and well what we have bought, and what we have learned, we will need to apply these benefits to continued exploration and exploitation of the exotic environment of space."[37]

Speaking at the dedication of Grissom and Chaffee Halls at Purdue University, Mueller, began by paying respect to the memory of the Apollo 1 astronauts (see chapter 6), like him, both Purdue alumni, and then spoke about the advance of technology through the ages, before addressing ways in which to eliminate poverty and improve human welfare – topics that were on everyone's mind in the wake of riots following the Martin Luther King assassination, which occurred the same day as the Apollo 6 launch. He reminded this audience that the money spent by NASA remained on Earth, and that "Not one dollar has ever been spent on the Moon." The space program had created jobs, products, and processes, as well as new companies and industries. But with national treasure increasingly going to support the Vietnam War and new social programs, Congress was cutting spending for space, and this led Mueller to tie investment in the space program to economic, technological, and educational development of the US. As he had argued elsewhere, he insisted "the space program is not in conflict with efforts to end poverty and improve human welfare. On the contrary, it contributes to the fundamental solution of these problems." Following this speech, an article in the *Washington Star* expressed outrage, writing "Mueller's optimism is understandable, even excusable," because his reputation was on the line. "But to an observer with no personal ax to grind there seems little justification for predicting [a] lunar landing within the next 18 months if the ordinary dictates of prudence and cautioned are followed."[38]

Continuing to speak publicly in support of human spaceflight, in back-to-back speeches at the end of May, Mueller said that Apollo 6 "revealed some technical

---

[37] Mueller, Aviation/Space Writers Association, Washington, DC, 5/1/68; for a discussion of the mission types, see Orloff and Harland, *Apollo*, 52.
[38] Mueller, Dedication of Grissom and Chaffee Halls, Purdue University, Lafayette, IN, 5/2/68; "Is Safety Taking a Back Seat in Project Apollo?" *The Sunday Star*, 5/19/68, GEM-83-6.

difficulties, [but t]hese are being corrected and we plan to overcome them shortly."
He expected the first piloted Apollo flights to take place before year-end, again
claiming, "We believe that we have a reasonable possibility of accomplishing the first
Moon landing before the end of next year." Yet, he added, "The lunar landing is not
a goal in itself . . . Scientific and technological capability constitutes a basic source of
national power." He spoke about Apollo's importance in international relations, and
the necessity to properly fund NASA. And again, "Any further reductions in this
budget would cripple our space program for years to come." Tying the space
program to economic growth, the balance of payments, and other earthly benefits,
he said it also provided many spinoffs. Competition with the USSR "is not just a
race to be 'first' in some specific achievement. It is a competition to be in the
forefront of the world's new technology . . . In the eyes of the world, the nation's
space program . . . is considered a measure of our ability to compete with a
formidable rival." Nonetheless, as the Soviet Union increased its investment in
human spaceflight, the US lessened theirs – although the Soviets already spent more
on space based on real purchasing power. Then he warned, "As a consequence, we
must face the fact that the Russians will again draw ahead of us during the second
decade of the Space Age."[39]

Mueller brought his senior staff together in mid-June, the first meeting of its kind
in over a year; though in the past he held such meetings more frequently. Much had
happened since the last meeting, so he recapped the major events and described the
analysis of Apollo 6 flight data "a fantastic detective story" which would lead to a
piloted flight on the third Saturn V (Apollo 8) later that year. In the meantime, the
first human flight aboard an Apollo spacecraft (Apollo 7) would take place in the fall
on a Saturn IB. He introduced new members of his immediate staff, beginning with
Mathews, no stranger to the team, who became his principal deputy in May. This
and other changes were intended to strengthen OMSF during the final year before
the lunar landing. But Mathews later told an interviewer that his promotion
occurred because Webb wanted him to keep an eye on Mueller, not merely to
strengthen the program. Before Mathews' appointment, Webb had arranged for
Edgar M. Cortright, Newell's deputy, to become Mueller's deputy. As Mueller
described it, this transfer was to help to deal with the advocates of space science.
However, Mathews claimed, "Mueller did not allow Cortright to penetrate his
program. He gave him all kinds of diversionary tasks." In February 1968 Webb told
Mathews, "I want you to take over as General Manager of OMSF. I don't want you
to push Mueller too hard. I want you to do it quietly." The administrator told
Mathews to "run things" and to "be a spy" for him, but Mathews felt uncomfortable
doing that, and was happy when Webb gave Paine the job of controlling Mueller.[40]

---

[39]   Mueller, Milwaukee, AC Sparkplug, Wisconsin Speech, 5/28/68, and Rotary Club,
     Madison, Wisconsin Speech, 5/29/68, GEM-46-11.
[40]   Mueller, OMSF Staff Briefing (notes), Washington DC, 6/13/68; Lambright, *Powering
     Apollo*, 191-192; Mathews Interview, Lambright, 10/4/91; Mueller Interviews: NASM, 5/
     1/89 and Slotkin, 6/8/10.

The budget submitted for FY 1969 kept the Apollo Program intact, while the Apollo Applications Program received limited funding. Putting a happy face on a difficult situation, Mueller called it a "deferment," not a "cancellation," and told his senior staff that the Apollo follow-on program still had a "substantial budget that will allow us to accomplish a great deal this coming year." However, he cautioned that Congress had not finished with NASA's budget, and there would be a hiring freeze until the final appropriations became known. Further cutbacks might be necessary, perhaps including forced reductions; though he insisted that people should maintain a "positive outlook on all of our programs [and] change from negative to positive thinking by emphasizing the fundamental value of Manned Space Flight." He said they should continue to move forward from the defensive position taken after the fire, and reminded them that they were working on the "greatest project man has ever undertaken."[41]

The Senate came very close to cutting a billion dollars from NASA's FY 1969 appropriations, and in June 1968 the agency's authorization bill came to just over $4 billion, with $253 million for the Apollo Applications Program. Webb still expected additional cutbacks before Congress finished, and used the post-Apollo program as "a surge tank" for Apollo, only authorizing month-to-month letter contracts that could be canceled at any time. The administrator limited NASA expenditures to $3.8 billion, which would mean the closure of the Saturn V production line. The Apollo Applications Program operated under conditions of austerity, and by fall of 1968 the agency had issued fifteen letter contracts covering different projects, a practice that was normally used only to get contracts started. The final appropriations came to just under $4 billion, which saved the Saturn V production line, albeit for only another year.[42]

Opposition to spending on the space program continued, and opponents in Congress formed a special subcommittee of the House Government Operations Committee which held a hearing to review the agency's support contracts, focusing in particular on the Boeing TIE contract in June. In prepared remarks, Bogart told the subcommittee that NASA had expanded Boeing's role in the aftermath of the fire to strengthen its technical integration and evaluation capability, starting with the arrival of new flight hardware at KSC. The agency had "no other realistic alternative," he claimed, because Boeing had highly qualified personnel with extensive experience immediately available, but in-house capability did not exist. NASA had considered other contractors, but found Boeing the best qualified to do the job and the least disruptive to existing contracts. He called the company "uniquely qualified" because it had performed similar work for NASA and the air force. And by using a support contractor, the agency was able to save about seven percent of total cost.[43]

---

[41]   Mueller, OMSF Staff Briefing (notes), Washington DC, 6/13/68.
[42]   Lambright, *Powering Apollo*, 195; Compton and Benson, *Living and Working in Space*, 83 and 104.
[43]   Boeing TIE, 6/27/68, SCP-81-2.

The special subcommittee held a second hearing on July 15, after investigators had discovered that the TIE contract costs had increased from an estimated $20 million in FY 1968 to $73 million in 1969. Finger and Bogart testified, with Mueller accompanying them to respond to specific questions. In answer to one question, Mueller used the pogo effect as an example of how Boeing assisted NASA, and, adopting his usual role of technical expert, went into considerable detail, leading one member to remark that he did not have to do that just tell us "it is complex," because everything in the space program is complex. But the members were more interested in the number of people working on the contract than in what the work entailed, which led Mueller to say, "if I may say so, the number of people that we determined as being necessary was the result of very careful evaluation of the tasks that needed to be performed." The chair, Porter Hardy, Jr., then said, "I haven't got anything . . . to demonstrate that there is any reasonableness in this contract . . . and I haven't got anything to indicate to me that NASA is really intelligently exercising some judgment in getting this thing done." Again Mueller tried to explain, saying that Boeing looked at all the interfaces in the system, causing Hardy to call "interfaces" a "$14 word," but a "good one." The committee remained unconvinced that the agency could not perform the work themselves. Despite congressional objections, the TIE contract remained in effect until the end of 1969.[44]

STAC met jointly with the Space Science and Technology Panel of PSAC in Huntsville, Alabama and in Huntington Beach, California on July 19 and 20. The meeting began at MSFC with a review of the core Apollo Applications Program, where Harold T. Luskin, who had taken over the program following Mathews' promotion, proposed modest budget extensions to restore some congressional cuts. He said that if the agency received additional funding, future flights could use a dry workshop because testing at Huntsville's neutral buoyancy test facility had shown the difficulty of assembling the wet workshop in zero-gravity. Then flying to Huntington Beach courtesy of the air force, the two advisory committees reassembled the next day at McDonnell Douglas to hear about the MOL, which one air force general described as "a seven-shot series of missions lasting 30 days each and not overlapping." In the battle for competing space stations, Huntsville, Houston and Langley discussed the results of their workshop studies and came up with price tags ranging from $4 to $8 billion. After McDonnell Douglas engineers pitched using an expanded version of the MOL in place of the planned orbital workshop, another general told the committees "as the Air Force sees it, [the MOL] will never be more than a 60 day vehicle and the development of a one year capability would be equivalent to developing a new vehicle."[45]

---

[44] "Investigation of the Boeing TIE Contract – House Committee on Government Operations 7/15/68, NASA Archives, RN 10689; Moritz to Long, 7/15/69, Boeing TIE, NASA Archives, RN 10689.

[45] Joint Meeting of STAC and the Space Science and Technology Panel of PSAC Huntsville, AL and Huntington Beach, CA, 7/19-20/68.

    After the joint meeting, Mueller discussed with STAC the projected schedule slips in the Apollo Applications Program resulting from budget cuts, warning that unless some decisions were made soon, the first flight of the orbital workshop would be delayed until 1971 and the Apollo Telescope Mount experiment might have to be canceled. The STAC members "voiced strong approval" of the orbital workshop, and were satisfied with the agency's progress, calling it a "promising, legitimate example of what man can do in space" and "essential for the understanding of man in space." STAC members supported funding the telescope experiment but expressed concern that it would compete for resources with the agency's scientific lunar and planetary exploration programs. Some argued that part of the funding should come from OMSF and not be wholly attributed to space science. As Mathews recalled, while problems in funding the Apollo Applications Program persisted, Mueller struggled with the scientific community over the "manned/unmanned dichotomy." The scientists did not think it necessary to fly the space telescope on a piloted mission, although they could not get money for it without assistance from OMSF. However, hoping to attract more support, Mueller tied these projects to the post-Apollo program.[46]

    As FY 1969 began in July 1968, Mueller had reasons for both optimism and pessimism. NASA remained a year away from the planned lunar landing, and he had the successful Apollo 4 and Apollo 5 tests behind him, but what he regarded as the "partial success" of Apollo 6 remained. Not until the flights of Apollo 7 and Apollo 8 would the agency know whether the space vehicle could deliver a crew to the Moon. The budget for Apollo was in reasonable shape, but the Apollo Applications Program budget was shrinking, and the Saturn V production line might have to close. Mueller continued to face personal pressure from the administrator, and was having difficulty maintaining support for his programs in the scientific community. There were issues to resolve before landing on the Moon, but he never gave up confidence in achieving the objective as he increasingly looked to a future filled with space stations and space shuttles.

---

[46]   Ibid.; Mathews Interview, Lambright, 10/4/91.

9

Orbiting the Moon

"'The triumph of the squares' – meaning 'the guys with crew-cuts and slide rules who read the Bible and get things done'."
Times of London quoting an unnamed NASA official speaking of Apollo 8,
January 10, 1969

NASA planned two piloted Apollo-Saturn missions in 1968: AS-205 (Apollo 7) to prove the Block II spacecraft ready for a lunar mission, and AS-503 (Apollo 8), the first piloted flight aboard a Saturn V to test the lunar module in low Earth orbit. Because of development delays, however, Apollo 8 would test the command and service modules in lunar orbit, and is remembered for the spectacular pictures of Earth rising over the lunar horizon. Mueller called the decision to fly Apollo 8 in lunar orbit "probably the most critical decision we made; certainly coming off of the 502 mission." The agency also planned two piloted Saturn V tests in the first half of 1969 – Apollo 9 and Apollo 10, to complete the learning phase of the Apollo Program, coming as it did between the first flight in deep space aboard Apollo 8 and the lunar landing on Apollo 11.[1]

On his way to Vienna to join Webb attending a United Nations conference on the peaceful uses of outer space in August 1968, Mueller gave a speech at a British Interplanetary Society meeting in London, making extended remarks about replacing the Saturn V with a reusable space shuttle. He had previously spoken about the role of space shuttles, but never in such depth or detail. "I believe that the exploration of space is limited in concept and extent by the very high cost of putting payloads into orbit, and the inaccessibility of objects after they have been launched ... Therefore, I would forecast that the next major thrust in space will be the development of an economical launch vehicle for shuttling between Earth and the installations, such as the orbiting space stations which will soon be operating in space." Space planners had been intrigued for many years by the idea of using space stations as supply bases and refueling points to allow space vehicles to transfer from

---

[1]   Mueller Interview, Slotkin, 6/8/10; Cortright, *Apollo Expeditions*, 10-1.

9-1   Apollo 8 (AS-503) – Earthrise 1968. (NASA photo)

low Earth orbit to higher orbits, or fly to the Moon or distant planets. With space stations and space shuttles, astronauts could install, checkout and maintain satellites without having to return to Earth for supplies. However, a space station would have to be routinely replenished and its crew exchanged, and existing space vehicles were too costly for that function. Mueller therefore argued for developing an efficient system – an economical space shuttle – to reduce the operating cost of gaining access to space by at least an order of magnitude. Yet to accomplish that required new manufacturing techniques and the development of new test procedures; plus volume production of components capable of withstanding repetitive use, as well as more efficient maintenance and handling methods. And he said the space shuttle would ideally operate like a large commercial air transport, and have a cockpit similar to a jet. After taking off vertically, it would land on a runway using an automated guidance system.[2]

The economics of the space shuttle remained to be determined, although over the previous decade technology had advanced from the Vanguard and Explorer, which had cost one million dollars per pound of payload placed in low Earth orbit, to the Saturn V which had reduced that cost by three orders of magnitude. The space shuttle should therefore be able to further reduce the cost by one or two orders of magnitude. However, such savings would require repetitive use. And while initial space shuttle designs reused all high cost components, the relatively inexpensive propellant tanks remained expendable. This idea had been studied for decades, the technology would be within the state of the art, and if the program were to begin in 1968 then the space shuttle could be ready in the late 1970s. Combined with the space station, the space shuttle would open up space for new industries, providing increased applications and benefits. And in conclusion, he said the "exploitation of the foreseeable techniques to their limit could result in truly commercial space transport being in sight by the end of the century ... The space shuttle is another step toward our destiny, another hand-hold on our future. We will go where we choose ... Man will never be satisfied with less than that." This speech received wide attention, and Mueller later called giving it overseas "a convenient way" to focus attention on the requirement to reduce the cost to access space and permit its exploitation. The impending completion of the Apollo Program dictated the timing, because as Apollo wound down NASA would require a replacement vehicle. And the election of a new president would mean losing a known supporter in the White House.[3]

Mueller realized that he had to involve the air force and the centers to increase political support for the development of the space shuttle. Payload requirements quickly surfaced as an issue, because the centers and the scientific community had lesser needs than the air force. He also found that just about anybody could say no to the space shuttle, but few could say yes, so he involved as many interest groups as possible in order to give them a stake in the outcome, which made planning more difficult since it complicated the ability to achieve consensus. He worked behind the scenes building support for several years, and finally got the air force and the centers to agree to replace their expendable launchers with reusable vehicles as a means of saving large amounts of money in the long term. However, he would face significant

[2]   Mueller, BIS, University College, London, England, 8/10/68.
[3]   Ibid.; Mueller Interviews: NASM, 5/1/89 and Slotkin, 2/12/10 and 6/8/10.

opposition, and spend much of his remaining time at the agency convincing doubters to support the development of a space shuttle. And as he recalled, "everybody wants to do their own thing, and who was going to depend upon NASA producing this marvelous new machine? But eventually they did, because everybody wanted a low cost access to space, and no one had a better idea."[4]

NASA conducted joint studies with the Aerospace Corporation representing the air force, and included the centers in their planning. Contractors pushed their own designs until Mueller said he "finally had to make the decision that this is where you want to go." However, the "constraints put on by the Air Force on the cargo bay size, and correspondingly then the liftoff weight, were very real," leading to significant discussions until they settled the configuration after getting everyone to understand the trade-offs. He favored a large cargo bay to carry space station modules, so he saw little downside in compromising with the air force.[5]

After the London speech, Mueller went on to the UN meeting in Vienna, where he adopted his usual role of NASA's technical expert and spoke about the contributions of human spaceflight to the understanding of space. He cited three examples – the ability of humans to live in space, to do useful work, and to create new technologies. In 1958, doubt existed about the ability of humans to survive in space because the impact of the weightless state and the effects of rapid acceleration and deceleration remained unknown. People asked if technology could be developed to sustain life. They wondered whether astronauts could accept confinement in a small spacecraft for long periods of time, and they worried about radiation hazards. Answering these questions became the first task, and experiments were begun immediately. Scientists remained doubtful, until in 1961 when the USSR launched the first human to fly in space, preceding the US by less than a month. Longer duration flights followed, but in 1965 doubt remained about surviving weightlessness for more than five or six days. NASA proceeded with Gemini despite these concerns, and proved the ability of astronauts to live in space for up to fourteen days. Over the twenty months of the Gemini Program, the agency achieved a number of important objectives. And the ten piloted Gemini missions established "without equivocation" the ability of humans to perform useful work in space. And the space program led to the development of technologies and products with significant spinoff potential. To conclude, he said, "In the exploration of space, man is indeed engaged upon a voyage of discovery which promises to return richer rewards than any expedition upon which has embarked."[6]

As preparations for the upcoming Apollo-Saturn missions gathered pace, the administrator had to decide about his own future, aware that a new president would take office on January 20, 1969 because Johnson had announced that he would not stand for re-election. Richard M. Nixon, the Republican nominee, would expect him

---

[4]   Mueller Interviews: Garber, 2/12/01 and JSC, 1/20/99.
[5]   Mueller Interviews: NASM, 5/1/89 and Slotkin, 6/8/10.
[6]   Mueller, UN Conference on the Peaceful Uses of Outer Space, Vienna, Austria, 8/14-15/ 68.

to resign, and as he had not remained on good terms with Vice President Humphrey, who represented the Democrats, Webb decided to leave the agency at the end of the Johnson administration. On September 16 he went to the White House to discuss several issues, and told the president of his personal plans. For some reason, Johnson decided to announce Webb's retirement immediately; he quickly called reporters into the Oval Office and told them that the administrator would leave NASA effective October 7, his sixty-second birthday. Webb wanted his deputy for his successor, and Johnson appointed Paine acting administrator; though it remained unknown whether he would get the permanent job. Sherrod later asked Teague who he would have favored for administrator at the time, and Teague replied "I was for both of them – Paine if they picked a Democrat, Mueller if they wanted a Republican," although he suspected that Johnson would not have appointed a Republican.[7]

Mueller was surprised by the timing of Webb's resignation. "I knew he was concerned because he was afraid that Apollo 11 might not work and he didn't want to go through another Apollo fire," he said. Adding, he "was probably the right person at the right time to maneuver us through that morass of Congress," praising both his political instincts and his use of agency programs to establish "some part of NASA in almost every key state in the union supporting some activity." This gave the agency visibility, and garnered influence with Congress, "so in that sense he was almost a genius." The administrator got "the key things done" and kept the space program going, although following the fire he clearly got "lost," Mueller said. Criticizing the bureaucracy that Webb created, he called his decision making process "weak," and noted that Webb changed directions "rather arbitrarily," requiring Mueller to "work around him sometimes."[8]

On October 8, three days before the launch of Apollo 7, Mueller spoke at a meeting of newspaper writers and publishers. Just a few days after the agency celebrated its tenth anniversary, and one day after Webb retired, the final steps were in position to bring the lunar landing to a successful conclusion. Journalists were already assembling at KSC and MSC to cover Apollo 7, and pool reporters went aboard the aircraft carrier *Essex* to witness the recovery. After enumerating the accomplishments of the space program since 1958, Mueller told them, "With these achievements behind us, we are moving into the manned phase of the Apollo Program with confidence that the hard work by thousands of people ... has made the Apollo/Saturn system as reliable as humanly possible." Apollo 7 would be the first flight test using the open-ended mission concept, allowing controllers to step through as many objectives as possible, and using telemetry to evaluate the readiness of the spacecraft to take the next steps in real time. And it would achieve as much as the first five piloted Gemini flights accomplished together. After launch and checkout in orbit, the Apollo spacecraft would separate from the S-IVB, turn around

---

[7]   Bizony, *The Man Who Ran the Moon*, 212-214; "Webb Quits as Head of Space Agency...," *NYT*, 9/17/68; Holmes (Jay) Interview, Sherrod, 6/28/72; Teague Interview, Sherrod, 4/1/70.

[8]   Mueller Interviews: Slotkin, 2/25 and 6/8/10 and Lambright, 9/21/90.

and simulate approaching to extract a lunar module from the top of the spent stage. It would then use its service propulsion system to move away in order to perform a rendezvous. Crew activities, system performance and ground support would be evaluated, and "if all goes well," he said, about eleven days after liftoff the crew would land in the Atlantic, south of Bermuda. He ended by pointing out that the flight plan for Apollo 8 would not be finalized until after the results of Apollo 7 had been fully analyzed and evaluated.[9]

On October 11 the Apollo 7 astronauts sat atop the Saturn IB, ready to lift off. The 1,225 telemetry measurements on Apollo 5 had been reduced to 720 for this flight. The Block II command and service modules included all of the improvements made since the fire – a new one piece hatch, a fire resistant interior, and the use of an oxygen-nitrogen atmosphere whilst at sea level. Overall, the changes had improved the workmanship in the capsule so that the crew could walk around "without blowing themselves up," as Mueller noted. Because of the failure of the J-2 engines on Apollo 6, technicians kept a close eye on the single J-2 of the second stage in the run up to launch. Then as the vehicle lifted off, controllers heard Schirra call the ride "a little bumpy," but a little later he shouted out, "She's riding like a dream." The J-2 engine performed well and achieved the desired orbit. The command and service modules separated from the second stage and turned around to perform the simulated docking with the phantom lunar module, but one of the four panels of the adapter section had only partially deployed and so the spacecraft was not able to approach as closely as intended. The *New York Times* quoted Phillips at the post-launch press conference saying that the astronauts "steered perfectly" while maneuvering of their spacecraft. (Mueller no longer attended the post-launch press conferences at KSC, because he left shortly after launch in order to follow progress in Houston.) Phillips went out of his way to praise Boeing's role, and the *Times* wrote, "Spokesmen from both NASA and the Boeing Company ... were cautiously proud as the craft moved into the first of 11 days in space." The paper's headline said "NASA Praises TIE With Consultant; Says Boeing Contract Paid Off in Shot's Success." It was an overstatement of Boeing's role, but most likely what the agency intended when releasing the story.[10]

The Apollo 7 crew came down with head colds, starting with Schirra, making him uncomfortable, grouchy and unwilling to cooperate with mission control when asked to turn on the TV camera. After a lot of persuasion, he finally relented. Yet despite their colds, the crew performed their other duties and successfully fired the service propulsion system a number of times to rendezvous with the second stage. At the end of the mission the astronauts refused to wear their helmets for re-entry because, with

---

[9]   Mueller, 9th Conference of UPI Editors and Publishers, 10/8/68; Mueller Interview, Slotkin, 6/8/10.

[10]   Bilstein, *Stages to Saturn*, 343-344; Mueller Interview, Slotkin, 6/8/10; DCR, Block II CSM and Delta DCR on CSM 101, 7/10-11/68, SCP-86-7; "3 On Apollo 7 Circling Earth In 11-Day Test For Moon Trip," and "NASA Praises TIE With Consultant," *NYT*, 10/12/68.

head colds, they were concerned that the pressure in the helmet would rupture their eardrums, and neither Slayton nor Berry could convince them to put them on. Schirra had already announced his retirement prior to the mission, so he was safe from administrative wrath, but neither of his two colleagues were permitted to fly in space again. And Mueller later said, "I guess [MSC] got the message across after a few of them retired." After completing almost eleven days in orbit, Apollo 7 re-entered the atmosphere and landed near the target point, where the *Essex* was stationed with a complement of reporters on hand to observe the recovery operation.[11]

When NASA officials met with PSAC on October 19, the acting administrator discussed key problems and future prospects, then turned the meeting over to Mueller who briefed these advisors about the status of the Apollo Program. He discussed the projected launch schedule, and outlined the primary objectives for each mission. The third Apollo flight in 1969, scheduled for mid-July, would be a mission-G designated Apollo 11, to be followed by two more lunar landings in September and November that year. The schedule showed four lunar landings in 1970 and three more in 1971. Although for the first lunar landing the astronauts would spend about twenty hours on the Moon, they would venture outside for only a few hours. During this time, as STAC had suggested, they would perform some basic geology observations and collect samples for return to Earth. Apollo 7 was mission-C, and if all went according to plan, Apollo 8 would be mission-D. It would be launched on a Saturn V and test the complete Apollo spacecraft in low Earth orbit. But the first lunar module rated for piloted flight would not be ready until early in 1969. This posed the dilemma of whether to proceed with Apollo 8 on time without the lunar module, or to wait until the lunar module became available. And it was decided to stick to the schedule. Mueller said the nominal plan was to test the command and service modules in low Earth orbit, *but* alternative mission profiles included flights in high Earth orbit, a circumlunar flyby, or a lunar orbital mission. Called mission-C' (pronounced "see-prime") since it would involve only the CSM, it would extend spacecraft operations to achieve additional objectives consistent with the results of Apollo 7.[12]

In prepared remarks at the press conference following the Apollo 7 splashdown on October 22, Mueller said, "The solid success of the Apollo 7 mission makes me confident that we will soon regain the momentum that characterized the manned phase of the Gemini Program." He discussed the open-ended mission concept that was designed to take advantage of real-time telemetry to maximize the benefits of each flight, and said preparations for Apollo 8 were accelerating. The first extended

---

[11]   Kraft, *Flight*, 289-291; "Schirra Pulls Plug, Won't Turn on TV," *Evening Star*, 10/12/68, GEM-52-9; Mueller Interview, Slotkin, 6/8/10.

[12]   PSAC Space Science and Technology Panel Meeting, 10/19/68, GEM-89-3; Mueller Interview, Slotkin, 6/8/10. The use of the prime symbol means a derivative of something, e.g., C' being a derivative of C. When used by engineers, C' means the next in a sequence following C, but not yet D.

option for mission C' involved flying at an altitude of several thousand miles to gain experience in navigation and communications at long distances. The second option, a circumlunar flight, would provide navigation, communications and ground support experience at distances out to the Moon. And a lunar orbital flight, the third and most adventurous option, would confirm the ability to use lunar surface landmarks, make measurements of lunar gravity, and verify navigation and control programs. But the decision about which mission to attempt was dependent on analyses of the Apollo 7 results, certification of solutions to problems on Apollo 6, and more ground testing. Apollo 8 had seven major reviews remaining, ending with the design certification and flight readiness reviews, where each system, subsystem and procedure would be certified before the space vehicle could be declared ready for launch.[13]

With Nixon's election on November 5, speculation began about who would be the next NASA administrator. The *Washington Star* called Schriever the most likely person because he campaigned for Nixon, and was "chiefly responsible for today's missile arsenal." But, the *Star* added, Paine "is not regarded as a candidate." Sherrod later reported that Nixon had offered Schriever the job, but he turned it down, as did several other campaign supporters. As Mueller said later, "That was when unrest was pretty high and it wasn't exactly a plum job." It was just before the lunar landing, and, "Nobody was willing to take on the task in case it didn't land properly," calling the job a high risk politically.[14]

In a speech Mueller gave shortly after the election, he quoted the president-elect saying, "I would have it clearly understood: That I consider this [Apollo] program as one of our national imperatives, that it must be supported at a level assuring efficient and steady progress." And as Mueller always said in his public speeches, "We believe that we have a very good possibility of carrying out President Kennedy's mandate." With Webb's departure from NASA, he felt free to add that after the initial lunar landings "we will want to return to fully explore the lunar surface. Using equipment already developed, we can establish permanent unmanned observation stations for continuing maximum scientific benefits." NASA, he indicated, would ask the new Congress to continue this "forward step in space exploration." And with all members of the congressional space committees re-elected, he felt that the prospects looked bright.[15]

II

In August 1968, while Mueller and Webb were in Vienna, Grumman officially slipped the lunar module schedule by three months. Every time that Mueller visited

---

[13]   Mueller, Remarks, Apollo 7 Press Conference, 10/22/68, GEM-46-3.
[14]   "Retired Air Force General May Head NASA," *Sunday Star*, 11/17/68, SCP-93-4; Teague Interview, Sherrod, 4/1/70; Mueller Interview, Slotkin, 6/8/10.
[15]   Mueller, 20th Century Club, Hartford, CT, 11/11/68.

Grumman they discovered another slip, he said, and this became a standing joke. But it was no laughing matter because NASA planned for Apollo 8 to dock with and test the lunar module in low Earth orbit. And when Mueller learned of this latest delay, he expressed concern that "the minute" he left the country the schedule slipped by a couple of months, which began the story of how Apollo 8 came to orbit the Moon on Christmas Eve 1968.[16]

With the lunar module behind schedule, NASA's budget at its lowest point in years, and Webb shaking up the agency, its morale reached a low point. And the looming delay in delivering the lunar module made it appear that meeting the lunar landing goal might not be possible. At a management council meeting on August 7, Low said that the lunar module would probably not be ready for the scheduled flight. A few days later, when he officially learned of the delay, he decided to recommend a test of the command and service modules in deep space, with the possibility of the spacecraft venturing to the Moon. He considered a circumlunar mission feasible and perhaps a necessary step before attempting the lunar landing. And he had another powerful reason – concern over what the Soviet Union might do in 1969. Intelligence reports said they could not land on the Moon, but could still upstage the US with a circumlunar mission. But the agency had valid technical reasons to orbit around the Moon as a means of resolving questions concerning lunar orbital flight mechanics, because when satellites went behind the Moon they emerged in unexpected places, as if the lunar gravitational field were irregular. And by flying around the Moon, NASA could obtain the information needed to make better orbital predictions. Low brought his idea to colleagues at MSC, and on August 9 he met with Gilruth, Kraft and Slayton. After discussing it, they liked the idea. But he had to persuade von Braun and Phillips, and in turn Mueller, Paine and Webb to accept the new mission. On that same day, Gilruth called von Braun and arranged to meet him in Huntsville that afternoon. Next he called Phillips, at the Cape for the day, to explain the proposal, and the general agreed to meet in Huntsville. The MSC group arrived at von Braun's office about two hours later, with Phillips, Hage, Debus and Petrone already there. Gilruth said that unless they waited three months they would have to fly the mission without a lunar module, and if they did that "we might as well swing around the Moon." With the Saturn V ready, the exact mission did not matter to von Braun, and he said the "risk difference between a Saturn V launch to Earth orbit and to go from there on to the Moon was a relatively small thing." Phillips later wrote, the "three-hour conference didn't turn up any 'show-stoppers.' Quite the opposite; while there were many details to be reexamined, it indeed looked as if we could do it. The gloom that had permeated our previous program review was replaced by excitement."[17]

---

[16]   "History," 3/13/69, Notes & Transcriptions, GEM-79-15; Paine Interview, *"Before This Decade Is Out..."* Ch. 2.

[17]   Lambright, *Powering Apollo*, 198; Kraft Interview, JSC, 5/23/08; Von Braun Interview, Sherrod, 8/25/70; Cortright, *Apollo Expeditions*, 9-3; Logsdon, *Exploring the Unknown*, 431; Orloff and Harland, *Apollo*, 56.

With Webb out of the country Paine served as acting administrator so Phillips called him, and he liked the idea. Although according to the general, "Paine reminded me that the program had fallen behind, pogo had occurred on the last flight, three engines had failed, and we had not yet flown a manned Apollo mission; yet 'now you want to up the ante. Do you really want to do this, Sam?'" Phillips responded, "Yes, sir, as a flexible mission, provided our detailed examination in the days to come doesn't turn up any show stoppers." The deputy administrator then said, "We'll have a hell of a time selling it to Mueller and Webb," which proved correct. As Phillips recalled, "A telephone conversation with Mueller in Vienna found him skeptical and cool. Mr. Webb was clearly shaken by the abrupt proposal and the consequences of possible failure." Paine sent a classified telegram to Webb through the US Embassy on August 15, emphasizing that the final decision would await the evaluation of Apollo 7. He asked for permission to replace mission-D with C' on Apollo 8, and indicated that he had agreement from Phillips, the center directors and all their key staff. Paine told Webb, "We recommend proceeding on this basis which is not [a] final commitment or decision on your part, but is [the] minimum cost insurance to maintain the option to fly such a mission in December of this year." He then added, mistakenly, "Technical and operations experts assign a lower overall degree of crew risk to this mission than to [the] previously scheduled 503 LM mission, although time exposure is longer." Mueller recalled, "I was surprised that Houston was pushing for it and very pleased actually, although I didn't say so . . . I didn't tell him [Webb] that. I said this is something that we have to think about. And so let's plan on a thorough review." And Webb said he wanted to "consider it but very, very carefully."[18]

Webb, Paine, Mueller and Phillips spoke the next day, and the general sent another telegram further explaining the plan. At the administrator's direction, Phillips discussed it in more detail with Mueller and Newell. Everyone agreed to the revised plan, and the agency scheduled an announcement to postpone mission-D and test the command and service modules on mission-C'. The draft press release sent to Webb said that the revised plan would "provide the maximum gain consistent with standing flight safety requirements in maturing the Apollo-Saturn V space system in Earth orbital operation. Studies would be carried out and plans prepared so as to provide reasonable flexibility in establishing final mission objectives after the flight of [Apollo 7]." The original crew for Apollo 8 was reassigned to perform mission-D on Apollo 9, and the crew for mission-E were brought forward. The administrator approved the changes, although he emphasized that mission-C' must fly one of three options depending on studies to be "carried out and plans prepared so as to provide reasonable flexibility in establishing final mission objectives."[19]

---

[18]   Von Braun Interview, Sherrod, 8/25/70; Lambright, *Powering Apollo*, 199; Cortright, *Apollo Expeditions*, 9-3; Paine to Webb, 8/15/68, Apollo 8, SCP-87-6; Mueller Interview, Slotkin, 6/8/10.

[19]   Phillips to Webb, 8/16/68, Apollo 8, SCP-87-6; Mueller Interview, NASM, 11/8/88.

On August 19, Phillips officially delayed mission-D until Apollo 9 and informed the center directors that Apollo 8 would be a piloted command and service modules mission designated C'. Continuing, he wrote, "The objectives and profile of the C-prime mission will be developed to provide the maximum gain consistent with standing flight safety requirements." He attached the press release that announced Apollo 8 would not carry a lunar module but would perform an alternative mission, the profile of which would not be defined until the results of Apollo 7 were known. When announcing the changes, he said that "circumlunar or lunar orbit were possible options," but threw the press off the trail by repeatedly saying "the basic mission is Earth orbit." As he later said, most of the media "missed the point," although *Space Business Daily* did note that December 16 to January 15 were optimum dates for a circumlunar mission.[20]

Mueller continued to question whether it would be worth the risk to fly to the Moon on the first piloted Apollo spacecraft launched by a Saturn V. We needed to be sure that we could do it, he said, "while Webb's attitude was, 'It doesn't make any sense.'" The fact that Webb did not tell them *not* to move forward gave Mueller time to conduct a thorough, in depth review of the proposal. However, he recalled, "I did not agree to go forward until this review was completed," about two months before the mission; "I was trying to make sure we really understood what it was we were doing."[21]

The USSR launched a spacecraft on September 15 which circumnavigated the Moon and returned to Earth on September 21. A mission "taken by many in NASA as proof that the intelligence they had received was sound," according to the *New York Times*. Were the Soviets planning a piloted circumlunar mission? As Mueller recalled, concern about what the Soviet Union might do "always plays a part in your decision making but I was adamant that wasn't going to be the reason we did anything. It was because we were ready. And I spent a lot of time making as sure as I could that we were ready and getting everybody else to do the same soul searching." He knew the USSR had the capability to fly a piloted circumlunar mission, though they could not land on the Moon. So it became a question of could they get to the Moon and back, and whether their space capsule was reliable enough to spend the necessary time in space. The CIA told NASA of an on-the-pad explosion of a Proton booster in July 1968, the same type of rocket that would launch a piloted circumlunar mission, that had killed one person and damaged the launch vehicle. Consequently, the Soviet Union's leadership decided they could not afford the risk of another accident, this time in deep space, which would be widely publicized. They made propaganda tradeoffs, and concluded a piloted circumlunar mission was too risky. And Mueller said, "I suspect they also doubted that we would really do

---

[20] Logsdon, *Exploring the Unknown*, 702-704 *Space Business Daily*, 8/21/68, Apollo 8, GEM-53-5.

[21] Murray and Cox, *Apollo*, 322-324; Mueller Interviews: Lambright, 9/20/90 and Slotkin, 6/8/10.

anything," because Apollo 8 surprised them as much as everyone else. With NASA recovering from the fire, "it was a little hard to believe" the agency would undertake a lunar mission, and indeed it was "hard for me to believe too." However, he explained, "we didn't know truly the capabilities of the Soviet system but we did know [a piloted circumlunar flight] was possible. And so we certainly talked about it but didn't consciously make the decision to try to one up them." On September 24, Mueller told Sherrod the chances of a lunar orbit mission on Apollo 8 "were only one in ten."[22]

On October 12, STAC discussed Mueller's plan to extend human spaceflight in the post-Apollo period based on NASA's budget submission for FY 1970. Hoping for a more aggressive Apollo Applications Program, and with Webb retired, Paine included funding for a space station in low Earth orbit by 1975 – although the submission still fell short of achieving the level that NASA facilities were designed to support, leaving options limited and the future looking "austere," according to Mueller. He told the advisory committee about additional studies conducted to better define the role of "man in space," although some members thought he should not focus solely on human spaceflight, while he argued otherwise. Phillips briefed STAC about mission-C' and outlined the three options under consideration. The members asked about NASA's decision making process, the risks involved, how much C' would contribute to the lunar landing goal, and "the consequences of failure." The general strongly supported the lunar orbit option, while Mueller "pointed out the great risk to the program as a whole in taking the second manned flight all the way to the Moon." The committee debated the advisability of flying Apollo 8 in lunar orbit and, after reviewing the pros and cons, it concluded that the major danger involved something going wrong while three days from Earth, versus a one hour return from low Earth orbit. After Phillips outlined why he supported the lunar orbital mission, Mueller pointed out "General Phillips has presented an optimistic picture in terms of technical capability," then he said "grave risks" existed to the program as a whole, not just Apollo 8, by flying a lunar orbital mission aboard the second piloted flight of the Block II spacecraft, and that "we have to face the fact that there is the possibility that it could appear to be a pernicious, risky venture where the propaganda payoff is the only gain." Flying the lunar orbital mission involved a difficult decision because the agency had a "complex and relatively immature vehicle," Mueller recalled. The paradox between the performance of AS-501 and 502 illustrated his point. And being three days instead of one hour away from Earth added risk, which the agency could accept if the spacecraft operated reliably for that period of time. Though based on the Apollo 7 results, he would conclude that the "risks were probably reasonable and acceptable." STAC supported the decision to go ahead with the lunar orbital flight, accepting the added risks. Yet Mueller still cautioned "if failure comes, the reaction would be that

22   "Soviet space missions," NASA, Goddard; "Soviet Seen Losing Chance to Best US on Moon Trip," *NYT*, 12/10/68; Mueller Interviews: Slotkin, 6/8-10/10; Skaggs to Slotkin, e-mail, 6/5/11.

'any fool would have known better than to undertake such a trip at this point in time.'"[23]

NASA planned to announce the Apollo 8 decision on November 10. Mueller listed the pre-decision steps, and the procedures planned to take place before the final decision would be made. And on November 2, he told Sherrod that the chances for a lunar orbit mission had increased to 50-50. The agency issued a press release noting that the three options remained under consideration, and quoted Mueller saying, "Our mission planning is based on careful assessment not just of the risks inherent in each mission, but rather on the objectives of minimizing the total risk of the complete set of missions required to carry out the lunar landing." It listed the remaining reviews, culminating in a NASA management review. The press release concluded, "The final decision on the Apollo 8 mission will be made only after a very careful assessment of the total risks involved against the progress that we can make towards a manned lunar landing." Mueller asked Bellcomm to conduct one final study of the mission software, and to calculate alternative trajectories and abort options in case of an emergency. He then sent Paine a decision schedule, containing three pages of Gantt charts showing the remaining reviews, briefings and decision points.[24]

Mueller continued to review the alternatives, and would not make the decision until he felt sure that the space vehicle was ready, the spacecraft had the capability to perform the mission, and it was safe to do. He eventually agreed it was safe to fly to the Moon and later claimed that he offered to personally fly on Apollo 8 with Phillips and Borman, the mission commander. As he saw it, he and Phillips did not need astronaut training because the mission profile did not require it, and they knew as much or more about the spacecraft as any of the astronauts. Furthermore, he would not ask someone to fly the mission if it was not safe enough to do it himself. Webb and later Paine would not hear of it, he said with a smile. But, he pointed out, whilst Webb was reticent about venturing to the Moon, "He might have let *me* go." We will never know whether this story is apocryphal, because there is no written record of this unusual request to support Mueller's personal recollection more than forty years later.[25]

Paine met with Mueller, Phillips, and the center directors on November 11, and he listened to a series of presentations about the C' mission. Phillips wrote in his notes, "Dr. Mueller indicated that this situation had been discussed with STAC, PSAC, DOD, and the Apollo executives group. He pointed out STAC members had made a penetrating review of the flight and clearly understood the risks. Their reaction was a positive one." The general continued, "PSAC was favorably disposed," while "DOD also generally favors the mission." However, Bellcomm recommended against it. In a memorandum to Mueller they wrote, "we find more tangible evidence of increased

---

[23] Minutes of the Meeting of STAC, KSC, FL, 10/12-13/68.
[24] "Decision Date on Apollo 8," 10/29/68, *NYT*, 14; Freitag Interview, Sherrod, 2/18/70; Draft Press Release, 10/24/68, Apollo 8, GEM-53-6; Mueller to Paine, Apollo 8 Schedule, 10/28/68, GEM-53-7.
[25] Mueller Interview, Slotkin, 6/10/10.

risk than increased gain" by flying Apollo 8 in lunar orbit, and the same gain could be obtained at lower risk in Earth orbit. Paine held a second meeting with a smaller management group, and then a third with just Mueller and Newell. After the last meeting he approved the lunar orbit mission with a launch date of December 21, with the implication that Apollo 8 would go around the Moon on Christmas Eve. As Mueller later admitted, the real reason to fly Apollo 8 as a lunar orbit flight "was to build back up the morale of our troops and to get everybody concentrated on getting everything right so that we could do that, and coming off the fire that was really the driving force ... I was enthusiastic about it ... for that very reason. Not that it got us any closer to landing on the Moon per se but it got us closer in terms of the morale of the troops and the motivation for getting it done ... Psychologically it was tremendous and important. But it would have been [a] disaster if it failed," making it "an interesting decision to make."[26]

Prior to the final approval of the mission, Phillips distributed a risk assessment questionnaire to twenty-six top NASA executives, requesting them to evaluate the probability of *crew loss risk* based on various mission profiles. (Interestingly, Gilruth was the only one not to return the questionnaire.) Normalizing the data to determine the relative risk of each mission profile, the survey results show on average, mission-C' performed in lunar orbit had *three times* the risk as the same mission in low Earth orbit. And Mueller assessed the risk as *lower than the average*. In his view, the lunar orbit mission involved less risk than mission-D by about twenty percent. He later said, human judgment, "when properly applied is somewhat better than statistics." To successfully go to and from low Earth orbit, he explained "everything has to work." Then, if you spend the same amount of time performing the mission in low Earth orbit as you would in going to and from the Moon, you have the same chance of something going wrong, because it is the time that is spent in space which creates the possibility of something going wrong. The only extra risk of the lunar orbit mission occurred because of firing the service propulsion system twice to enter and leave orbit around the Moon, as compared to once in low orbit in order to return to Earth. However, the agency planned to test the service propulsion system several times on mission-D and also did so on the way to the Moon in order to verify and calibrate it prior to attempting the lunar orbit insertion maneuver. Consequently, both missions presented similar risks. Although, he added, "The consequences [of failure] would have been different," and going out to the Moon and back "adds some complexity."[27]

[26]   Logsdon, *Exploring the Unknown*, 430 and 704-706; Bellcomm, C-Prime Mission Decision, 11/7/68, Apollo 8, SCP-89-1; Mueller Interview, 6/10/10.

[27]   "Apollo: Looking Back," Caltech, Pasadena, CA, 5/18/71, SDC Speeches; Crew Risk Assessment, 11/18/68, Apollo 8, GEM-54-8; Mueller Interview, Slotkin, 6/8/10.

III

Speaking in San Francisco just days after the presidential election, Mueller said that NASA's schedule for Apollo 8 called for launching three astronauts on a 63 hour flight "to the vicinity of the Moon, and possibly into lunar orbit for 20 hours." The decision to fly to the Moon on the next flight occurred "only after a thorough assessment of the total risks involved and the total gains to be realized in this next step toward a manned lunar landing." Spacecraft ground testing proved satisfactory, and reviews of the previous flights led to the decision that the Saturn V was ready for human flight into deep space. However he cautioned, risk existed in any spaceflight, although "we feel that [in] taking this deep space flight now we will gain engineering and operational information that will ultimately reduce the total risk of a manned lunar landing during the next year." And based on a complete checkout of flight systems during the mission, a decision would be made to either go around the Moon or into lunar orbit. Apollo 8 would prove the flight worthiness of the spacecraft, demonstrate the communications network, provide data to explain lunar gravitational anomalies, and scout possible landing sites. In addition to the engineering benefits, he pointed out, "history will note – that this will be the occasion when man will first fly, truly free of the gravity of his mother planet." Achieving the lunar landing would not be "an end in itself, but when two American astronauts return from the surface of the Moon, we will mark the accomplishment of our greatest national peacetime goal." Because of the space program, he said, the US had "the newest, largest and most sophisticated national plant, equipment and skills which we have ever had outside of wartime ... American industry has demonstrated its resourcefulness and its capability – it has in effect, created its own renaissance." He concluded optimistically that after the initial lunar landing "we will want to return to fully explore the lunar surface."[28]

Then speaking at the National Space Club later that month, Mueller described the open-ended nature of the Apollo 8 mission, with "plateaus" followed by "decision points," in order to provide "maximum crew safety together with maximum forward progress." The mission trajectory permitted the spacecraft to swing around the Moon and automatically return to the Earth (referred to as the "free return"). And he said, "If all systems are 'go' when Apollo 8 reaches the Moon's vicinity, a decision will be made to perform 10 lunar orbits." Yet, while a circumlunar flight would significantly advance the Apollo Program, "Lunar orbit will be even more valuable." Tests and rehearsals would continue right up to liftoff, and not until then would Apollo 8 be launched. As usual, he argued the case for the Apollo Applications Program, and told the space club audience, "New goals must be set, or we will lose the impetus we have thus far developed." Although he added that repeated delays in

---

[28] NASA Press Release 68-199, Apollo 8 Moon Orbit Flight, 11/12/68, GEM-53-9; Murray and Cox, *Apollo*, 322-324; Mueller, Commonwealth Club of California, San Francisco, CA, 11/15/68.

funding had already caused much to be "lost for lack of a firm and forward thrusting policy for our manned space program beyond Apollo." Yet despite the nation's other problems, "aggressive creation and utilization of new technology have been vital forces in making this land, in less than two hundred years, the greatest in the world." And the only thing standing in the way of this "forward thrusting national space program," remained the lack of sufficient funding.[29]

To preclude the problems experienced on Apollo 6, engineers and technicians carefully rechecked the AS-503 launch vehicle, and after a two day quality and reliability review, they deemed it ready to go. MSFC then conducted one final quality check, and corrected some minor problems before approving the flight. The vehicle went through a simulated countdown in early December and then a pogo suppression test on December 18. Final ground tests took place three days before liftoff, and on December 21 the Apollo 8 mission took off from KSC. Performing as planned, the pogo suppression system damped out longitudinal vibrations, and stage separation was nominal. The S-II stage performed flawlessly and the S-IVB achieved the initial orbit. The J-2 engine reignited on the second orbit, putting Apollo 8 on a translunar trajectory. As the spacecraft set off for the Moon, the mission commander suffered a brief bout of "space sickness". The *New York Times* called the launch "flawless," and said Apollo 8 was heading for a lunar rendezvous on Christmas Eve.[30]

Two days later, on December 23, 1968, the spacecraft reached the point in space where the gravitational pull of the Moon overcame that of Earth, making the three astronauts the first humans to break free of Earth's gravity. On Christmas Eve, Apollo 8 slipped behind the Moon, creating the tensest time of the mission because it lost radio contact for thirty-three minutes. The suspense was lifted when, precisely on time, Lovell's garbled voice was heard telling mission control that they had achieved lunar orbit. Mueller watched "with baited breath," but said he knew "that wasn't the end point. Apollo 11 was the end;" and when Apollo 8 came from behind the Moon it was the beginning of the end. The crew sent dramatic pictures of Earth from lunar distance, and took detailed photographs of the Moon's surface. Because it was the holiday season, some people attached religious significance to the mission, which the astronauts aided by unexpectedly reading verses from the *Book of Genesis*. Mueller later called that "very impressive. In fact I think it impressed the whole world ... A very excellent reading and apropos and it helped the program. It won support for the program."[31]

On December 25, shortly after midnight Houston time, the service propulsion system kicked the spacecraft out of lunar orbit and put it on a trajectory for return to

---

[29]   Mueller, NSC, 11/26/68;"A NASA Aide Sees Industry in Space," 11/27/68, *NYT*, 44.

[30]   Murray and Cox, *Apollo*, 366-367; "3 Astronauts Speed Toward Moon Orbit ..." *NYT*, 12/22/68.

[31]   Murray and Cox, *Apollo*, 328; "Small World," 12/24/68, and "3 Men Fly Around the Moon Only 70 Miles from Surface," *NYT*, 12/25/68; Kraft, *Flight*, 299; Mueller Interview, Slotkin, 6/8/10.

9-2   Samuel C. Phillips at Kennedy Space Center during Apollo 8's launch, December 21, 1968. (NASA photo)

Earth, and the capsule landed safely two days later in the Pacific within sight of the aircraft carrier *Yorktown*. President Johnson phoned the astronauts from the Texas White House, while President-elect Nixon watched on TV from a vacation home in Key Biscayne, Florida. The public reaction "was sort of astonishment that we were able to do it," Mueller later observed.[32]

In an interview published in the *Times* of London on January 10, 1969, Mueller said, "No unmanned flight could ever have duplicated the work performed by the Apollo 8 astronauts ... There is no machine that begins to equal the abilities of man's brain." The *Times* called Mueller, "A slim bespectacled and energetic 50-year-old," but with analysis of flight results remaining to be done, he would not comment on the few flight anomalies other than to call them minor. The *Times* quoted an unnamed NASA official who called Apollo 8, "'the triumph of the squares' – meaning 'the guys with crew-cuts and slide rules who read the Bible and get things done'" pulled off the difficult mission. However, the paper said, "the Soviet Union will 'inevitably' move ahead of the United States," because NASA funding had

---

[32]   Murray and Cox, *Apollo*, 368; "World's Leaders Hail Apollo Feat," *NYT*, 12/28/68; Mueller Interview, Slotkin, 6/8/10.

declined by ten percent per year for the last three years. And they quoted Mueller saying, "If the present trend continues ... we will inevitably be out of the manned spacecraft business."[33]

Mueller later called the decision to put Apollo 8 into orbit around the Moon very important; not the most important, but a turning point. "In a real sense every decision is the most important decision you make ... [but] Apollo 8 in my view was the most difficult mission we undertook. Maybe Apollo 11 is the second most difficult. Those first tests and the first Apollo, all-up test of Saturn [were] the most important. So those are all key events and if any one of them had failed we would have been set back by a substantial amount. Just as the Apollo fire set us back a year."[34]

After spending about $20 billion of the taxpayer's money, Mueller explained to the New York Society of Securities Analysts at the end of January 1969, just what the money bought and the expected return on investment. Recalling the challenge of Sputnik in 1957, he said people questioned whether the US led technologically and expressed concerns about national security. The country reacted by strengthening its educational system, investing in industry, developing new management systems and concepts needed to achieve and verify the reliability of complex space equipment, resulting in the training of hundreds of thousands of people across the nation. After mounting "the most ambitious science and engineering program in man's history," he said, more remained to be done; though Apollo 8 "renewed the world's confidence in our strength and vitality as a technological leader." He called the reaction to this flight "almost overwhelming." The news media around the world praised it "as a witness of the vitality of capitalism, and a demonstration of the technological, organization, and scientific skill of the United States." The Apollo 8 flight achieved a "demonstrated reliability of 99.9999 percent – a phenomenal level" as he called it, with only five non-critical parts failing out of five million. The flight to the Moon required only one mid-course correction of four feet per second on a vehicle traveling at thirty-six thousand feet per second, representing "another facet of perfection" due to "phenomenal precision" of the spacecraft guidance and control system. However, he noted, "We must realize that space flight is still in an embryonic stage." Yet, in 1969, within the target date set in 1961, and "well within the lower range of the costs estimated at that time," the US fully expected to land astronauts on the Moon and return them safely to Earth with the first samples of lunar material. Discussing the future, he argued that space stations would be platforms for scientific research and manufacturing, with significant commercial potential. Although the first space station would not be ready until 1975, they would be used for many years. The cost of putting objects into low Earth orbit remained expensive but, he said, to "make space activity the commonplace utility it will eventually be, we are going to have to reduce the cost of transport by orders of magnitude," requiring the development of reusable space shuttles of various types. And he noted that NASA had already awarded study contracts to get industry's best ideas for a space shuttle that would operate in low Earth orbit. Then predicting commercial markets for reusable shuttles, he added, "the things we have developed ... today will create the products that people are going to be building, selling, and using ten years from now."[35]

A few days later at a conference of defense contractors in Cocoa Beach, Florida, Mueller expanded on those remarks, saying Apollo 8 demonstrated NASA's "third generation capability in the most advanced and thoroughly successful space mission to date." He described progress during the first decade of the space age, equating it to the early years of aviation after the Wright brothers' first flight. And he said the time had come to consider the next generation of low cost space transportation – space shuttles moving between the Earth's surface and orbiting space stations and from low Earth orbit to lunar orbit and back. During the second decade of the space age, he said, "we need to further develop and strengthen that productive and cooperative working relationship between NASA and industry and university scientists." And he pointed out that more than ninety percent of the agency's funds were contracted to industry. Yet, space related employment had dropped by almost half from its peak, whereas the Soviet Union continued to build their space program "with the clearly expressed intentions of attaining preeminence in space." He called NASA's greatest problem the preservation of the knowledge that it created during the first decade of the space age, cautioning that the people, infrastructure, technology, and knowledge developed must not be allowed to disperse, since once they were gone they would be difficult and expensive to reassemble. Nevertheless, other national priorities forced NASA to operate near its breakpoint of approximately $4 billion per year, and if its funding were to fall below that level then it could not "hold together our hard-won capabilities and utilize them effectively in critical programs." However, the approved agency budget only allowed "modest investments aimed at reducing costs of future space activities." But, he argued, the agency had reached the point where it became necessary to "spend money to save money ... And it is time to start planning the next NASA center, not on Earth – but in space – a major national orbiting platform where extremely valuable work can be performed."[36]

Mueller explained that alleviating poverty would require the creation of additional wealth, and said, "NASA's space exploration programs ... create new jobs and new opportunities that will not merely alleviate poverty but get at its root." He called conditions in the cities a national disgrace, but added "as we tackle these grave social ills, we must also continue to forge ahead in other areas." It would be "an international tragedy if America were to turn back now from its forward thrust

---

[35] Mueller, NY Society of Securities Analysts, NY, NY, 1/28/69.
[36] Mueller, NSIA, KSC, FL, 1/30/69.

in space at the end of this truly astonishing first decade." Rather, he argued, the US should seek ways to "bring space-age management capabilities" to bear on the nation's other problems. Again raising the specter of the Soviet Union regaining preeminence in space, he asked for help to "develop a better public understanding and support for the promise of space as the nation watches our Apollo Program reach its dramatic conclusion."[37]

Nixon established a new Space Task Group in February 1969 to recommend goals for the future of spaceflight. Chaired by Vice President Spiro T. Agnew, the task group expected to issue its report the following September. Its members included Paine, the former NASA deputy administrator and new secretary of the air force Seamans, and STAC member and presidential science advisor DuBridge, and it had several other administration officials in the role of official observers. The task group met for the first time in March, whereupon Paine supported Mueller's expanded post-Apollo space program, DuBridge expressed concerns about its impact on the effort to reduce federal spending, and Seamans argued for a compromise. Agnew, an avid supporter of an expanded post-Apollo space program, favored Mars exploration, but unfortunately he had little influence within the new administration.[38]

Just before the launch of Apollo 9 in early March, Mueller said that Apollo 10 might land on the Moon. An advocate of doing the most possible on each mission, he explained that if all went well performing mission-D in low Earth orbit, then the next flight could either be a test of the lunar module in lunar orbit as mission-F, or a lunar landing attempt as mission-G. But regardless of which mission the agency selected, the July 1969 target for attempting the first lunar landing would remain. As Phillips recalled, the rationale for mission-F in May "was hotly debated ... Here we would be, 50,000 feet above the Moon, having accepted much of the risk inherent in landing. The temptation to go the rest of the way was great." And as Sherrod wrote, "Mueller initially saw no point in going to the Moon a second time without touching down. However, for this one time the LM wasn't completely adapted to the task [it weighed too much for the ascent stage to be able to perform a rendezvous after lifting off from the Moon, meaning that it was unable to land] and the program management decided they were not ready for the big step."[39]

Apollo 9 would be the first mission to include a production lunar module. Gemini veterans McDivitt and Scott served as commander and command module pilot, while rookie Russell L. Schweickart became the lunar module pilot. It took off on the morning of March 3 to spend ten days in low Earth orbit flying mission-D. The launch vehicle performed as planned, and after the CSM separated from the S-IVB

---

[37]   Ibid.

[38]   McDougall, *The Heavens and the Earth*, 421; "Beyond Apollo, plans for the exploration of space from the age of Apollo 1959-1979," Portree.

[39]   "Apollo 9 To Decide Moon Trip," *Washington Post*, 3/1/69, GEM-83-7; Logsdon, *Exploring the Unknown*, 431; Cortright, *Apollo Expeditions*, 9-3.

stage it turned around and retrieved the lunar module. Mueller described the lunar module as "the first true spacecraft," because it was designed for flight only in the environment of space. Stowed with its legs folded in the adapter at the top of the rocket, engineers had been concerned about whether the adapter that supported it could handle the dynamic loads of the Saturn V launch, whether the lunar module could handle the vibrations, and whether the mechanism to extend its landing legs would jam due to flight loads. After the successful launch, all appeared to be going well. When Schweickart disclosed that he had "space sickness", flight surgeon Berry told him to take one of the medicines carried on board. Nonetheless, the two hour EVA scheduled for March 4 had to be delayed a day. The next day, despite his weakened state, Schweickart insisted he could perform his assignments. He transferred to the lunar module with McDivitt to conduct a standup EVA in the open hatch and demonstrate the Apollo Portable Life Support System backpack that was to be used on the lunar surface. However, the test had to be cut short, and the rest of the EVA in which he would have performed an external transfer to the command module to demonstrate this emergency procedure, was canceled.[40]

Meanwhile, on March 4, Nixon appointed Paine as NASA's administrator, making him "one of the highest ranking officers of the Johnson Administration to be retained," according to the *New York Times*. The paper reported, "Dr. Paine's permanent appointment comes at a time when the space programs' accomplishments are spectacular and its budgetary problems are particularly severe. He is expected to play a vigorous role in continuing the former and trying to improve the latter." Mueller said Paine got the job because "they couldn't find anybody else." And he claimed he did not to want to be administrator, insisting that he "didn't even think about it," never lobbied for it, and never discussed it with anyone. "I was too busy doing what I was doing," he said, and "at that time I was more interested in new challenges" outside of NASA. Planning to leave the agency after the successful lunar landing, unless somebody actually asked him to take the administrator's job, he would not lobby for it. Teague told him it would be difficult to increase NASA funding after Apollo, and that without strong support from the president, the space program could not succeed. So he insisted "it wasn't something that I really wanted to do, particularly since the decision they made not really to do robust space program after Apollo."[41]

---

[40]   NASA Press Release 69-3, 1/8/69, Apollo 8, GEM-54-13; Bilstein, *Stages to Saturn*, 368; Cortright, *Apollo Expeditions*, 10-3; "Apollo Spaceship, In Orbit, Links Up To Lunar Module," *NYT*, 3/4/69; Mueller Interview, Slotkin, 6/9/10; Person to Rowsom, 7/10/75, "Getting It All Together," GEM-195-4.

[41]   "Paine Named by Nixon . . ." 3/4/69, *NYT*, 18; Mueller Interviews: Slotkin, 2/25/10 and 6/8/10.

IV

The day that Paine received his permanent appointment, Mueller testified before the House space committee in support of NASA's budget. As the Apollo Program neared its goal, he reminded the members that the original program objective involved "the development of the <u>capability</u> for carrying out manned space flight activities in all areas between the Earth and Moon including landing men on the Moon and returning them safely to Earth." He called the first landing "the historic beginning of lunar exploration" to demonstrate this national capability. Yet the FY 1970 budget of $3.7 billion represented a reduction in the funding necessary to continue the program: a 7.5 percent drop from FY 1969, and the fourth consecutive year of declining space budgets. "Because of the relentless pressure of other national needs, the erosion of national capability for manned space flight continues apace," he said, and reminded the members the budget eliminated the Apollo Program and the Apollo Applications Program in 1972. He lectured the committee about the "need for progress in space," and again reminded them that the goal of Apollo began with Soviet competition, and its stimulating effects on national development led to significant benefits. While the "direct economic impact of the space program is quite obvious," he said, "[n]ot so obvious, but of perhaps greater importance in the long run are the basic contributions of the program in bringing together people from all disciplines ... and causing them to work together toward a common goal – the long term survival of men in space." He pointed to economic development near the NASA centers, and said the "economic effects of the space program are spreading throughout society. Everywhere, it has widened our horizons. Its energizing force permeates our economy. It is a seedbed for invention, a stimulus to higher productivity and a task-master for precision and reliability ... Our national growth and every facet of our life on Earth is affected, enriched and improved by the adventure and the fact of space flight."[42]

While the FY 1970 budget called for additional lunar landings, Mueller said it did not provide for the "meaningful utilization of the remaining Apollo space vehicles." The Apollo Applications Program budget funded five Earth orbital flights beginning in 1971, but they would end in 1972. The future of human spaceflight depended on how much launch costs could be reduced. Fortunately, the state of technology created opportunities to develop reusable space vehicles that would lead to substantial reductions in the cost of spaceflight. With reduced cost of access, the use of outer space would increase to the point that it would be "less expensive to use man than to design machines to replace him," and he spoke of cost reductions by a factor of ten or more. Looking toward the mid-1970s and beyond, he called for establishing a semi-permanent space station that would start out small and be expanded by attaching modules. Funding in FY 1970 would permit planning, but cost effective access to space required reusable space shuttles with low operating

---

[42]   Statement of Mueller before the House Space Committee, 3/4/69.

costs. However, the budget request represented a significant decline, and by June 1970 the program would be down to one-third of its peak of personnel. Reporting his testimony, the *St. Louis Post-Dispatch* said, "Because of the relentless pressure of other national needs, the dissipation of the national capability for manned space flight continues apace."[43]

As Apollo 9 continued circling the Earth, on March 10 Mueller addressed the National Research Council's annual engineering division dinner – the first sitting member of the National Academy of Engineering to be invited to speak at the dinner. He gave immediacy to his remarks by announcing that the crew of Apollo 9 was engaged in photographing areas of the Earth's surface using a multispectral camera, pointing to its direct application to the collection of Earth resources data. He spoke of studies showing that it would be possible to reduce the cost of launching payloads to low Earth orbit by orders of magnitude, and repeated that astronauts had advantages over machines in space because they possessed "a very wide-band set of sensors for acquiring information ... [with a] built in memory and computer that cannot yet be matched," and had "a remarkably versatile capability for action." These factors made them valuable for operations in space, when feasible. He predicted that putting humans into space "within the next decade ... will become almost as inexpensive and commonplace as sending up an airplane today." And he then spoke about the contributions astronauts would make to Earth applications – leading to immediate benefits instead of waiting for years to develop automated equipment. Thus, he explained, the "recycle time for a versatile manned station, could be cut down to days if the required equipment is onboard, or a few months if new equipment must be brought up at the next resupply of a long duration space station." Semi-permanent space stations would permit maintenance and repair of satellites in orbit, reducing costs significantly. But astronauts could do more than routine work, they could also deploy and calibrate large systems, and their ability to operate as sensors in their own right opened opportunities to view Earth for science and applications. They could perform as technicians, engineers, or scientists, and by "dynamic, synergistic reactions to all that he observes, and everything that he touches, we can expect not only to exploit what we have already learned in this new dimension, but also to continue to discover by his exploration of the unknown." All this, he concluded, could be achieved by establishing a low cost space transportation system that could become operational six years after its approval.[44]

After five days in space, McDivitt and Schweickart pulled the lunar module free of the command module and flew it as a separate spacecraft for the first time, testing the decent engine as they moved about 113 miles away from the Apollo spacecraft. Then they jettisoned the descent stage and successfully fired the engine of the ascent stage to rendezvous with the command and service modules, finally docking after about six hours of independent operations. And after a little more than ten days in

---

[43]   "NASA Aims at Planets in '70s," *St. Louis Post-Dispatch*, 3/14/69, GEM-83-7.
[44]   Mueller, NRC Division of Engineering Annual Dinner, Washington, DC, 3/10/69.

space, Apollo 9 re-entered the atmosphere and landed in the Atlantic on March 13, having successfully completed all of the mission objectives.[45]

As the initial lunar landing mission neared, Mueller increased his emphasis on planning for the post-Apollo period, and the future of human spaceflight. Ideas about what NASA could do following the initial lunar landings jelled, forming the basis for the agency's integrated long range plan which Paine would present to the president's Space Task Group in July. And Mueller increasingly spoke about his return to industry, though not until the achievement of the lunar landing goal, just days away.

---

[45]   Bilstein, *Stages to Saturn*, 368; Kraft, Flight, 302-306; Logsdon, *Exploring the Universe*, 431.

10

A railroad in space

"The decision facing us is that of taking the next step."

Mueller, November 1969

In a speech delivered shortly after Apollo 9 landed, Mueller said that NASA must preserve its scientific and technological base and the management capability that the space program had created. However, the real challenge beyond Apollo would be the next generation of low cost spaceflight systems. As a result of the Vietnam War and domestic unrest, the nation stood at a crossroads. Without adequate funding, much of the investment in human spaceflight would disappear. The agency needed a budget of at least $4 billion per year to properly use its existing capabilities to carry out major new space programs, and modest additional investment would reduce the cost of future space activities. As he explained, "When people ask why we should spend money on space instead of on poverty programs, they overlook the fact that these are two different questions that need to be sorted out and addressed separately." To accumulate the resources to alleviate poverty requires the creation of additional wealth. However, the productivity generated by advancing science and technology creates new jobs and opportunities "that will not merely alleviate poverty but get at its root." And he said it would be a tragedy to turn away from space at the end of a "truly astonishing first decade." Rather, "we should be asking ourselves how we can find new ways to bring space-age management capabilities to bear on our other problems."[1]

Questions about the need to fly AS-504 (Apollo 10) became a topic for discussion at STAC's meeting in late March 1969. Phillips told the advisory committee, "NASA appears fully prepared to support the current dates for Apollo 10, Apollo 11, and any subsequent launches that may be necessary to assure a landing mission by the end of the year." Hage then told STAC that while experience with the command and service modules continued to be "excellent," additional flight experience with the lunar module remained desirable because it had only flown in space once with a crew

---

[1]   Mueller, Annual Kiwanis Dinner, Milledgeville, Georgia, 3/21/69.

on board. This led to extensive discussion of lunar orbital navigation and the need for more testing of the Moon's gravity field. Experience on the three previous Apollo-Saturn missions led the agency to reject the need for another Earth orbital flight. And while NASA could proceed directly to a lunar landing without another lunar orbital mission if it waited until the lighter weight lunar module became available, Phillips advised against it to assure "safety and success." The STAC members then expressed support for flying Apollo 10 in lunar orbit as mission-F, rather than postponing it in order to attempt the first lunar landing as mission-G.[2]

STAC chair Townes had also chaired Nixon's space transition team, and characterized the transition team's report as saying, "NASA was here to stay and that its budget should not be changed violently." The transition report supported building a space shuttle following the Apollo Applications Program, although it recommended deferring the other elements of the space transportation system. Nonetheless, Mueller did not give up on a more extensive post-Apollo plan, and repeated some of the arguments used in his recent congressional testimony in support of it, pointing out that while "we are probably staying even with the Russians," they would move ahead of the US without sufficient investment. Due to budget cuts, utilization of facilities remained subcritical, and the agency planned to end Saturn V production. He called the space station "a logical and necessary step in a national space program, as a way-station to the Moon, as practice for planetary missions, and a step in reducing the costs of space operations." And of course it would advance space science. However, the Space Science Board's chair and STAC member, Hess, objected to Mueller trying to justify "a manned program on the basis of space science," and indicated that "if given the choice scientists would probably prefer to spend money in other ways."

After reviewing the economics of the space transportation system, Mueller urged proceeding with both the space station and the space shuttle. Townes disagreed, and restated the transition team's recommendation to build the space shuttle but defer the space station to a later date. Paine joined the STAC meeting at that point, and asked for an outline of alternative goals for the post-Apollo period. He asked the advisory committee's view of the space station, because decisions had to be made shortly. This led to a wide ranging discussion of missions and goals, with Mueller advocating the adoption of a major new goal in space to avoid day-to-day budget oscillations. Consequently, Townes asked for a Bellcomm study of the "real justification for 'man in space,'" and this became the rationale for what would be the agency's integrated long range plan for the post-Apollo period. The first steps towards such a long range plan had actually been taken as early as 1967. With planetary exploration his ultimate goal, Mueller began with the space station. He had to determine what the agency could do with it, and what its long term significance would be. To be useful, NASA needed multiple space stations in different orbits. But based on projected utilization, the agency could not justify such

---

2   Minutes of STAC, NASA HQ, Washington, DC, 3/22-23/69.

a plan unless the space stations became nodes in an overall space transportation system. Mueller set out to justify building these space stations, and since moving between orbits required energy he proposed using them as refueling points. To commercialize space, he realized that NASA had to reduce costs, primarily by developing a less expensive means of accessing low Earth orbit. Initially he sought cheap expendable launchers, the so-called "big dumb boosters," although he noted, "[i]t soon became clear ... there's a certain cost per pound of material, and that cost per pound is fixed depending upon the structural components that you use." He spent time figuring out how to change the equation so that the cost per pound did not determine the cost of the vehicle and therefore the transportation costs; this led naturally to the conclusion that the cost could be significantly reduced by reusing the launch vehicles – and the more reusable they were, the lower was the cost per pound. For completely reusable vehicles, the cost of transportation approached the cost of the fuel and supporting infrastructure. While it would take money to develop reusable vehicles, spreading development cost over a large number of flights would reduce the cost per flight, which was what really mattered. So, he said, "that simple minded set of concepts led to a fully reusable system" with a small supporting infrastructure.

And a key to this concept was to fly it like an airplane. The space station needed an economical method for resupply, and as studies progressed the need for a more far-reaching target emerged, an overall system with nodes that did not require taking everything into space each time they launched from Earth. A space station in low Earth orbit became the first node, with a second node around the Moon serviced by inter-orbital transfer vehicles and lunar landing vehicles, "like a railroad in space," he said. A node around Mars allowed for the transfer from Earth to Mars orbit, and it did not require much more energy than transferring from low Earth orbit to lunar orbit. This system reinforced the conclusion that keeping costs down would require a fully reusable space vehicle.[3]

After Webb gave Newell agency-wide responsibility for program planning, he created a planning steering group with twelve working groups and representatives from each program office. All these groups dealt with human spaceflight in one way or another. But in the end, Newell did not produce an integrated agency-wide plan and, as Mueller remarked, "We watched that go on for some time," and it did not yield any results "primarily because of the compromises necessary to get agreement." Paine recognized that Newell's planning efforts would not build a workable plan; and Mueller said it soon became "pretty obvious that Homer's plan wasn't a plan." It was really seven or eight different plans that were not integrated. "They were planning for everybody, but not very coherently." And that is when Mueller set out to define an integrated plan that addressed what "everyone wants to have done and design it in a modular fashion so that you can back down from it." Mueller used the

---

3   Ibid; Mueller to Senior Staff, MSFC, Huntsville, AL, 12/8/69; Mueller Interviews: NASM, 5/1/89 and JSC, 1/20/99.

requirements which Newell's working groups had listed, and then he devised the most economical way to meet those needs. Mueller used Bellcomm and the advanced planning groups at the centers to build the plan. STAC "challenged me," he said, "to try to find a way of implementing a program that would meet all of the objectives of the unmanned space programs." With the objectives identified, he set out to "find some economical way of doing everything anybody had thought of." Consequently, he developed "as complete a plan as you could," which "covered all of the bases."[4]

In mid-April, Mueller held an off-site meeting for the OMSF directors at KSC, where he estimated that implementing an integrated long range plan would require a budget of about $5 billion per year for the next six years, though it limited Saturn V flights to three per year and required that Apollo spacecraft be refurbished and reused. This amount exceeded NASA's budget plan, but he called the increased cost "reasonable," and proposed establishing a joint NASA/air force program office in order to allow the agency to spend more to develop the shuttle than NASA could get in its budget alone. He wanted MSFC to lead the design effort, again expressing preference for Huntsville over Houston. However, Mathews argued for involving multiple centers, because shuttle aerodynamics required different skills and he reminded Mueller of Newell's dissatisfaction with OMSF's planning because it did not contain enough space science. That is when Mueller decided the integrated plan should include more options covering space science, because, he said, "that is the only thing we haven't tried." And "we should talk about science first and then the incidental hardware." The agency required a plan that tied everything together, and Mueller built such a program, pulling all the other plans together into one plan. "In other words," he said, "it would be smart to review the planning that is going on to see how much of it we could support with our integrated program . . . [and] integrate our program with the Agency's program."[5]

Bellcomm produced the first drafts, reviewed them with Mathews and Mueller, and presented a draft capitalizing on lunar mission capability that would use existing technology to build two new systems: a space station and a low cost logistics vehicle to make the space station economically feasible. Their proposal emphasized that the system should be fully reusable to minimize cost, with several different vehicles operating between the Earth's surface and low Earth orbit, between low Earth orbit and lunar orbit, and between lunar orbit and the lunar surface. They envisioned four space stations operating in low Earth orbit, synchronous Earth orbit, lunar orbit, and on the lunar surface. The orbital workshop of the Apollo Applications Program would be replaced in the mid-1970s by a semi-permanent space station holding up to six astronauts for two or more years in autonomous operation. Space station modules would be joined to form larger structures capable of holding additional

---

[4]   Notes from. . .PSAC Presentation, 5/10/67, GEM-89-3; Mueller Interviews: Sherrod, 5/20/73, Slotkin, 6/9/10 and NASM, 5/1/89; Launius and McCurdy, *Spaceflight and the Myth of Presidential Leadership,* essay by Joan Hoff; Presentation from Mueller to NASA Staff, 12/4/69.

[5]   OMSF Directors Retreat, KSC, 4/14/69, GEM-97-14 and GEM-79-15.

crew. The space shuttle would carry a crew of up to twelve, and a nuclear stage would use a space-tug as its crew compartment. The Saturn V would remain the workhorse of the system, launching space station modules and the nuclear shuttle into low Earth orbit. This plan went through a number of iterations before OMSF published it on July 10, calling it an *Integrated Program of Space Utilization and Exploration for the Decade 1970 to 1980*.[6]

Meanwhile, NASA continued to prepare for the eight day Apollo 10 mission to test the lunar module in lunar orbit. The lunar module would separate from the CSM and enter an elliptical orbit with its low point at an altitude of 50,000 feet, but it did not have the ability to land. It would then jettison the descent stage. The ascent stage would carry 63 percent of the maximum fuel load in order to represent the weight that it would have at that altitude after lifting off from the Moon in order to conduct a rendezvous. The crew members, all Gemini veterans, were led by Stafford, with Cernan as the lunar module pilot and Young as the command module pilot. Like Apollo 9 – although fortunately without any bouts of "space sickness" – the mission achieved all of its objectives and proved once and for all the viability of lunar orbit rendezvous. Their landing in mid-Pacific on May 26 cleared the way for Apollo 11 to attempt mission-G in July.[7]

Mueller wrote of Apollo 9 and Apollo 10, "Fortunately we had very few problems on both missions. Three of the reasons for that good fortune were the extensive ground testing to which every subsystem was subjected, the simulation exercises which provided the crews with high fidelity training for every phase of the flights, and the critical design review procedures ... fundamental to the testing and simulation programs." However, flight software remained a problem on the critical path as this pushed the state of the art in software development. Because of its criticality, Mueller personally led a software review group which, over a nine month period, developed techniques to resolve software issues that could have delayed the lunar landing.[8]

STAC members went to KSC to witness the Apollo 10 launch, along with the members of the Space Science and Technology Panel of PSAC. The two advisory committees met jointly, where Mueller presented the integrated long range plan and described the work of the joint space shuttle task force. He emphasized that NASA had worked directly with DOD to manage the space shuttle program, and that the air force participated in all space shuttle planning and reviews. Although DOD had an interest in the shuttle, they had no requirements for NASA's space station. (In fact, the air force's MOL was on the verge of cancellation, its intended function

6   "Integrated Manned Space Flight Plan," B. T. Howard, 4/18/69, GEM-74-7; "An Integrated Program..." 7/10/69, GEM-74-13; "NASA Integrated Program..." Executive Summary, 7/22/69, GEM-74-14; Paine to distribution, 5/29/69, Space Task Group, GEM-99-7.
7   "Apollo Flight Tests, SA-505: Apollo 10," apollosaturn.com. Mueller Interview, Slotkin, 6/9/10; Logsdon, *Exploring the Unknown*, 431-432.
8   Person to Rowsom, 7/10/75, "Getting It All Together," Apollo Program, GEM-195-4.

having been superseded by automated satellites.) Describing the integrated long range plan, Mueller told the advisory committee members that it was designed to achieve low cost transportation in cislunar space. Space stations in Earth and lunar orbit would be transfer points for low cost reusable space shuttles, designed to support science, technology and applications. They would become the "building blocks for manned planetary missions," and the short lived orbital workshops would be superseded by space stations that had a ten year life span and could accommodate from six to twelve astronauts. Unfortunately, he said, budget cuts had delayed the start of work on the workshop for the Apollo Applications Program by at least a year.[9]

At an internal review of the post-Apollo program on May 27, Mueller confronted a number of difficult problems compounded by the limited funding in FY 1970. The Saturn IB wet orbital workshop weighed too much, its design limited the space that was available for stowage, and it experienced significant developmental difficulties. Replacing it with the dry workshop would resolve most issues, and after looking at the alternatives he agreed to the switch, but set the ground rule that it must carry the same experiments as the wet workshop, and the funding, schedule, ground support requirements, and Saturn V production costs would have to remain unchanged. In early June, with Paine's agreement he proposed using one of the Saturn V boosters nominally allocated to the Apollo Program to launch the dry workshop, and then to send up the Apollo Telescope Mount experiment on the second flight of the Apollo Applications Program, thus eliminating one Saturn IB, and resolving a number of key issues. And as NASA's William C. Schneider wrote in his history of the program, "In July 1969, after a period of intensive study of the pros and cons, it was decided to convert to the 'dry' workshop configuration." Consequently, when Mueller left the agency, the AAP had "a solid technical base with good prospects of meeting cost and schedule targets."[10]

II

Mueller returned to London on May 30 to address another meeting of the British Interplanetary Society. Just six weeks before the launch of Apollo 11 he said, "For the first time in eons of man's history, he will set foot upon another body in our solar system. Miles traveled are no longer a measure of meaning, but three days will be required to reach the destination where mankind will enter upon his greatest adventure." He credited "men of all ages, and of all nations" for the discoveries and developments of hundreds of scientists, thousands of inventors, and hundreds of thousands of technicians and workers which would lead to the "final victory." The

---

[9]   Minutes of STAC, KSC, FL, 5/19-20/69; Mueller Interview, Slotkin, 6/9/10.
[10]   Schneider to Mueller, 6/16/69, GEM-49-14; Schneider to Mueller, History of AAP, 12/5/ 69, GEM-50-1.

groundwork laid would result in bases on the Moon, space stations in low Earth orbit, and space shuttles taking passengers, crew, and cargo to bases in space and on the Moon. Much progress had been made defining the new space transportation system, "rather more than I had hoped," he said. And study contracts that were expected to be completed by year-end would lead to a space vehicle providing transportation to and from Earth orbit at a cost more than one order of magnitude lower than existing capabilities. Interesting new designs for reusable equipment had evolved, though the development of new technologies remained necessary to make the space shuttle fully operational. "If we persevere, I believe that a space shuttle can be operation within seven years. And within these seven years many more of the nations of the world will have begun to express their aspirations in space activity ... The consequences of the synergistic effects of all these elements are truly beyond our ability to forecast."[11]

Back in Los Angeles in mid-June, Mueller spoke to a group of electrical engineers about the role of computers in the space program, a topic of particular interest to him. He wrote several articles on the subject, calling computers "one of the few basic elements upon which our whole space program is constructed." During Mercury, the entire computer complex supporting the program performed one million calculations per *minute*, and the systems supporting Apollo did almost fifty times that many – approaching one million instructions per *second*.[12] Computers, he said, would do "more and more work to help solve some of the massive problems of our time ... [and] increased computer speeds are another of these characteristics which will contribute appreciably to greater efficiency." Then he noted that within a month NASA would make the first attempt at a lunar landing. And with its accomplishment, "the manned space flight program will have, in effect, graduated to a new phase of its development ... [a] more mature segment of our total operation." The Saturn V would continue as a major component of the space program through the 1980s, although to make it more cost effective it had to be simplified in order to reduce complexity and thus cost. That meant reducing the number of interfaces between the space vehicle and the ground, and placing more responsibly on internal systems to perform automatic checkout. Space stations in low Earth orbit would also reduce the cost of spaceflight by limiting the number of launches needed for a specific investigation. A major cost of space operations involved logistics, which could be reduced by using lower cost vehicles between Earth and low orbit. However, those vehicles would have to be fully reusable, because routine transportation between Earth and points in space would be an ongoing requirement. Therefore, a low cost reusable system would be a necessity. This system could be built with existing technology, but required new approaches to the

---

[11] Mueller, BIS, London, England, 5/30/69.
[12] In measuring computer processing speeds, IBM's fastest computer *chip* in 2010, the "z-Enterprise 196," processed 50 billion instructions per second, or 50,000 times the speed of the whole *Apollo* computer complex. *TG Daily*, September 6, 2010.

preparation of vehicles for launch. While 20,000 people worked to checkout and launch the Saturn V, the space shuttle had to simplify that. Major breakthroughs in electronics would provide the tools, but to "realize the economies which are inherent in our new electronics advances, we need to arrange that each subsystem, each black box, is self-checking."[13]

Mueller reviewed his integrated long range plan with Paine on June 24. The plan assured that human spaceflight would continue after the completion of Apollo, and it would achieve objectives for science, applications, and engineering in an integrated fashion by eliminating what he called the "line between manned and unmanned space flight." The first strategy involved minimizing new development, limiting cost, and living within tight budgets for several years. The second required full reusability. Although that increased near term costs, it would deliver long term savings. And reusability had to extend not only from Earth to orbit, but include orbit to orbit shuttles and vehicles going from Earth orbit to the lunar surface. A space station in polar orbit around the Moon could place a crew sixty miles above any point on the surface. A landing vehicle would allow astronauts to touch down, conduct extensive explorations, and return to the safety of the space station when the orbital plane was realigned fourteen days later. However, he pointed out, "We haven't eliminated unmanned satellites by any matter of means," because there remained many things where economy and other considerations required their use. And while this plan did not include human exploration of the solar system, he said the capabilities discussed would be able to be "directly applicable to such manned planetary activities," and as such represented new ways to achieve interplanetary missions at low cost once initial investments were made.[14]

And the integrated plan did not assume a commitment to the next phase until the completion of the prior one, giving it the advantage of flexibility. The first phase involved extending Apollo capabilities while building a base in lunar orbit. Then, assuming the discovery of something interesting, the third phase would address lunar surface exploration. Next would be planetary exploration, although Mueller pointed out that during the 1970s this would be done with robotic satellites in advance of human missions in the 1980s, with particular attention being paid to Mars. He called low Earth orbit the primary locus for performing space science and applications. This would begin with the Apollo Applications Program and would be followed by the space shuttle and space station. This plan called for $6 billion per year, as compared to under $4 billion in NASA's FY 1971 budget request. The question, Mueller told Paine, "is should we go ahead and press for a return to a $6 billion national space program in view of all the social problems and student unrest ... [or] recommend to the President that the politic thing to do here is to sort of think of pressing gradually down to $4 [or] $3.5 billion and to cut our cloth to fit." Asked by

---

[13]   Mueller, IEEE, Los Angeles CA, 6/18/69.

[14]   Transcript of Mueller's remarks, Program Planning meeting with Mr. Paine, 6/24/69, GEM-47-1.

Paine if he had underestimated the short range budget problem, Mueller replied "My inclination ... is that we should press very hard for a return to the $6 billion level. The space program is something that is very solidly rooted now in the American dream ... and to turn it back in favor of some dubious activities in other areas would in the long run not be something the President should do."[15]

On July 3 Mueller spoke at the Canaveral Press Club, where he announced that the agency had just completed a "very successful" countdown demonstration test of Apollo 11 and planned to launch on July 16. Quickly changing subjects, he said the real question concerned the future of spaceflight after Apollo. Based on the existing schedule, the Apollo Program would end in 1972. The president's Space Task Group had studied what should be done after Apollo, and this had caused him, "and a number of my friends ... [to spend] a great deal of time in the last several months [at] work on the basic question of what would make a sensible long term program." He discussed the reusable space shuttle and the space station, telling these reporters, "I am enthusiastic about it." He then explained that the cost involved in developing that capability would be comparable to that of the Apollo Program as a whole. Questions ranged from the impact of the follow-on program on NASA staffing, through to the characteristics of the new equipment required. Sidestepping a question about future planetary exploration, he explained, "we don't have any plans specifically addressed to going to manned planetary flight," but the agency continued to conduct studies and "one of the attractive things about this shuttle is that once you get that kind of shuttle system built it is also applicable in a follow-on sense to planetary travel." However, he refused to state the goal of sending astronauts to Mars.[16]

As planned months earlier, Apollo 11 took off around 9:30 a.m. on July 16, carrying astronauts Armstrong, Aldrin and Collins – all Gemini veterans. Inside the spacecraft, as the crew watched their instruments and displays, the noise was not as loud as they expected, and the ride was softer than the Gemini-Titan. Staging was nominal and the third stage achieved the desired orbit. When ready for translunar injection, the J-2 engine of the S-IVB restarted and sent Apollo 11 to the Moon. As he stood there, Mueller thought, "Is it going to be going up? Which of the thousands of things that could go wrong are going to go wrong?" Nothing did – but as he later said in relation to Apollo 13, "it could have gone wrong just by one simple mistake." Then, he smiled, and laughed to release the tension of the moment.[17]

Millions of people around the world watched the Saturn V take off, and saw the astronauts beam color TV pictures from space showing the receding Earth. The vice president held a press conference at the Cape, telling reporters that he supported a new national goal of landing humans on Mars by the end of the century. However,

---

[15]   Ibid.
[16]   Mueller, Press Conference, Canaveral Press Club, FL, 7/3/69, GEM-47-2.
[17]   Mueller Interviews: Slotkin, 6/8-9/10; Gilruth Interview, NASM, 3/2/87; Logsdon, *Exploring the Unknown*, 432-433; Bilstein, *Stages to Saturn*, 369-372.

10-1   George E. Mueller during the Apollo 11 launch, July 16, 1969. (NASA photo)

10-2   George E. Mueller after the Apollo 11 launch, July 16, 1969. (NASA photo)

according to the *New York Times*, Senate majority leader Michael J. Mansfield "promptly ruled out further space efforts until problems here on Earth are solved." In the days following the launch, Mueller got his share of publicity. Newspapers praised the people behind Apollo 11, frequently mentioning Mueller, Phillips and the center directors. The *St. Louis Globe-Democrat* and the *St. Louis Post-Dispatch* published long articles quoting their favorite son, as did the *Washington Post*, and the *Chicago Tribune*, along with the *Times* of London and innumerable other newspapers and magazines.[18]

After the launch, Mueller's colleagues threw a surprise fifty-first birthday party for him, which he called the most rewarding experience he had at NASA. "So many people appreciated the work I had done," he said. "There was a fair amount of camaraderie at that time. Everybody was up and working together as a real team." Returning to the hotel, the marquis read "Happy Birthday." And as he recalled, "Sam [Phillips] and the troops organized all that." It was "a high point, no question. You know I had no idea that that was going on. I did of course have an idea that the

---

[18]   Various, Newspapers & Magazine Articles, 7/1-8/22/69, GEM-83-8.

launch was on my birthday. And I've always had a suspicion that Sam delayed it for a day so it was on my birthday but I don't know that ... It was great."[19]

Three hours into the mission, Collins separated the command and service modules from the S-IVB and performed the maneuvers necessary to retrieve the lunar module. The crew had selected the names *Columbia* for the command and service modules and *Eagle* for the lunar module. The translunar coast passed without incident, and the spacecraft entered the desired lunar orbit. At about four in the afternoon Eastern Time on July 20, *Eagle* landed on the Moon. Armstrong's heart rate, which was normally 70 to 75 beats per minute, jumped to 110 when *Eagle* started its decent, and hit 156 at touchdown. Forty-five minutes later it remained in the 90s. As Mueller put it, Armstrong had good reason for his heart to race because, "The landing of Apollo 11 'had everyone scared to death.'" It had come close to being a disaster, and "There wasn't much cheering in Mission Control because of that. It was not near to being catastrophic as many thought, but we did come close to losing the mission."[20]

Mueller called the landing "a fascinating time." He sat in a small room at the side of mission control, "listening and listening and listening and thinking, 'Are they going to be able to land?'" The actual landing took place very quickly, and he said, "they landed by the time all of the worries got organized in my mind. But that was a traumatic moment." The final phase was particularly tricky because it occurred in the so-called "dead man's box", where the vehicle was too low to abort because it would hit the surface before staging could be successfully achieved. For Mueller, however, liftoff of the Saturn V created more tension because, he explained, with the "whole stack there if anything goes wrong you've got a major problem ... You would blow up half the Cape." The crew spent about two and a half hours outside the lunar module collecting soil and rocks, planting a US flag, unveiling a plaque, and receiving a call from the president. Then they returned to the lunar module to try to get some sleep before preparing for liftoff. They remained on the surface for about twenty-one hours. Then they flew *Eagle* back to *Columbia*, where Armstrong and Aldrin rejoined Collins, who had spent the day orbiting the Moon in solitude.[21]

III

Following the lunar landing, the *New York Times* published a long article by Mueller about the integrated long range plan. "The triumph of Apollo ... is only a beginning – it has given us the confidence to dream those impossible dreams and the knowledge required to make those dreams reality," he wrote. Covering nearly a full page of

---

[19] Mueller Interviews: Sherrod, 11/19/69, NASM, 5/1/89 and Slotkin, 6/9/10.

[20] "Astronauts Land on Plain...," and "Armstrong, Neil A's Pulse Climbs," *NYT*, 7/21/69; Mueller Interviews: Sherrod, 11/19/69, Slotkin, 6/8-10/10, and JSC, 8/27/98.

[21] Logsdon, *Exploring the Unknown*, 434.

newsprint, he described plans for lunar and planetary exploration using the new space transportation system. And concluding his narrative, he said, "we will use two major new developments in the realm of manned space flight that are within our grasp today, a long-duration manned space station and reusable transportation between Earth and space, within space itself, and out to the Moon." In numerous interviews at the time, he also spoke about the future of human spaceflight, not about Apollo 11. He wanted people to focus on the future, not the present, because he said "there was a great tendency to think that Apollo 11 or 12 or so was an end to the space program and I wanted to try to get that thought turned around. It's just the beginning of the space program."[22]

After *Eagle* docked with *Columbia*, and following another rest period, the crew began their return to Earth with a successful burn of the service propulsion system. The trajectory brought them back to Earth on July 24, where they and forty-four pounds of lunar material landed in the Pacific near the aircraft carrier *Hornet*. In Houston for the splashdown, Mueller joined in the celebration at the end of the mission. Nixon on a round-the-world trip that was part of the cover for Henry A. Kissinger's secret trip to Beijing, greeted the astronauts aboard ship, later calling Apollo 11 "the most exciting event of the first year of my presidency," and claimed to lament the end of the Apollo Program because it lacked popular support. However, according to Mueller, the new president did not back human spaceflight; he used it for his own purposes but did not support it because "it was not invented here, wasn't invented on his watch."[23]

Paine planned to send congratulations to the "the top dozen" people who had contributed directly to the successful lunar landing, and asked Mueller and Phillips for lists. Mueller provided *thirteen* names; most of them expected ones, though he included a few surprises (Table 10-1). He wrote his list by hand, but not in order of their contributions, he said. Reviewing a copy of the list forty years later, he called them "a remarkably capable crew that got involved. We were lucky having the right people at the right time I guess."[24]

However, he said, "there were another hundred that I could have named ... Freitag was our interface with Congress and he was important. And of course Bob Seamans probably doesn't get as much credit as he should and he contributed to things ... So twelve was far too few." He said the whole staff in OMSF deserved recognition, called Berry "a key player," and said John A. Hornbeck, the president of Bellcomm, was "an important character." Although "other guys at North American were very important ... Ralph Ruud should have been on that list," as should Joseph Gavin of Grumman. "As I say the list was too short. And you always

---

[22] "In the Next Decade: A Lunar Base, Space Laboratories and a Space Shuttle," *NYT*, 7/21/69; Mueller Interview, Slotkin, 6/9/10.

[23] Mueller Interview, NASM, 5/1/89; Nixon, *Memoirs*, 394 and 428-430.

[24] Mueller and Phillips from Paine, 8/6/69, Apollo 11, GEM-58-1; Mueller Interview, Slotkin, 6/9/10.

Table 10-1: Top "twelve" contributors to the Apollo Program (August 8, 1969) [25]

Phillips	"the best manager we ever had."
Low	"a great organizer and manager."
Shea	"A near genius in terms of engineering and really one of the leaders."
Rees	"A great guy and one of the key players in getting the Saturn V built, working."
Von Braun	"an extraordinary engineer … more than a manager – he was a leader."
Gilruth	"important as a father figure … or else we would have fired him."
Debus	"The kind of guy you have to have in the launch business."
Bergen [William B.]	"He came into North American and rescued" the CSM.
Stoner [George E.]	"Probably one of the best brilliant engineers [Boeing] had."
Richard [Ludie G.]	"One of the very good engineers we had [at MSFC]."
Rudolph	"He was … a major contributor."
Kraft	"He did a tremendous job with organizing the support activity."
Petrone	"He really was the one that drove through the construction of the facilities at the Cape. He and Kurt were a rare combination."

make mistakes when you put it this way. But I wouldn't take back any of the names I had." Then he added, "I could have put Bennie Schriever on that list."[26]

Mueller addressed the National Space Club once again on August 6, with the three center directors at the head table. After expressing appreciation for the work of the whole team, he noted that nine more Apollo flights were scheduled, and he expected to find "new and exciting things as we go forward that we will want to continue to use the extended Apollo capabilities to explore more sites on the lunar surface." He predicted space stations would become laboratories, refueling facilities and terminals for space shuttles traveling between the Earth and lunar orbiting space stations, and hoped that the payload launch cost of $1,000 per pound on the Saturn V would be reduced by one or two orders of magnitude. After describing the elements of the integrated long range plan, he said that, by establishing the Space Task Group, Nixon "recognized the need for a decision on our future in space," although he added "it remains for the people … to determine whether mankind will make the 'Great Leap' into the planets."[27]

Speaking at another IEEE meeting in Los Angeles in early September, after acknowledging those who made the lunar landing a success, he quickly turned to the "two key concepts" needed for future space programs, reusability and commonality. NASA's long range plan would eliminate the distinction between human or robotic spaceflight, because, he said, "They in effect become a single program aimed at lunar exploration, Earth orbital applications, and planetary exploration." Because

---

[25]   The comments are those Mueller made after seeing the list for the first time in forty years, except for von Braun, which comes from: Mueller, *Memorial Tribute,* 12/77, Huntsville, AL.

[26]   Mueller Interview, Slotkin, 6/9/10.

[27]   Mueller, NSC, Washington, DC, 8/6/69.

missions to the Moon would find many "new and exciting things," NASA would need more equipment to continue lunar exploration. And with a reusable lunar landing vehicle, astronauts could travel from an orbiting space station to any point on the Moon. Using cislunar space for the benefit of people on Earth, the agency could establish a space station in low Earth orbit with reusable space shuttles providing an economic means of traveling to and from the Earth's surface. With such a system, he said, NASA could develop the capability for planetary exploration, including missions to Mars. And between 1979 and 1989, using the lessons of Apollo, he predicted that over a hundred people would live on space stations and a lunar base, with a fleet of shuttles facilitating resupply, satellite repair and maintenance. If the agency moved rapidly, the first crew could leave for Mars by 1981; and by 1989, he said, "I would expect that men would have established their first colony on the Moon and the first permanent base on Mars. And they would be beginning to fly out to the further planets."[28]

The president's Space Task Group met during the spring and summer of 1969 to develop plans, using NASA's budget for FY 1971 as their baseline. Mueller briefed them on the integrated long range plan in May. However, he later said Nixon used this effort to contain NASA, because he did not want the vice president or the Space Task Group to "get politically out of hand and get its own momentum going." And domestic turmoil made it inopportune "to launch a new wonderful space program," as Mueller described it. The agency lacked congressional support, the war drained resources, and riots took place on college campuses and in major cities. Planning for future space activity was not "the first topic" on the new administration's agenda, though he said "it might have been the best possible thing they could have done." Teague told him "the realities of the situation are that I can't get you more money," and while Mueller said that he "had good enough rapport with Congress," the budget bureau remained a stumbling block, focusing on the next year's budget, ignoring the longer term, and lacking interest in space except to find out what it cost.[29]

The integrated long range plan became the basis for NASA's official submission to the Space Task Group in September, which Paine named *The Post-Apollo Space Program: Directions for the Future: America's Next Decades in Space*. This plan projected budget increases that would peak in FY 1976 at more than $10 billion, before falling to just under $8 billion per year into the 1990s. It included everything in Mueller's plan, and more. On September 15 the Space Task Group issued their report, which said, "We have concluded that a forward-looking space program for the future of this Nation should include continuation of manned space flight activity. Space will continue to provide new challenges to satisfy the innate desire of man to explore the limits of reach." Not wishing to create a crash program they

---

[28] Various, Apollo 11, 9/2-11/69, GEM-58-5; Mueller, IEEE Council Meeting, Los Angeles, CA, 9/5/69.

[29] Mueller Interview, NASM, 11/8/88.

nevertheless concluded, "NASA has demonstrated organizational competence and technology base, by virtue of the Apollo success and other achievements, to carry out a successful program to land man on Mars within 15 years ... [and] a manned Mars mission should be accepted as a long-range goal for the space program." Listing a series of program objectives emphasizing commonality, reusability and economy, the report recommended three alternative program plans, each with different budgets. Nixon did not announce his decision until shortly after Mueller left NASA, but from the time that he received the report until then, the agency's budget faced cutbacks that would negate the impact of the Space Task Group's proposals.[30]

At the end of September, Mueller spoke to the senior staff at KSC, describing the agency's long range plan. He argued that reducing the cost of human spaceflight would require a quick turnaround of launchers, with a high in-service availability. "Essentially," he said, "we're trying to get in the space transportation system to about the same kind of operating mode as the airlines have." And he noted, "One of our objectives is to explore the solar system," with the design of the space transportation system permitting the start of planetary exploration by the end of the century. The plan called for building four new space vehicles by the 1980s and a planetary excursion vehicle later on. The first step remained the orbital workshop. When fully operational, the space shuttle would eliminate expendable launch vehicles, and by the 1980s as many as 300 people could live and work in space on a continuing basis. He insisted that this could be done within existing budgets, simply by increasing the returns on the dollars spent. And if the administration adopted the plan, then the number of "people devoting their time to it here on Earth will remain roughly constant over the next 20 years," although they could expect the return on investment would increase "by at least a hundred times." NASA had adopted the plan for the next two decades, and the Space Task Group accepted it with some modifications, delaying parts for a year and decreasing some activities planned in the out years. However, the Space Task Group's plan included all of the elements contained in NASA's plan, which created "a great deal of future for the nation and for everybody in NASA."[31]

Mueller later explained that organization was the secret to these plans. He had established the OMSF organization so that the interfaces were limited and well defined. The idea of program offices at headquarters and project offices in the centers had already been adopted throughout NASA, making it easier to pursue an integrated long range plan. But the principal audience for his integrated long range plan was not the Space Task Group. The development of the plan had begun "long before ... [in] response to the planning activities, rather than vice versa." His motivation was to exploit what they had accomplished in Apollo, and provide a

---

[30]   Mueller Interview, NASM, 5/1/89; "Post-Apollo Space Program: Directions for the Future," 9/69, GEM-75-2/3; "Report of the Space Task Group 1969," hq.nasa.gov; Logsdon, *Exploring the Unknown*, 436; McDougall, *The Heaven and Earth,* 421.

[31]   Mueller, Speech Transcript, KSC, FL, 9/30/69.

meaningful follow-on program. And he designed it to pull all of NASA together into a single integrated program.[32]

Mueller had spoken about leaving NASA for quite some time, although in August he became more public about his desires. Sending letters to friendly CEOs of major corporations, he explained that he was seeking an operational role, although he never directly asked for a job. Rather, he invited "advice," but several firms immediately expressed interest and made job offers. Soon he had a number of discussions going, and by mid-October had received several concrete offers, including one from Roger Lewis, the CEO of the General Dynamics Corporation. As Mueller considered his options he debated the pros and cons of each, and spoke with friends. By the time he finished his search, he had several large books filled with notes and letters. He finally opted for General Dynamics. On October 13, he sent Paine a handwritten letter of resignation, saying that "pressing personal reasons require that I leave NASA in the immediate future." Once he had made up his mind, he wanted to quickly begin his new career, and requested to be relieved of duty on November 10, and to terminate employment on December 10. He closed by writing, "I regret leaving the agency and my many friends, believe me that I shall do all that I can to support the space program in the future." Paine asked if he would remain as deputy administrator, but he "told him I didn't think that was a good idea."[33]

IV

Mueller continued his involvement in several technical societies while at NASA, speaking at meetings and writing articles for magazines, although he did not serve on committees or accept any official position in them. A fellow of many organizations, he remained closest to the IEEE which he first joined in the 1940s and the AIAA, of which he had been a member since the 1950s. Then in 1969 he became a member of the AIAA's committee on international cooperation. This committee worked closely with the IAF and the IAA, the groups which organized the annual International Astronautical Congress, which he had also attended since the late 1950s. The AIAA appointed him as a delegate to the International Astronautical Congress in Mar del Plata, Argentina, and he traveled to South America shortly after submitting his resignation in October, but before it became public knowledge. He gave the keynote address at this congress, describing the Apollo Program and NASA's integrated long range plan. At a plenary session, the delegates elected him one of four IAF vice presidents, together with representatives from Belgium, the USSR and Argentina. He later remembered these congresses as "one of the places where there was some interchange of information" with Soviet delegates, and said they "were likely to be a little freer with what they were willing to talk about than they were anywhere else ... so

---

[32] Ibid.; Mueller Interview, NASM, 5/1/89.

[33] Employment Records, 12/15/67-10/15/69, GEM-72-12/17; Mueller Interview, Slotkin, 6/8-10/10.

in that sense it was a useful forum." He recalled talking with Vladimir A. Kotelnikoff, vice president of the Soviet Academy of Sciences, who told him about their shuttle, saying "we have a program that is at least as good as yours.'" Kotelnikoff and Mueller, Cold War rivals, became friends when both later served as IAA officers.[34]

Mueller went on a speaking tour in Europe following his trip to Argentina, and during one of his last speeches as a NASA official he addressed the International Air Transport Association in Amsterdam in the last week of October, delivering the same messages as he had all year. He said, "Never before has man been in a position to make a conscious decision to follow a path that would change the future of all mankind." And he asked, "Are we to choose the path that leads to men colonizing the solar system, or will we turn back?" If the path is forward, he said that it should be an international effort. He attributed the lunar landing on schedule and at the lower end of original cost estimates to all-up testing, which allowed the third Saturn V to fly astronauts on the first lunar orbital flight. He spoke of studies to select the best designs for a fully reusable space shuttle, which he expected to fly in six or seven years. However, there were areas of technology requiring additional research. He did not know how much the space shuttle would cost, but estimated $6 billion to develop a prototype. Then after describing the space transportation system he said, "The kinds of resources involved are similar to those applied to the Apollo Program." And he called it the "key to the program ... It is literally man's key to the solar system."[35]

In November, Mueller wrote an article for *Astronautics & Aeronautics* magazine, simply named "An Integrated Space Plan, 1970-1990." He said, "We stand today on the verge of what I believe is one of the great decision points in the history of the planet," one of the decisions that would "change the future for all mankind." With the achievement of the lunar landing goal, "we can decide to use or not to use this potential for the benefit of all man on Earth and to expand the scope of man's activities to make him truly a citizen of the solar system." He spoke of practical benefits, scientific advances, and utilization of the Moon. However, he said, "The decision facing us is that of taking the next step – the step of creating a low cost transportation system" to permit humans to explore the solar system and develop the means to use space for scientific, technological and economic applications. The obstacle standing in the way of achieving that objective remained the high cost of spaceflight, but "in this time of intense completion for available funds, it is fortunate that we have learned enough from the Apollo experience to permit us to design a system that will vastly reduce costs and thus increase the amount of space flight benefits for each dollar." In one of his last attempts to gain support for human spaceflight while at NASA, he wrote, "this integrated program reflects a new strategy adopted in response to opportunities to greatly increase the rate of program

---

[34] Harford to Mueller, 9/24/69 and 10/31/69, AIAA, GEM-49-5; Mueller Interview, Slotkin, 2/24/10. Mueller joined the Institute of Radio Engineers while at Bell Labs and the American Rocket Society at Space Technology Laboratories. Both groups were predecessors of the IEEE and AIAA, respectively.

[35] Mueller, IATA, Amsterdam, 10/23/69.

output per dollar through substantial reductions in cost. This strategy places increased emphasis on scientific experiments, on practical applications and on the exploration of the solar system." In conclusion he wrote, "The program for the 1970's can be accomplished at a cost comparable to the investment in NASA programs of the 1960's."[36]

The *Washington Post* broke the news of Mueller's resignation on November 1, writing that he would resign when he returned from Europe, and said his resignation had become an open secret in Washington. In addition to running human spaceflight, the newspaper said Mueller had served as NASA's chief lobbyist on Capitol Hill, and had been in line for the deputy administrator's position before Paine arrived. The Associated Press spread the story. Publishing it in Europe, the *International Herald Tribune* said Mueller "had been passed over twice for the deputy administrator job," and "he had been talking about leaving for months." Perhaps because he was on the continent when the story broke, newspapers from Finland to Spain published articles about his departure.[37]

When he returned to the United States, Mueller submitted a second letter of resignation, and the agency publicly announced it on November 10, to take effect on December 10. He accepted the position of vice president for corporate development at General Dynamics, and during his last days at NASA sent letters turning down the other job offers. Why did he choose General Dynamics? Mueller said, "I always was impressed with the work at General Dynamics . . . I had been involved with them in the ballistic missile program so I knew many of the engineers down there. And it just seemed to me that it was one of the more forward looking companies." He said Roger Lewis, "spent a great deal of time talking to me trying to convince [me] about the future with General Dynamics. I suspect if some of the other companies had been equally persuasive we would have had a different outcome." Shortly after announcing his resignation, Mueller privately told Sherrod, "I've done my duty by the government." Working for NASA had been more fun than industry, even although his job had become "an all-consuming task." More than forty years later he candidly described his decision to leave the agency this way: "we had landed on the Moon and I had done what I set out to do and . . . it didn't look like there was any great future in NASA because the decision had already been made politically to . . . shut Apollo down. And that was most of NASA they were shutting down." And, he wryly noted, "There's hardly anything you could do that aces the Apollo Program." The combination of circumstances which had maintained the support needed to complete the Apollo Program no longer existed. Had Kennedy lived, he may have been able to change the decision to land on the Moon, decide not to go because of other priorities, or travel to the Moon but not land. That *might* have been possible, although after he made the commitment to land on the Moon the country accepted the idea. Then a gunman assassinated him and Johnson continued Apollo

---

[36] "An Integrated Space Program, 1970-1990, *Aeronautics & Astronautics*, AIAA, Nov. 1969, GEM-49-6.

[37] "Mueller Quitting, Space Job," *Washington Post*, 11/1/69, GEM-83-10.

in his honor. Mueller called it "very powerful in the sense that the goal was established by almost acclamation and then the president got killed. So [the] 'acclamators' were constrained to make it happen in [his] memory." In one sense, the memory of Kennedy as a motivator for the Apollo Program "was fortuitous," he said. "[T]he one thing that is important in Washington ... is to have a firm set of goals with milestones and indications that you are achieving those goals [because] then you can continue through some adverse circumstances. But without that overlapping and overriding need you can get diverted and accomplish nothing."[38]

Mueller flew to KSC to view his last Saturn V launch as a NASA official. The agency prepared Apollo 12 as a mission-H, a "minimalist" landing that, in Shea's words, would show that Apollo 11 was not a "random success." Aboard were two Gemini veterans, Conrad as the mission commander and Gordon as the command module pilot, together with rookie Alan L. Bean as the lunar module pilot. Mueller had a seat in the operations management room to watch the takeoff. Across the water separating the launch pad from the VIP viewing stands, with a press pass from the AIAA, your author mingled with NASA's guests and the other news media. It was dawn and white floodlights dramatically illuminated the Saturn V. The now familiar large digital clock sat in front of the viewing area, ticking off the time until launch. About an hour before the launch, a large marine corps helicopter landed. Nixon stepped out and walked down the line of visitors shaking hands with everyone. The weather was overcast and there were showers of rain. The launch clock ticked down, and when it reached zero, after a slight delay the five F-1 first stage engines ignited. The Saturn V slowly lifted off the launch pad in a vivid display of fire and color. About fifteen seconds later the sound of the engines reached the viewing area with a loud roar. The rocket moved into the overcast sky and entered the low cloud cover at 800 feet, with the glow of the engine showing through. Then it looked like lightning hit the rocket, and this was later confirmed.[39]

The decision by launch director Walter J. Kapryan to proceed with the launch in such conditions was based on the fact that there were no thunderstorms in the area, but as the vehicle went from about 6,000 feet to 13,000 feet inside the thick clouds, electrical discharges occurred. The crew saw a bright light as lightning struck the vehicle and discharged down the hot plume of exhaust to the ground, and alarms and warning lights went off inside the spacecraft and electrical equipment dropped out. The NASA investigation concluded that the structural bonding designed for such discharges held in place, but the electrical "systems that were affected had previously exhibited sensitivity to transient voltage conditions during ground testing." From his position Mueller could not see the lightning, although he heard the thunder clap. As he recalled, the space vehicle "ran a conducting path from the ground to where it was," in effect becoming a lightning rod. Although it only lasted a second, "That was

---

[38]   Mueller to Paine, 11/10/69, Employment Records, GEM-72-17; Mueller Interviews: Sherrod, 11/19/69, Slotkin, 2/25 and 6/9-10/10, JSC, 8/27/98 and NASM, 11/8/88.
[39]   Slotkin, unpublished recollections; Orloff and Harland, *Apollo*, 52.

10-3   Apollo 12 lightning strike, November 14, 1969. (NASA photo)

a long second." It was "pretty tense." However, he went on, "Pete Conrad was a remarkably good engineer as well as pilot. So he didn't panic and that was good because a lesser person might have pushed the escape button." That was one of several launches where the astronauts could have opted to trigger the escape system, but as Mueller said, "It's a lot safer waiting actually." He considered Apollo 12 "the most dangerous launch we had until you got to Apollo 13 when the [oxygen] tank blew up [in space] . . . That lightning strike could have really destroyed that mission. And it is probably because we did so much work after the Apollo fire to re-do all of the wiring that it survived. To my amazement we got a direct lightning strike and nevertheless everything recovered and it was able to go to the Moon and back." Although the power supply of the spacecraft dropped out and its inertial platform tumbled, the Saturn V had its own guidance and control system and the rocket kept on going, proving the success of redundancy.[40]

---

[40]   "Preliminary Findings of the Lightning Incident During The Apollo 12 Launch," GEM-59-14; Mueller Interview, Slotkin, 6/9-10/10; Remarks by Mueller to NASA Senior Staff, MSFC, Huntsville, AL, 12/8/69.

Sherrod interviewed Mueller one more time before his resignation took effect. They were both in Houston on November 19, and met while waiting for the second EVA of the mission. Thinking Mueller would "let his hair down" now that he had announced his resignation, the journalist wrote of his disappointment. "Mueller is one of the most puzzling characters I have met in NASA . . . Nobody denies that he is brilliant. Even his most severe critic, Chris Kraft, admits that. But he certainly never made himself loved." Sherrod wrote that "reliable reports" maintained that Paine "eased him out." The reporter's first question concerned the scientist versus engineer controversy, which was "what everyone talks about nowadays." Mueller responded that scientists were uneasy because of the decline in university research money over the past few years. "Scientists are also naïve. They are accustomed to seeking funds in the universities, not in government. They think 'If we can only cut it out of NASA' without realizing if one gets cut all get cut." He discussed the inside politics of the engineering versus science controversy and said, "In addition, there was a let-down after the Moon landing. Many people found it hard to readjust," and that led to some scientists "blowing off steam" about what they considered the lack of support for science at the agency. Mueller also claimed the relationship between center directors improved during the previous six years. When he first arrived they were territorial and found it difficult to work together; although that changed, he insisted. The fire brought people together and, he said, "I'm pleasantly pleased by relations nowadays between OMSF and the centers." He attributed the success of Apollo to the program management system, and the use of all-up testing which reduced cost and schedules. And he pointed out "the only way you can control schedules is to control money." He also credited the Apollo executives group, though Sherrod noted Paine disbanded it as soon as Mueller left the agency. Free of other duties, Mueller went out to the aircraft carrier *Hornet,* to be on hand when the Apollo 12 crew landed in the Pacific on November 24.[41]

Mueller spoke to his senior staff in Washington on December 4. Having less than a week before his departure, he said, "I think quite clearly we have accomplished a great deal in the past year and in the past six years." If they had not succeeded, he said, "then this would have been a very different time indeed (laughter) . . . Yet in many ways they would have been simpler and more straight forward." Leaving NASA came with difficulty, although he thought it a "fairly good point in time to leave. We finished Apollo . . . [and] laid out a plan for the future." He explained "there is hardly a better time to make a decision to leave," because it would take seven or eight years to implement the new plan "and it is time for somebody else to have the opportunity to work on this kind of exciting challenging thing." He recognized that when a plan was laid out there was a good chance it would change, but he called it "a framework and an approach that is a viable one . . . [capturing] the imagination of enough people so that it ought to go forward." The plan was well on its way toward implementation, especially the "first and most vital link in the whole

---

[41] Mueller Interview, Sherrod, 11/19/69.

operation and that is the space shuttle." He argued that by building the shuttle it became essential to have a space station, and predicted that the different parts of the space transportation system would fall into place. "So in a real sense," he said, "we had better get that first increment going and implement it. The rest will follow just as day follows night." As he observed in a similar talk in Huntsville the same week, in addition to the plan gaining support from NASC, PSAC, the air force, and DOD, "strangely enough" the Bureau of the Budget also supported it "in principle."[42]

A basic assumption of the integrated long range plan was that the space shuttle and the space station would be built concurrently. The Apollo Applications Program would continue until the next phase began in the mid-1970s, and Mueller said, "it is essential that we do so ... So I am urging that we fully fund the space station and space shuttle in this next year in order to be sure that we do early on the kinds of studies and the kinds of trade-offs that will permit us to be certain later that we can carry out the program." But he cautioned that its success depended on commonality and reusability. Although engineers could always design perfect hardware to achieve a specific goal, the success of the plan depended on not sub-optimizing for single tasks but instead building systems to perform multiple missions. The space shuttle was designed for flexibility. He said it was "just a cargo plane" with a capacity to deliver 50,000 pounds of cargo into low Earth orbit. Similarly the space-tug and nuclear shuttle designed to do several things, would not be thrown away after each use. Also, "there is much to be said for the common use of subsystems as well as systems." In his final speech at MSFC, Mueller said, "I hope the Saturn Workshop is going to lead to ... a better understanding of how we design the space station module." Because of these efforts, future generations would find life on other planets – not because we find indigenous higher life forms, but because humans will be colonizing those planets. Explaining, "I expect that somehow or another during the course of the next several years the imagination of the public will be caught up with [the idea that] men ought to go to Mars ... Imagine the first step of men on Mars. It is just unbelievable that we would not take that step since we have the technology available to do so."[43]

Along with the other recognition of Mueller's work, Senator Smith praised him in a speech that she gave on the Senate floor in which she said his leaving NASA was "the end of the beginning of the amazingly successful man-in-the-Moon program." According to the senator, "to him must go the primary recognition for the tremendous achievement of that program." The nation owed him a "deep debt," she said, and "Dr. Mueller had a rare knack for translating and relating the high technical and scientific space program into words and phrases and presentations that even we non-technicians and non-scientists on the Senate Aeronautical and Space Sciences Committee could understand ... He could make even me understand." And

---

[42] Senior Staff OMSF, Washington, DC, Mueller, 12/4/69; Senior Staff MSFC, Huntsville, AL, Mueller,12/8/69.

[43] Senior Staff OMSF, Washington, DC, Mueller, 12/4/69; Remarks by Mueller to NASA Senior Staff, MSFC, Huntsville, AL, 12/8/69.

she closed by saying, "I really do not think that anyone will be able to fill the shoes of Dr. Mueller."[44]

V

Having the support of Vice President Agnew did not prove sufficient to keep human spaceflight on track, because he did not carry much weight in the White House. In March 1970 the Nixon administration scaled back the recommendations of the Space Task Group, deferring the space station, though they opted to build the space shuttle. As Mueller recalled, "there were riots going on all over the country and spending money on space was regarded to be as taking money away from the poor people who needed it on Earth. So if you can put a man on the Moon, why can't you take care of the riots in New York City, or feed all those hungry people who aren't working." So, while the space program provided many rewards, there were no constituents in space. The space program had a "fair number" of supporters, although they were not as influential as the people opposing the war and promoting social welfare programs. "It was an exciting and terrible time," he said. The president used the space program while not actually supporting it. "Nixon had decided he was not going to be known in history for his space activities . . . [And] he wasn't going to make that his legacy." However, Mueller added, "It sure would have been a better legacy than [the one] he had . . . The lesson one learns from that is that no president is going to try to make the legacy of the preceding president more than his own. And so Nixon had to create his own legacy and he wasn't going to spend money keeping . . . Johnson's legacy going." When the president officially responded to the Space Task Group's report in March, he said: "Space expenditures must take their proper place within a rigorous system of national priorities. What we do in space from here on in must become a normal and regular part of our national life and must therefore be planned in conjunction with all of the other undertakings which are important to us." As Mueller recalled, "I knew that there would be sustained support in Congress," albeit on a "fairly level" basis. It would not be a sprint to Mars, but a marathon stretched out over a longer period of time. But in the end, the war, domestic problems, high inflation, and a president with little interest in space doomed NASA's integrated long range plan.[45]

Mueller continued to advocate building the fully reusable space transportation system, while the nation lost the opportunity which briefly existed in 1969. Human spaceflight would not be Nixon's program, and he would not spend political capital supporting it. Ironically, Mueller concluded, Apollo's very success contributed to the lack of support for the post-Apollo program because Congress and the president

---

[44]   "Retirement of Mueller from NASA," *Congressional Record-Senate*, 12/2/69, Smith, 1966-1969, GEM-43-10.

[45]   McDougall, *The Heaven and Earth,* 421; Mueller Interviews: NASM, 5/1/89 and 11/8/88 and Slotkin, 6/8-9/10; Logsdon, *Exploring the Unknown*, 436.

thought "we had done everything" worth doing, and it took several years before they recognized the need for a follow-on to Apollo. Yet despite its popular appeal, the administration remained reluctant to proceed with the space shuttle. Finally, they worked out an internal compromise to build it at a reduced cost. Congress cut the space shuttle budget before work began and, Mueller argued, "every time you cut a budget, particularly before you start, you end up with a situation that is unattainable." NASA agreed to proceed with a reduced budget, which forced a switch to a partially reusable shuttle, and unfortunately it failed to recognize the financial impact of doing so. Mueller told James C. Fletcher, his first boss at Space Technology Laboratories, who in 1971 succeeded Paine as NASA administrator, it would be "a disaster and he ought not to do it," but Fletcher "decided to go forward with it." Mueller considered it a mistake to accept a reduction in development funding which would essentially reduce the capabilities of the vehicle and raise its operating costs. Fletcher "would have been better off just not doing it," he argued. Sticking with expendable launchers and using the Saturn V would, "in the long run," have been a better solution. So once Nixon got all of the publicity he could from the space program, he abandoned it and moved on to other priorities.[46]

---

[46]   Mueller Interview, Slotkin, 2/23/11.

Epilogue

"Man needs frontiers."

Mueller, July 17, 1969

In December 1969, Mueller joined General Dynamics, a company whose financial difficulties were compounded by serious problems with several of its aircraft development contracts. As federal spending for defense and space R&D declined, revenue fell and in 1970 the company showed a loss. A hands-off manager, Roger Lewis hired Mueller to help solve some of the company's technical problems, and on December 11 the new vice president for corporate development reported for work at General Dynamics headquarters in New York City. On several occasions during his short stay at the company he met with Teague, and in mid-April 1970 he shared his ideas about the administration's plans for human spaceflight. Writing in support of space shuttle funding, he told Teague, "this single development is the key to all of our future space activity, both manned and unmanned ... Reusability is a prime factor. At least 100 round trips into space will replace one time use on a one way trip of all present launch equipment." He called the space shuttle key to the nation's defense, and to the control and utilization of space. And, he argued, "the space shuttle will save billions of dollars ... It will be a barrier to technological surprise. It will be an effective shield for our national security."[1]

After leaving NASA, Mueller continued to go to the Cape to witness Saturn launches. However, no longer an agency official, he sat in the VIP/Press stands across the river from the launch pad, which he referred to as "the boon docks." Speaking at a space congress held in nearby Cocoa Beach before the launch of Apollo 13, he criticized plans to build the partially reusable space shuttle, saying, "the original space agency plans called for the development of a reusable spacecraft to greatly reduce the cost of getting men and supplies in and out of orbit ... It doesn't make sense to build a train where you make the engine reusable but where

---

[1]　GD Press releases, 12/1/69, GEM-122-9; "Henry Crown, Industrialist Dies...," *NYT*, 8/16/90; Mueller Interview, Slotkin, 6/9/10; Mueller to Teague, 4/15/70, GEM-111-5.

you throw away the freight cars each time you use it." To use "a rocket that dropped into the sea after each launch would cause the whole effort to lose its significance."[2]

Perhaps he missed leaving the agency, although probably not when Apollo 13 ran into trouble on its way to the Moon. But he said the problems had nothing to do with the design or construction of the spacecraft, they occurred due to human error. And, "we had looked [at] all of those contingencies before Apollo 11 and my immediate reaction was well there are enough things they can do so they will get back." As he explained it, "Some guy left a heater on when it was supposed to be off and ... [this subsequently] caused an explosion." Nonetheless, in a letter to his mother dated April 28, he wrote, "As you can guess, I spent some very tense hours and a few very sleepless nights, but now that they are back we have learned why it all happened. I was down at the Cape for the launch ... It was a very strange feeling to be there strictly as a spectator for the first time!" Yet from the tone of his letter, it appears that he adjusted to life outside of NASA, or at least that is what he wanted his mother to believe. In the fall of 1970, Roger Lewis reorganized General Dynamics. He put six divisions under Mueller, promoted him to senior vice president, and announced that the reorganization would take effect in December. With expanded responsibilities, Mueller returned to his whirlwind ways. He visited plants, reorganized divisions, met with customers and members of his management team, and established a new reporting structure. To support his efforts, he brought Skaggs from NASA to New York as his director of management and operations. In all, he assembled a staff of ten to manage eighty percent of the company. However, just as he began his new job, Henry Crown, who owned a majority interest in General Dynamics and controlled the board of directors, shook up the top management. On October 22, David S. Lewis, the former president of McDonnell Douglas Corporation, replaced Roger Lewis (no relation) as CEO. Caught totally unaware, Mueller had a new boss. And as it turned out, this Lewis, a hands-on manager, did not want anyone between him and most of the company. As Mueller remembered, "that was pretty clearly going to be a real problem," and it only "took Dave Lewis [a while] to decide he wanted to have me disappear."[3]

As it turned out, a board member of the System Development Corporation visited Mueller and asked if he would be interested in leading that firm at around the time Lewis told him "well you really ought to think about getting another job." He knew that SDC had developed the software for the North American air defense system, and thought highly of their work, but did not know many details about the company. Unlike his job search prior to leaving NASA, this time he did not cast a wide net to evaluate his employment options; he liked the company and they liked him, though he did not know much about their business or financial condition. And after meeting with the SDC board in early February, they offered him the position of

---

[2] Mueller Interview, Slotkin, 6/9/10; "Confidence Voiced in Space Program," AP, and "Importance Cited of Shuttle System," *The News Courier*, 4/24/70, GEM-111-5.

[3] Mueller Interview, Slotkin, 6/9/10 and 2/22/11; George to Mums, 4/28/70, GEM-111-5; Mueller to Skaggs, 9/17-30/70, GEM-111-10; Skaggs to Slotkin, e-mail, 6/3/11.

chairman, president and CEO. It was a tempting opportunity, albeit a much smaller company than General Dynamics. However, after his recent experience he wanted to be his own boss. And as he recalled, it was a "challenge" and "quite an awakening."[4]

Then on January 27, 1971, the White House surprised Mueller by announcing that he would receive the National Medal of Science, becoming just the fifteenth engineer so honored. William D. McElroy, director of the National Science Foundation which administered the award, called it well deserved: "The National Medal of Science is the Nation's premiere recognition for distinguished science achievements, and your work is richly deserving of this honor." Mueller later learned that he received the award because two former Bell Labs colleagues, Edward E. David, Jr. and Harald T. Friis, the former previously Nixon's second science advisor, appreciated his work at NASA. At the White House ceremony on May 21, the president presented the medal, reading a citation saying: "For his many individual contribution to the design of the Apollo system, including the planning and interpretation of a large array of advanced experiments necessary to insure the success of this venture into a new and little known environment."[5]

With an enhanced reputation, Mueller arrived in Santa Monica, California to lead SDC with a staff of 2,000 software developers. His task was to make the small not-for-profit company a commercial success. As he later explained, taking over and converting SDC to a profit making company became "a greater challenge than trying to get Houston and Marshall working together." And his first task involved building a team, because, he said, "There are no more dedicated individuals than computer scientists." However, a few days after arriving he asked himself "what have I gotten myself into?"[6]

Mueller promptly began recruiting and hiring new managers with profit making and marketing experience to augment the company's technical capabilities. As he said, "We knew we had a solid technical management ... Now all we needed were people who knew how to make profit." He also brought in a number of key associates from NASA and the air force, and one of his first hires was Skaggs, who would earn a reputation at the company for action. He hired other talent to augment the old hands on board, and by March 1972 he had a number of executives with computing and aerospace backgrounds at SDC. He also recognized and promoted some existing SDC managers, mainly to fill technical and administrative management positions. The combination of SDC veterans and the new hires resulted in an eclectic group of people at the top of the company; and as Mueller recalled, SDC was "a fairly eclectic company."[7]

---

[4] Mueller Interview, Slotkin, 6/9/10.

[5] "National Medal of Science," NSF, nsf.gov; Drury to Colleagues, 4/23/71, NSF, GEM-213-5; McElroy to Mueller, 2/2/71, GEM-112-3; Mueller Interview, Slotkin, 6/9/10; National Medal of Science, 1/28-5/21/71, GEM-113-15.

[6] Mueller Interview, Slotkin, 6/9/10.

[7] Baum, *System Builders,* 165-167; Mueller Interview, Slotkin, 6/9/10; Mueller to Employees, James B. Skaggs Elected SDC President, 1/29/81, GEM-126-9.

Mueller continued his involvement with the civilian space program as a consultant to NASA, although SDC did not win significant business with the agency until after he retired. He served as a member of the Air Force Systems Command Advisory Committee for two years when Phillips became its commander in 1973, and joined the Air Force Studies Board, of which he remained a member until his retirement from SDC in 1983. These voluntary assignments required major time commitments, but it made good business sense because the air force was SDC's main customer. He also became a member of the National Security Agency Scientific Advisory Board, advised the CIA, and joined the Defense Communications Agency Scientific Advisory Group. In addition, he contributed to a National Research Council study on Space Solar Power.[8]

Interested in helping to resolve the energy crisis in the 1970s, Mueller became a member of the congressional Office of Technology Assessment's Energy Advisory Committee, reviewed the Carter administration's energy plans, and advised Congress about the proposed National Energy Act, nuclear proliferation, application of solar energy, recovery of oil, and other energy issues. Working with the Office of Technology Assessment he evaluated plans for various projects in the late 1970s, and in the early 1980s participated in a congressional assessment of the space station as a member of the Civilian Space Station Advisory Panel.[9]

Throughout the 1970s, Mueller attended the annual International Astronautical Congress and chaired the 1976 congress in Anaheim, California. The event attracted astronauts and cosmonauts – including the crews of the recent Apollo-Soyuz Test Project – and space agency heads from around the world, along with almost one thousand space scientist and engineers from forty-three countries. The meeting turned out well, enhancing Mueller's role with the IAF and its US member, the AIAA. As a result of this success he agreed to take on leadership roles at subsequent congresses. In 1978, AIAA's members elected him president for 1979-1980. And he attacked this volunteer job like everything else he did, using the position to crisscross the nation, attending meetings, making speeches, and contributing to their monthly magazine *Aerospace America*. In July 1979 he wrote in an editorial, "It was ten years ago this month that man first stepped on the Moon. That exciting moment was the culmination of the most remarkable and sustained engineering program in history." He congratulated those responsible, and proposed using the Apollo 11 anniversary "to stimulate a resurgence of interest in our space program." Also around that time, he published another article in *Aerospace America* with the message that "America relinquished its commanding post-Apollo potential and now awaits the shock of

---

8   Correspondence, Thomas O. Paine, 1981-1982, GEM-128-15; Space Station Task Force, 9/2/82, NASA, GEM-201-12. (In 1986, three years after Mueller retired, SDC won a role as the software integration subcontractor to Rockwell International, amounting to $800 million over fifteen years on the Space Transportation Systems Operations Contract.)

9   Mueller Interview, Slotkin, 6/9/10; Mueller Testimony, OTA, Washington, DC, 6/14/76, SDC Speeches.

another major Soviet achievement to create public backing for major space missions."[10]

On July 17, 1979, ten years and one day after the Apollo 11 launch, Mueller addressed an AIAA meeting, saying "the success of Apollo richly rewarded the American people." He said that the reception given to the Apollo 11 crew upon their return to Earth was comparable to Charles Lindbergh's triumphant return to the US in 1927. Yet in the ten years since that lunar landing, the nation "drew back from the promise of space," and he found this discouraging. He believed the pullback to be temporary, though the country remained depressed in the aftermath of the Vietnam War, a stagnant economy, and record inflation. However, he said that a new space program could once "again provide a sense of purpose for all of us … Space is man's manifest destiny. By accepting this destiny, we reaffirm our uniqueness as a species, and we open the door to unimagined potentialities." He expanded on this theme in Huntsville, in a lecture honoring von Braun, who died in 1977. Quoting an article that his colleague wrote in 1949 forecasting the exploration of Mars, he said that if von Braun had thought of it, he would have said "Space exploration is our manifest destiny." He spoke of Fredrick Jackson Turner's frontier thesis, and added, "I am not constrained by professional niceties of the historian, I am perfectly willing to take that extra step and extend Turner's thesis universally. Man needs frontiers … [If] a civilization draws back from a frontier because of fear of the unknown, it will inevitably decay." He called space "the greatest frontier of all and it offers unlimited potential for mankind." Following this speech, the *Huntsville Times* quoted Mueller saying, "Only when we gain the knowledge from the exploration itself can we gain a glimmer of the ultimate impact of that exploration. Thus, support of any true exploration must be an act of faith."[11]

II

Mueller rebuilt SDC with Skaggs' help, and ten years after becoming a profit seeking company it had increasing revenues and profits. Meanwhile, the not for profit System Development Foundation that owned most of the company stock was eager sell its shares. Although the foundation had converted small amounts of equity into cash by several transactions over the years, it still owned two-thirds of the company. In 1979, Mueller recommended that SDC's senior management seek a merger partner because attempts to go public had not been successful.[12]

---

[10] IAF, GEM-173 to 175; Draft Editorials, AIAA, 1979, Writings, GEM-241-14; Correspondence, Apollo 11, 7/16/74- 3/20/79, GEM-192-10.

[11] Mueller, AIAA Antelope Valley Section, CA, 7/17/79, and "Manifest Destiny," University of Alabama, Huntsville, AL, 7/18/79, SDC Speeches; "Ex-Space Program Leader Calls Exploration a 'Manifest Destiny," *Huntsville Times*, 7/19/79, Newspaper & Magazine Articles, GEM-215-11.

[12] Baum, *System Builders*, 259-285; Mueller to distribution, Draft Criteria, 1/16/80, GEM-154-2.

Fletcher, who had stepped down as NASA administrator in 1977, sat on the Burroughs Corporation board of directors, and in early 1980 he contacted Mueller to suggest that Burroughs buy SDC. Mueller turned this over to the foundation, and Burroughs began discussions to acquire all of SDC's shares. A cash offer for $98 million was made on August 27, and the deal was closed on January 5, 1981. The foundation received $66 million in cash for their portion, nine times the value of SDC's equity when Mueller had arrived in 1971. SDC had almost $200 million in revenues, more than four times the amount in 1971, and during ten years under his leadership the company had grown, diversified and become profitable. So at 63 years of age, he became chairman and CEO of SDC, operating as a Burroughs subsidiary, and Skaggs became its president and chief operating officer.[13]

MIT's Charles Stark Draper, 80 years of age in 1981, stepped down as president of the IAA and handpicked Mueller as his replacement, naming him acting president in December. At that time, the academy did little more than organize sessions at the annual International Astronautical Congresses, publish a scholarly journal, and honor members with diplomas. Mueller wanted to breathe new life into the organization; no mean task considering it had limited assets with greater liabilities. In 1982 Mueller officially succeeded Draper and, like his predecessor, remained president for many years, finally stepping down at the age of seventy-seven in 1995. After his election in 1982 Mueller reinvigorated the academy, making it more dynamic with an expanded and active membership. He selected new candidates to fill leadership roles, and grew the membership to include some of the foremost workers in the field of astronautics. Being president of the IAA gave him new speaking platforms, and over the next thirteen years he frequently spoke about the future of space exploration as he traveled the world. He expanded the academy's reach by conducting meetings jointly with national academies of sciences, improved its finances, and began holding standalone specialized conferences. He enlivened the academy, decreased the average age of the academicians, and accomplished most of the objectives that he set himself. However, as his wife Darla once told your author, his involvement with the IAA is "only a footnote" in his extremely successful life.[14]

In 1982 the original SDC had a revenue exceeding $263 million and a pre-tax profit of almost $23 million. Financial results continued on plan in 1983, but later in the year, shortly after his sixty-fifth birthday, Mueller retired from the company. At that point in his life he did not want to run another firm. When he married Darla J. Schwartzman (née Hix) in 1978, she had two pre-teen children, and the new Mueller family moved to Santa Barbara into a residence that overlooked the Pacific. After retirement Mueller devoted himself mainly to his new family, but he remained involved in outside activities and worked with the University of

---

[13]　Mueller Interviews: Slotkin, 2/22/10, 6/8-9/10; Mueller to SDC Board, 8/27/80, GEM-154-3; Baum, *System Builders*, 259-285.

[14]　Mueller Interview, Slotkin, 2/24/10; Draper to Mueller, 12/3/81, GEM-177-9; Speeches, IAA, 3/86-5/91, GEM-268-9. Mueller appointed Jean-Michel Contant and Arthur L. Slotkin as IAA co-secretaries.

California at Santa Barbara in human system research, a project which occupied him for several years.[15]

In 1986, in the wake of the *Challenger* accident and the report of the Rogers Commission, Fletcher returned as NASA administrator and awarded the National Academy of Public Administration a contract to study the effectiveness of agency's management. NAPA tapped Phillips to lead a study group which included Mueller, Mathews, Skaggs, Lilly, and other NASA and industry Apollo alumni. At one of their first meetings, Mueller made several observations consistent with his approach to program management during the Apollo Program. Notes from September 23, 1986 quote him saying, "Any ability to trace decisions and updates between centers is now purely coincidental ... NASA must clearly define interfaces and program information transfers and linkages." He expressed concern that the agency had "lost sight of the concept of roles and missions," allowing the center directors freedom to select the work they wanted to do. And, he argued, "NASA needs to develop institutional – and program – based strategic plans with a focus that goes beyond next year's goals." He continued working with the NAPA group, studying the agency's organization and management until they issued their final report in early 1988, which said that NASA needed to improve its approach to policy development and recommended separating development of new programs from day-to-day operations. This project gave Mueller a close look at the agency once again, and he visited its major facilities, receiving briefings about the agency's organization and management. Unlike other studies of its kind, this one had an impact when NASA relocated the International Space Station Program Office from Houston, the "lead-center," to a centralized program office in Reston, Virginia. Another of Mueller's ideas involved commercializing the space shuttle in order to "offload it from NASA and put it into commercial operation." He discussed this at headquarters and in Houston, recalling "they supported the idea which was surprising to me." However, in the end, nothing came of it.[16]

In July 1994, the nation celebrated the twenty-fifth anniversary of Apollo 11 and Mueller participated in some events commemorating the mission. Newspapers and magazines published articles looking back to the first lunar landing and forward to planetary exploration, including a mission to Mars. While many of these articles highlighted the accomplishments of the astronauts, and referred to von Braun and several of the others who contributed, very few mentioned Mueller. The nation had nearly forgotten the man who successfully managed human spaceflight during the race to the Moon. In a letter that Mueller wrote for publication at this time, he reminisced about the first lunar landing, and credited nine men with the accomplishment: Phillips, von Braun, Rees, Debus, Petrone, Gilruth, Low, Rudolph,

---

[15]   Mueller Interview, Slotkin, 2/24/10.

[16]   Correlation Meeting – 9/23/86, NASA Management Study Group, GEM-298-1; "Phillips, Samuel C. to Mueller, 3/8/88, "Effectiveness of NASA Headquarters," GEM-297-2; Space Station Program Office Status, December, 1987, GEM-302-3; Mueller Interview, Slotkin, 2/22/11.

and Shea in that order (all of whom he had included on his list of thirteen in 1969). He wrote about the Apollo executives group and STAC, and concluded by saying, "These are only a few of the 250,000 men and women who worked directly on the Apollo Program. Every one of them is a hero in their own right."[17]

Involved with a number of startup companies throughout the 1970s, 1980s and 1990s, Mueller began his association with the Kistler Aerospace Corporation in February 1995, when Robert Citron and Walter Kistler visited him in Santa Barbara and asked him to join their board of directors. He told them that he had no interest in becoming a board member unless they made him CEO, something that surprised them. He did not want to get involved with Kistler's new K-1 launch vehicle without being given control of its design. After learning that Kistler shared his dream of building a fully reusable launch vehicle, he saw the K-1 as an opportunity to develop "that first and most important link from the Earth's surface to [low] Earth orbit," he said. And while he had remained busy since retiring from SDC, he felt ready to return to full time work once again. After Kistler and Citron had thought his proposal over, they agreed to make him CEO.[18]

So in April 1995, at the age of seventy-seven Mueller became CEO of Kistler Aerospace Corporation, initially signing on for three years, which stretched to eight. For a modest salary, he found himself the head of the startup company, and relocated to Kirkland, Washington. The design team that he found when he arrived consisted of "an interesting group of individuals," but according to Mueller, none of them could design a reusable space vehicle. He realized they had problems because a major part of the development involved building new engines, although he said "when we began to test the engines they couldn't control the thrust accurately enough to be able to fly it." And after trying for some time, they could not stabilize the engines. Then things "got more interesting," he said, and he spent a year trying to make the original design work before replacing the designers.[19]

The new design team included people that Mueller felt "really understood this work." One of the first people he brought in was Aaron Cohen, former director of the Johnson Space Center. In turn, Cohen brought Henry O. Pohl, an expert in engine design who had retired as chief engineer for the International Space Station. Cohen and Pohl helped to attract others, and Mueller hired Myers, the man who replaced him when he first left the agency and who had retired as deputy administrator of NASA in 1989. With this help, Mueller said, "we began to really try to understand what Bob Citron and Walt [Kistler] had put together and decided that it was unlikely to succeed. So we took off and tried to do something that would work." Back in the space launch business, Mueller set out to do what he first planned at NASA, to build a completely reusable launcher because he

---

[17] Apollo Program, Twenty-Fifth Anniversary, GEM-296-1; Names he left off the 1969 list: Bergin, Stoner, Richard and Kraft.

[18] Gottfried to Mueller, 2/4/95, GEM-292-6; Mueller Interview, Slotkin, 6/10/10.

[19] Correspondence, 1995, GEM-292-6; Mueller Interviews: Slotkin, 6/10/10 and 2/22-23/11.

believed the development of the communications satellite business finally justified the need.[20]

In 1996, Kistler Aerospace Corporation had discussions going with a number of potential investors, strategic partners and contractors, and development and testing of the K-1 was set for completion by 1999. The company won its first commercial order for ten launches as part of Motorola's Iridium venture from Space Systems/ Loral worth more than $100 million, which helped in their search for additional funding. Mueller spent time on design work and attending design reviews, while Kistler's founders pursued funding. His strategy involved using major aerospace companies as subcontractors in order to assure investors and customers of the company's viability. For example, in 1997 he hired Lockheed Martin to build the K-1's fuel tanks. And as they progressed towards initial launch date, Northrop Grumman invested $30 million, with a promise of a similar amount in 2000 "if the company proves it has the financial wherewithal to proceed with its first launch test," according to the *Los Angeles Times*.[21]

The design of the K-1 did not present major problems, but getting the funding to build it did. "The real problems we have encountered," Mueller said in 1999, "are not the design, because we had a pretty competent crew of designers ... probably the best design team in the country." The challenge involved building something reusable that was also simple and inexpensive to operate. Because, "in the long run it's the cost of operations that determines the success or failure of any program," he explained. He anticipated using a ground crew of about sixty people, with a turnaround time of nine days. And to do that, they had to look at each part of the vehicle to insure that it was simple, foolproof, and reliable. A sophisticated K-1 health monitoring system would inform them what needed to be done after each flight in order to get the vehicle ready for the next mission.[22]

The company raised $400 million in capital and was well into development when hit by the Asian financial crisis in 1997. Motorola's Iridium venture went bankrupt in 1999, and many of Kistler's foreign investors backed away. Since most of the money came from overseas, mainly from Asia and the Middle East, that spelled trouble, and everything started to fall apart. Without funding to finish the K-1, the company was impacted by marching army costs, which sapped the remaining funds. Although they eventually raised a total of about $700 million, much of it was wasted in supporting the staff while seeking additional funds, or was paid off as subcontractor overheads. But for these delays, Mueller believed they could have finished the K-1. They had planned to start flight testing in 1998 and conduct their first commercial launch the following year, but that timetable proved unachievable

---

[20] Mueller Interviews: Slotkin, 2/22/11 and JSC, 1/20/99.

[21] Mueller Interview, Slotkin, 2/22/11; "Lockheed Gets Kistler Rocket Tank Contract," 6/ 18/97, "Coming Along at Kistler," 9/11/98, and "Northrop Takes $30-Million Equity ..." 3/19/99, *LA Times*.

[22] Mueller Interviews: JSC, 1/20/99 and Slotkin, 2/22/11.

without additional funds. As Mueller put it, Kistler Aerospace Corporation "always were on the brink of success," and "we came close to succeeding."[23]

Unable to recover from its financial difficulties, the company filed for bankruptcy protection in July 2003. Mueller tried to shepherd them through reorganization, and kept the developers working. In 2004 he finally stepped down as CEO, although he remained chairman and chief vehicle architect. Despite all the money raised, in the end the company fell $100 million short and could not raise additional funds. Mueller remained with Kistler until he got sidelined with health issues that kept him out of action for several years. Nonetheless, he insisted, "the whole secret of a successful launch system is to reuse the parts." In 2011, he still believes current approaches using expendable vehicles are too expensive, and it will take at least another generation of space vehicles before the commercial space transportation business will succeed.[24]

III

The Introduction stated two objectives: first to describe Mueller's contributions to human spaceflight, and second to provide a narrative of how he managed Gemini, Apollo and post-Apollo programs at the same time. He applied system engineering to program management to manage human spaceflight. He defined system engineering as a "discipline which involves all of engineering ... applied to a particular system." And he regarded system management as a "structure for visualizing all the factors involved as an integrated whole" – in other words, it was the application of system engineering to program management. He considered his greatest contribution to Apollo to be the program management system which he introduced, a modified version of the air force system program office methodology. With Phillips's help, he imposed matrix management on the Apollo Program. The Apollo Program Office created dual reporting between the headquarters program management organization and project management offices at the centers. Each program and project office consisted of five functional organizations – the five box or GEM box organization – which included system engineering, program control, testing, reliability and quality assurance, and flight operations. The project offices remained in the centers, but the matrix separated the institutional responsibilities of the center directors from the program responsibility of the program directors, creating a decentralized system with centralized authority. At the top, Mueller reorganized the management council, and used it to manage the programs, with the program directors reporting to it. Separate from the program management system,

---

23 "Kistler Aerospace Corporation Restructures..." Press Release, 7/15/03; "Private Rockets," *Discover*, 4/99; Mueller Interviews: Slotkin, 2/22/11 and JSC, 1/20/99.

24 "Kistler Aerospace Corporation's K-1," GlobalSecurity.org; Mueller Interview, Slotkin, 2/22/11; "Kistler Aerospace Files for Chapter 11 Bankruptcy Protection," 7/23/03.

Mueller treated "external affairs," by which he meant relationships outside of NASA, as part of the overall management system, and became adept at working with the politics of the space program.[25]

The Gemini Program achieved all of its objectives, and by flying in space for fourteen days on Gemini VII "removed all doubts" about astronauts and equipment performing satisfactorily during the time needed to accomplish the lunar landing. During Gemini, NASA developed operational techniques for spacecraft rendezvous and docking critical to the success of Apollo. But only on the final mission, Gemini XII, did Aldrin prove that with proper training and equipment, astronauts were able to perform useful work while spacewalking. Mueller's main contribution to Gemini consisted of applying management pressure to complete it on time and below the cost estimates that had been established before he arrived. A major change he introduced involved converting Gemini contracts to incentive fee based on cost and schedule, a technique he also applied to Apollo. In the final accounting, Gemini cost about $200 million less than the estimate for its completion when he joined the agency in September 1963, and despite being two quarters behind at that time it was finished on schedule. Because Gemini was further along when he arrived, he was unable to implement all of the facets of the program management system he applied to Apollo. Nonetheless, the parts of it that he put in place contributed to Gemini's success. However, he never failed to give Mathews full credit for its success.[26]

In September 1963, Mueller said, "Without much improved management ... we will not achieve the lunar goal prior to 1972-1975 ... at a cost of $35 billion or more." But using the program management approach from the air force ballistic missile program, he set out to impose "the right set of working relationships between the centers and hold them there ... long enough so that communications could grow." In addition to the program management system, he used other management innovations. After receiving the Disher-Tischler study, he realized that he would have to change the traditional step-by-step approach to flight testing, and imposed all-up testing, a method first used in the Minuteman Program. Like most ideas, he did not invent all-up testing, but he applied it efficiently and used it to speed up Saturn development and cut years off of the schedule. Announced two months after arriving, all-up testing of the Saturn V began in 1968 with the flight of AS-501 (Apollo 4) after overcoming a one year delay caused by the disastrous AS-204 (Apollo 1) fire. Following the successful first Saturn V flight test, NASA faced the unsuccessful flight of AS-502 (Apollo 6). Officially a failure because it achieved only four of five major objectives, the second Saturn V was a successful flight test precisely because it revealed serious problems in the booster. After correcting these deficiencies, the agency flew AS-503 successfully and sent Apollo 8 to orbit the Moon.[27]

---

[25] Mueller Interview, Slotkin, 9/9/09; Mueller, Joint AIAA/CASI Meeting, Montreal, Canada, 7/8/68.

[26] Mueller, Gemini Conference, Houston TX, 2/1/67.

[27] "Organizational Meeting," 9/4/63, O&M, GEM-84-11; Mueller Interview, 9/9/09.

Concurrency, another management innovation brought to NASA from the air force ballistic missile program, involved the parallel development of systems; and when Schriever led the air force ballistic missile program he developing multiple missiles and major subsystems in parallel in order to open options. Cost and schedule considerations prohibited developing concurrent boosters for the Apollo Program, but Mueller used concurrent development of major subsystems at NASA to guard against "show stoppers". Another special consideration borrowed from the air force led him to establish the two industry executives groups to improve communications between the agency and its major contractors. He kept Congress informed through monthly briefings, and at the suggestion of his colleague from Bell Labs Charles H. Townes, he established the Manned Space Flight Science and Technology Advisory Committee to improve relations with the science community and serve as a sounding board for his ideas. Composed of some of the most prominent space scientists, medical doctors and engineers in the nation, Mueller credited STAC with major contributions, but he faced opposition from scientific community throughout the Apollo Program. These "special considerations," as he called them, contributed to the success of Apollo by improving communications.

Mueller established an advanced programs organization to develop a post-Apollo program as part of the original reorganization of the Office of Manned Space Flight in November 1963. Initially calling the post-Apollo program the Apollo Extension Systems, NASA later renamed it the Apollo Applications Program and still later, as its focus tightened, the Skylab Program. Planning for the post-Apollo period took place from the day that he arrived, but not until Apollo-Saturn flight testing began in 1967 did he focus on what he called the space transportation system. As the agency experienced budget pressures, Mueller narrowed the focus of post-Apollo planning. However, after Webb's departure he returned to planning and talking more broadly about the future of spaceflight. In a speech that he made in London in August 1968, he unveiled ideas about building a reusable space shuttle to travel between Earth and a space station in low orbit. Further planning led to a system that included multiple space stations which would be serviced by different types of space shuttles, including a nuclear powered interplanetary shuttle. Mueller's promotion of the space shuttle, another idea he borrowed from others, earned him the sobriquet "father of the space shuttle." However, he did not believe the space shuttle should stand alone – it was to be part of an elaborate space transportation system, like a "railroad" in space; and to be cost effective, it required to be fully reusable.

My second objective in writing this book was to interweave the story of Mueller's work on Gemini, Apollo and the Apollo follow-on programs into a single narrative, just as he lived them on a day-to-day basis; a goal which I will leave to the readers to verify.

Over the years Mueller met many people who not only impacted his career, but remained friends and played important roles in his professional life. He said that he never planned his career, "it just happened." He changed jobs, going from Bell Labs, to Ohio State University, to Space Technology Laboratories, to NASA. And after NASA he led the System Development Corporation until retirement. He kept himself busy and useful in retirement, devoting time to government and industry

E-1 George E. Mueller receives the Lifetime Achievement Trophy from the Smithsonian National Air and Space Museum, April 2011. (NASM photo)

boards and committees, and spent an enormous amount of effort rebuilding the International Academy of Astronautics, stepping down as president at age seventy-seven only to return to full time employment with Kistler Aerospace Corporation. He said none of these moves were planned, "it just was a set of circumstances that led to that being the logical thing to do at that time." Reflecting in 2011, he said, "I never would have imagined I'd be in aerospace, although when I started as an undergraduate, I was thinking in terms of aeronautical engineering." He did not get involved in the space business until he arrived at Ramo-Wooldridge/Space Technology Laboratories, but it came to dominate his life. He called his careers "very interesting," all of them.[28]

In April 2011 the Smithsonian National Air and Space Museum in Washington recognized Mueller with the 2011 Lifetime Achievement Trophy for contributions to human spaceflight. At ninety-two years old, Mueller said, "Looking at it today, one of the fundamental drives for humanity is spreading from this small Earth into the rest of the solar system and the rest of the galaxy – and from there into the universe." In accepting the award, he turned even more philosophical, and said, "I believe that men are going to live and work in space, and are going to explore and colonize the Moon as a stepping stone to establishing an outpost and then a colony on Mars ...

---

[28] Mueller Interview, JSC, 1/20/99.

As we build this new civilization and become citizens of the solar system I believe we will be building a better life for all men and, at the same time, building the capability required to men to go to the stars."[29]

---

[29]   Mueller Interview, Slotkin, 2/22/11; Mueller, Smithsonian Speech, 4/20/11.

Bibliography

PRIMARY SOURCES

Library of Congress, Manuscript Division, Washington, DC.[1]

The Papers of George E. Mueller (Indicated as GEM in footnotes).
The Papers of Samuel C. Phillips (SCP).
The Papers of Thomas O. Paine (TOP).

Interviews[2]

Alphabetical by interviewee last name (Indicates reference in footnotes)

Apollo Program Oral History, July 21, 1989, NASA Headquarters Archives Historical Collection, Washington, DC, Reference Number 18924, (Apollo Oral History).

Atwood, J. Leland, National Air and Space Museum, Martin Collins, Washington, DC, February 15, 1988, January 19, 1989, August 25, 1989, January 12, 1990, and June 25, 1990, (NASM).

Burdett, James Robert, National Air and Space Museum, Washington, DC, Martin Collins, July 19, 1989 and January 10, 1990, (NASM).

Cohen, Aaron, Lloyd Swenson, NASA Headquarters Archives Historical Collection, Washington, DC, Location XII/D/5, January 14, 1970, (Swenson).

Dannenberg, Konrad, Tom Ray, NASA Headquarters Archives Historical Collection, Washington, DC, Reference Number 0438, April 6, 1973, (Ray).

---

[1] To minimize space in footnotes, I have adopted the short hand as follows: Collection-box number-folder number. For example GEM, box 100, folder 2 is listed as GEM-100-2.

[2] Interviews are listed by name of interviewee, interviewer and date of interview as follow: XYZ Interview, Interviewer, date. For example Mueller interview conducted by Slotkin on February 22, 2011 would be listed as: Mueller Interview, Slotkin, February 22, 2011.

Dannenberg, Konrad, 50th Anniversary of First Launch at Cape Canaveral, Kennedy Space Center, Interviewed by Roger Launius, et al. July 25, 2000, (Launius).

Disher, John H. Disher, Robert Sherrod, NASA Headquarters Archives Historical Collection, Washington, DC, Reference Number 0471, April 15, 1971, (Sherrod).

Disher, John H., Ivan Ertel, NASA Headquarters Archives Historical Collection, Washington, DC, Reference Number 0471, January 27, 1967, (Ertel).

Draper, Charles Stark, Robert Sherrod, NASA Headquarters Archives Historical Collection, Washington DC, Reference Number 13286, March 11, 1971 and May 17, 1971, (Sherrod).

Freitag, Robert, Robert Sherrod, NASA Headquarters Archives Historical Collection, Washington, DC, Reference Number 12973, June 11, 1969, (Sherrod).

Freitag, Robert, Robert Sherrod, NASA Headquarters Archives Historical Collection, Washington, DC, Reference Number 13286, November 4, 1969, February 18, 1970, (Sherrod).

Freitag, Robert, W. Henry Lambright, NASA Headquarters Archives Historical Collection, Washington, DC, Reference Number 07106, June 27, 1991, (Lambright).

Fulton, James G. (R-Pa.), Robert Sherrod, NASA Headquarters Archives Historical Collection, Washington, DC, Reference Number 13286, September 15-16, 1969, (Sherrod).

Gilruth, Robert, National Air and Space Museum, Washington, DC, Martin Collins, March 2, 1987, (NASM).

Hodge, John D., W. Henry Lambright, NASA Headquarters Archives Historical Collection, Washington, DC, Reference Number 07106, October 5, 1991, (Lambright).

Holmes, Jay, Robert Sherrod, NASA Headquarters Archives Historical Collection, Washington, DC, Reference Number 12973, February 9 and 11, 1972 and June 28, 1972, (Sherrod).

Hull, Harrison, Robert Sherrod, NASA Headquarters Archives Historical Collection, Washington DC, Reference Number 13286, April 6, 1970, (Sherrod).

Kline, Ray, W., W. Henry Lambright, NASA Headquarters Archives Historical Collection, Washington, DC, Reference Number 7106 February 7, 1992, (Lambright).

Kraft, Christopher C., Robert Sherrod, NASA Headquarters Archives Historical Collection, Washington, DC, Reference Number 12973, July 27, 1972, (Sherrod).

Lilly, William, W. Henry Lambright, NASA Headquarters Archives Historical Collection, Washington, DC, Reference Number 7106, July 18, 1990, (Lambright).

Logsdon, John M., W. Henry Lambright, NASA Headquarters Archives Historical Collection, Washington, DC, Reference Number 7106, November 7-8, 1990, (Lambright).

Low, George M., Robert Sherrod NASA, Headquarters Archives Historical Collection, Washington, DC, Reference Number 12973, November 7, 1969, August 12, 1970, June 21, 1972, July 5, 1972, September 7, 1972, (Sherrod).

Lunney, Glynn, NASA Johnson Space Center Oral History Project (JSC), Houston, TX, February 8, 1999.

McCurdy, Howard, W. Henry Lambright NASA Headquarters Archives Historical Collection, Washington, DC, Reference Number 7106, October 25, 1991, (Lambright).

Mathews, Charles W., W. Henry Lambright, NASA Headquarters Archives Historical Collection, Washington, DC, Reference Number 7106, October 4, 1992, (Lambright).

Mathews, Charles, Robert Sherrod, NASA Headquarters Archives Historical Collection, Washington, DC, Reference Number 12973, February 17, 1970, March 20, 1973, (Sherrod).

Mathews, Charles, W. David Compton, NASA Headquarters Archives Historical Collection, Washington, DC, Reference Number 01729, July 15, 1975, (Compton).

Mettler, Ruben, NASA Johnson Space Center Oral History Project (JSC), Houston, TX, April 7, 1999.

Moritz, Bernard, W. Henry Lambright, NASA Headquarters Archives Historical Collection, Washington, DC, Reference Number 7106, February 19, 1992, (Lambright).

Mueller, George E., Mary Bubb, May 22, 1969, Meeting Notes and Transcriptions, GEM- 79-9, (Bubb).

Mueller, George E., Martin Collins, National Air and Space Museum, Washington, DC, January 12, 1987, April 30, 1987, July 27, 1987, February 15, 1988, May 2, 1988, June 22, 1988, November 8, 1988, (NASM).

Mueller, George E., Stephen Garber, NASA Headquarters Archives Historical Collection, Washington, DC, Reference Number 33476, February 12, 2001, (Garber).

Mueller, George E., NASA Johnson Space Center Oral History Project (JSC), Houston, TX, August 27, 1998 and January 20, 1998.

Mueller, George E., W. Henry Lambright, NASA Headquarters Archives Historical Collection, Washington, DC, Reference Number 7106, September 21, 1990, (Lambright).

Mueller, George E., Howard E. McCurdy, NASA Headquarters Archives Historical Collection, Washington, DC, Reference Number 6722, June 22, 1988, (McCurdy).

Mueller, George E., Putnam, NASA Headquarters Archives Historical Collection, Washington DC, Reference Number 1522, June 6, 27, and October 4, 1967, (Putnam).

Mueller, George E., Robert Sherrod , NASA Headquarters Archives Historical Collection, Washington, DC, Reference Number 13287, November 19, 1969, April 21, 1971, August 19, 1971, September 1971, March 20, 1973, (Sherrod).

Mueller, George E., Arthur L. Slotkin, June 15-16, 2009, September 9-10, 2009, February 24-25, 2010, June 8-10, 2010, February 22-23. 2011, (Slotkin).

Mueller, George E., Paul P. Van Ripper, Cornell University, NASA Headquarters Archives Historical Collection, Washington, DC, Reference Number 16203, December 7, 1966, (Van Ripper).

Myers, Dale, Robert Sherrod, NASA Headquarters Archives Historical Collection, Washington, DC, Reference Number 13287, March 17 and 31, 1970, (Sherrod).

Naugle, John, W. Henry Lambright, NASA Headquarters Archives Historical Collection, Washington, DC, Reference Number 7106, November 7, 1991, (Lambright).

Phillips, Samuel C., National Air and Space Museum, Martin Collins, February 23, 15, 1988, January 16, 1989, August 23, 1989, and September 8, 28, 1989, (NASM).

Potate, John, Tom Ray, NASA Headquarters Archive Collection, Washington, DC, Reference Number 01729, June 6, 1972, (Ray).

Ramo, Simon, NASA Johnson Space Center Oral History Project (JSC), Houston, TX, April 6, 1999.

Ramo, Simon, Martin Collins, National Air and Space Museum, Washington, DC, January 25, 1998, June 27, 1998 (NASM).

Rees, Eberhard, Robert Sherrod, NASA Headquarters Archive Collection, Washington, DC, Reference Number 13290, February 6, 1971, (Sherrod).

Scheer, Julian, Robert Sherrod, NASA Headquarters Archive Collection, Washington, DC, Reference Number 13290, September 11, 1969, (Sherrod).

Schriever, Bernard A., NASA Johnson Space Center Oral History Project (JSC), Houston, TX, January 15, 1998 and April 16, 1999.

Schriever, Bernard A., Jane Butler, NASA Headquarters Archives Historical Collection, Washington, DC, Location I/J, April 15, 1999, (NASA).

Seamans, Robert, NASA Johnson Space Center Oral History Project (JSC), Houston, TX, November 20, 1998.

Seamans, Robert, Paul P. Van Ripper, NASA Headquarters Archives Historical Collection, Washington, DC, Reference Number 16203, December 6, 1966, (Van Ripper).

Shea, Joseph F., Eugene M. Emme, NASA Headquarters Archives Historical Collection, Washington, DC, Reference Number 02011, May 6, 1970, (Emme).

Shea, Joseph F. Shea, Donald Neff, NASA Headquarters Archives Historical Collection, Washington, DC, Reference Number 0211, February 12, 1969, (Neff).

Shea, Joseph F., Robert Sherrod, NASA Headquarters Archives Historical Collection, Washington, DC, Reference Number 13287, May 6, May 16, 1971, and March 10, 1973, (Sherrod).

Shea, Joseph F., NASA Johnson Space Center Oral History Project (JSC), Houston, TX, August 26, 1998.

Teague, Olin, Robert Sherrod, NASA Headquarters Archive Collection, Washington, DC, Reference Number 13290, April 1, 1970, (Sherrod).

Thompson, Floyd L., Robert Sherrod, NASA Headquarters Archive Collection, Washington, DC, Reference Number 13290, September 10, 1969, (Sherrod).

Ullberg, Alan D., Robert Sherrod, NASA Headquarters Archive Collection, Washington, DC, Reference Number 13290, February 4, and August 22, 1969, (Sherrod).

Vogel, Lawrence W., Robert Sherrod, NASA Headquarters Archive Collection, Washington, DC, Reference Number 13290, October 23, 1969, (Sherrod).

von Braun, Wernher, Robert Sherrod, NASA Headquarters Archive Collection, Washington, DC, Reference Number 13290, November 19, 1969, (Sherrod).

Webb, James E., Robert Sherrod, NASA Headquarters Archives Historical Collection, Washington, DC, Reference Number 13290, August 2 and 8, 1968, November 15, 1968, June 8, 16 and 18, 1969, April 28, 1971, (Sherrod).

Webb, James E. and Thomas O. Paine, Robert Sherrod, NASA Headquarters Archives Historical Collection, Washington, DC, Reference Number 13287, October 7, 1971, (Sherrod).

Yarymovych, Michael, Charles D. Benson, February 2, 1976, NASA Headquarters Historical Collection, Washington, DC, Reference Number 02908, (Benson).

Note: All NASA JSC oral histories in his book are available on line at the following address, http://www.jsc.nasa.gov/history/oral_histories/participants.htm, *accessed June 20, 2009.*

OTHER PRIMARY SOURCES: NASA HEADQUARTERS ARCHIVES HISTORICAL COLLECTION, WASHINGTON, DC

White House Documents (By date):

"Statement by James C. Hagerty," The White House, July 29, 1955, NASA Headquarters Archives Historical Collection, Washington, DC, Reference Number 12377.

Dulles to Hagerty, October 8, 1957, NASA Headquarters Archives Historical Collection, Washington, DC, Reference Number 12401.

"Statement by the President: Summary of Important Facts in the Development by the United States of an Earth Satellite," October 9, 1957, NASA Headquarters Archives Historical Collection, Washington, DC, Reference Number 12400.

"Discussion at the 339th Meeting of the National Security Council, Thursday, October 10, 1957, NASA Headquarters Archives Historical Collection, Washington, DC, Reference Number 12400.

Eisenhower to Killian, November 22, 1957, NASA Headquarters Archives Historical Collection, Washington, DC, Reference Number 12306.

Eisenhower to Johnson, January 21, 1958, NASA Headquarters Archives Historical Collection, Washington, DC, Reference Number 12401.

"White House Press Release," March 27, 1958, NASA Headquarters Archives Historical Collection, Washington, DC, Reference Number 012376.

"Long-Range Ballistic Missiles – A History," October 21, 1958, White House Central File, NASA Headquarters Archives Historical Collection, Washington, DC, Reference Number 012405.

Eisenhower to Swenson, August 5, 1965, NASA Headquarters Archives Historical Collection, Washington, DC, Reference Number 12377.

NASA Documents:

Hugh L. Dryden, "The National Aeronautics and Space Administration," presentation to the Air Force Association, September 26, 1958, NASA Headquarters Archives Historical Collection, Washington, DC, Reference Number 18105.

Minutes of the Science and Technology Advisory Committee for Manned Space Flight (STAC), 1964-1969, NASA Headquarters Archives Historical Collection, Washington, DC, Reference Numbers 2358 for the years 1964-1969 and 19971 for 1965.

Mueller Speeches: (unless otherwise indicated Mueller speeches are in NASA Archives, Mueller speech collection, listed by date of speech)

"Manned Space Flight – Where Do We Stand?" American Institute of Aeronautics and Astronautics, Washington, DC, November 6, 1963.

"Introduction by Dr. G. E. Mueller," NASA Saturn Launch Vehicle Program Review, November 9, 1963, NASA Headquarters Archives Historical Collection, Washington, DC, Reference Number 17369.

"Webb Review, ADD-2 Introduction Discussing Saturn Program Schedule," NASA Saturn Launch Vehicle Program Review, November 9, 1963, NASA Headquarters Archives Historical Collection, Washington, DC, Reference Number 17369.

"Dedication of Satellite 6, Los Angeles International Airport," Los Angeles, CA, November 10, 1963.

Technology Club of Syracuse, Syracuse, NY, February 1, 1964.

"Scientific and Engineering Manpower Requirements in the Manned Space Program," AIAA, Los Angeles, CA, February 3, 1964.

California Institute Associates, Los Angeles, CA, February 6, 1964.

"Engineers' Week Dinner," Los Angeles, CA, February 21, 1964.

Fourth National Conference on the Peaceful Uses of Outer Space, Boston, Massachusetts, April 29, 1964.

"The Manned Space Flight Program," 10th Annual Meeting of the American Astronautical Society, New York, NY, May 5, 1964.

"Space Research – the Implications for the Earth," Honors Convocation of the College of Engineering, Wayne State University, Detroit, Michigan, May 20, 1964.

"The Space Program and the National Economy," 70th Annual Convention of the Pennsylvania Bankers Association, Atlantic City, NJ, May 27, 1964.

Annual Banquet of the Aviation/Space Writers Association, Miami, FL, May 28, 1964.

Commencement Address, Missouri School of Mines and Metallurgy, Rolla, MO, May 31, 1964.

Apollo – The Challenge to Telemetering, Annual Banquet of the National Telemetering Conference, Los Angeles, CA, June 3, 1964.

Commencement Exercises, New Mexico State University, University Park, NM, June 6, 1964.

Address, NASA-University Conference, Washington, DC, July 7, 1964.

"Address before the Professional Group on Antennas and Propagation of the IEEE," International Hotel, Kennedy Airport, New York, NY, September 23, 1964.

Presentation before the Senior Management of the Manned Spacecraft Center, Houston, TX, October 5, 1964.

"Man's Role in Man-Machine System in Space," American Institute of Aeronautics and Astronautics Luncheon Third Manned Space Flight Meeting, Houston, TX, November 4, 1964.

Address before the National Association of Real Estate Boards, Los Angeles, CA, November 12, 1964.

National Editorial Writers Conference, Cocoa Beach, FL, November 13, 1964.

Joint Meeting of the Columbus Sections of ASME, IES and AIAA, Columbus, OH, January 21, 1965.

Technology Club of Syracuse, Syracuse, NY, February 1, 1965.

Committee of 100 of the Greater Titusville Chamber of Commerce, Cocoa Beach, FL, March 12, 1965.

Colloquium, Department of Electrical Engineering, University of California, Berkley, CA, March 16, 1965.

Second Space Congress, Cocoa Beach, FL, April 6, 1965.

News Conference, Manned Spacecraft Center Mission Control Announcement, Houston, TX, April 19, 1965.

School of Electrical Engineering, Purdue University, Lafayette, IN, April 22, 1965.

Lunar Exploration Symposium, Marshall Space Flight Center, Huntsville, AL, April 26, 1965.

Space Medicine Branch, Aerospace Medical Association, New York, NY, April 28, 1965.

"Apollo Extension Systems, Opportunities for Advanced Space Applications," American Astronautical Society, Chicago, Illinois, May 5, 1965.

SAE Aerospace Fluid Power Systems & Equipment Conference, Los Angeles, CA, May 19, 1965.

Fifth National Conference on the Peaceful Uses of Space, and St. Louis Bicentennial Space Symposium, St. Louis, Missouri, May 26, 1965.

Eighth National Symposium, Society of Aerospace Material and Process Engineers, San Francisco, CA, May 27, 1965.

National Aerospace Education Council, Washington, DC, June 24, 1965.

First Annual Rudolph Bannow Memorial Address, Bridgeport, CT, June 28, 1965.

Annual Convention, American Trial Lawyers Association, Miami Beach, FL, July 27, 1965.

"Some Applications of Apollo," XVI International Astronautical Congress, Athens, Greece, September 14, 1965.

Hartford Rotary Club, Hartford, CT, October 4, 1965.

Annual Joint Conference on School Management, Columbus, Ohio, November 10, 1965.

51st Annual Meeting, National Dairy Council, Washington, DC, January 24, 1966.

Annual Symposium on Reliability, San Francisco, CA, January 26, 1966.

Annual Convention of the American Association of School Administrators, Atlantic City, New Jersey, February 12, 1966.

Business Council, Washington, DC, February 17, 1966.

Treasurers' Club, Columbus, OH, February 23, 1966.

Draft of Speech for Margret Chase Smith, March 3, 1966, Smith, Margret Chase, 1966-1969, NASA, GEM-43-10.

Statement of George E. Mueller before the Committee on Science and Astronautics, House of Representatives, March 11, 1966.

Astronomers of the University California at Los Angeles, Los Angeles, CA, April 29, 1966.

Presentation to the President's Science Advisory Committee, Identification of Goals and Missions for Future Manned Space Flight, May 5, 1966, Speeches, April 29-May 28, 1966, NASA, GEM-46-2.

Annual Meeting, Aviation/Space Writers Association, New York, NY, May 26, 1966.

Joint Meeting of the Texas Radiation Advisory Board and the Texas Atomic Energy Commission, Freeport, TX, May 28, 1966.

National Space Club, Washington, DC, August 16, 1966.

Acceptance Ceremonies for Spacecraft 12, St. Louis, MO, September 1, 1966, Speeches, 1 September 1966-12 December 1966, GEM-46-3.

Acceptance Ceremonies for Gemini XII Agena Target Vehicle, Sunnyvale, CA, September 2, 1966, Speeches, 1 September 1966-12 December 1966, GEM-46-3.

First Annual NASA-Industry Logistics Management Symposium, George C. Marshall Space Flight Center, Huntsville, AL, September 13, 1966.

Dedication of the Science Center, Cedar Crest College, Allentown, PA, October 19, 1966.

75th Anniversary Conference, California Institute of Technology, Pasadena, CA, October 25, 1966.

Wisconsin Chamber of Commerce, Milwaukee, WI, November 2, 1966.

Lectures at the University of Sydney Summer School, Sydney, New South Wales, Australia, January 10-11, 1967.

Institute of Electrical and Electronics Engineers, Washington, DC, February 20, 1967.

Statement to Committee on Aeronautics and Space Sciences, United States Senate, February 27, 1967.

Explorers Club, Annual Open House, New York, NY, April 1, 1967.

Statement of George E. Mueller," Oversight Subcommittee of the Committee on Science and Astronautics, House of Representatives, April 11, 1967, Apollo Program – Apollo 204, 7-29 April 1967, GEM-62-12.

George E. Mueller, draft speech delivered to the House Subcommittee on NASA Investigations, April 10-12, 1967, *Apollo* AS-204 Fire, NASA Headquarters Archives Historical Collection, Washington DC, Reference Number 31579.

Eleventh National Symposium, Society of Aerospace Materials and Process Engineers, St. Louis, MO, April 19, 1967.

American Power Conference, Chicago, IL, April 26, 1967.

Thirteenth Annual Meeting, American Astronautical Society, Dallas, TX, May 3, 1967.

George E. Mueller, Statement, Subcommittee on NASA Oversight, Committee on Science and Astronautics, House of Representatives, May 10, 1967.

American Institute of Aeronautics and Astronautics, New Orleans, LA, May 16, 1967.

National Capital Section, American Institute of Aeronautics and Astronautics, Washington, DC, June 6, 1967.

Statement before the Senate Committee on Acronautical and Space Sciences, George E. Mueller, June 12, 1967, Apollo Program – Apollo 204, 5-12 June 1967, GEM-62-2.

Texas Society of Washington, DC, Washington, DC, October 2, 1967.

Reading Branch, Royal Aeronautical Society, University of Reading, England, Speeches, 17 October 1967, NASA, GEM-46-10.

Physics Colloquium, Harvey Mudd College, Claremont, CA, December 12, 1967.

Economic Club of Detroit, MI, February 12, 1968.

Chamber of Commerce, Denver, CO, March 8, 1968.

Aviation/Space Writers Association, Washington, DC, May 1, 1968.

Dedication of Grissom and Chaffee Halls, Purdue University, Lafayette, IN, May 2, 1968.

AC Sparkplug, Milwaukee, Wisconsin May 28, 1968, Speeches, 12 February -29 May, 1968, GEM-46-11.

Rotary Club, Madison, Wisconsin Speech, May 29, 1968, Speeches, 12 February -29 May, 1968, GEM-46-11.

Joint AIAA/CASI Meeting, Montreal, Canada, July 8, 1968.

British Interplanetary Society, University College, London, England, August 10, 1968.

Dr. George E. Mueller, Introductory Session of the United Nations Conference on the Peaceful Uses of Outer Space, Vienna, Austria, August 14-15, 1968.

World Affairs Council, Pittsburgh, PA, September 12, 1968.

Tenth Anniversary Dinner of NASA, Cape Kennedy Area Chamber of Commerce, October 8, 1968, Speeches, 8 October-15 November, 1968, GEM-46-3.

Ninth National Conference of United Press International Editors and Publishers, October 8, 1968.

20th Century Club, Hartford, CT, November 11, 1968.

Remarks to Commonwealth Club of California, San Francisco, CA, November 15, 1968.

National Space Club, Washington, DC, November 26, 1968.

New York Society of Securities Analysts, New York, NY, January 28, 1969.

National Security Industrial Association, Kennedy Space Center, FL, January 30, 1969.

Statement before the Committee on Science and Astronautics of the House of Representatives, Washington, DC, March 4, 1969.

National Research Council Division of Engineering Annual Dinner, Washington, DC, March 10, 1969.

Annual Kiwanis Dinner, Milledgeville, GA, March 21, 1969.

Address before the Michigan Engineering Society, Grand Rapids, Michigan, April 19, 1969, Speeches, 28 January-18 June 1969, GEM-47-1.

Student Seminar, California Museum of Science and Industry, Los Angeles, CA, April 30, 1969.

British Interplanetary Society, London, England, May 30, 1969.

Address before the 1969 IEEE Computer Group Conference, Minneapolis, MN, June 18, 1969.

"Transcript, Dr. Mueller's remarks," Program Planning meeting with Mr. Paine, June 24, 1969, Speeches, 28 January-18 June 1969, GEM-47-1.

Press Conference, Canaveral Press Club, Cape Canaveral, FL, 3 July 1969, Speeches, June 24-July 3, 1969, GEM-47-2.

Statement to the Press, July 24, 1969, Speeches, 21-24 July 1969, GEM-47-3.

National Space Club, Washington, DC, August 6, 1969.

IEEE Council Meeting, Los Angeles, CA, September 5, 1969.

Speech at KSC (Transcript), John F. Kennedy Space Center, FL, September 30, 1969.

Twenty-fifth Annual General Meeting, International Air Transportation Association, Amsterdam, The Netherlands, October 23, 1969.

Press Conference, Canaveral Press Club, Cape Canaveral, FL, 3 July 1969, Speeches, 24 June-3 July 1969, GEM-47-2.

XXth International Astronautical Congress, October 6, 1969, Mar del Plata, Argentina, Speeches, 6 August-23 November 1969, GEM-47-4.

Norwegian Polytechnical Society, Oslo, Norway, November 4, 1969, Speeches, 6 August-23 November 1969, GEM-47-4.

Presentation to NASA Senior Staff, Office of Manned Space Flight, Washington, DC, December 4, 1969.

Presentation to NASA Senior Staff Wives, Office of Manned Space Flight, Washington, DC, December 4, 1969.

Remarks to Senior Staff, Marshall Space Flight Center, Huntsville, AL, December 8, 1969.

Address before the American Society of Civil Engineers, Houston, TX, April 16, 1970, General Dynamics, Speeches, GEM-112-1.

Commencement Address, Benjamin Franklin University, Washington, DC, June 26, 1970, General Dynamics Speeches, GEM-112-8.

Beverly Hills Rotary Club, Beverly Hills, CA, April 20, 1970, General Dynamics Speeches, GEM-112-9.

National Telemetering Conference, Los Angeles, CA, General Dynamics Speeches, GEM-112-14.

XXIst International Astronautical Congress, Constance, West Germany, Eugene Sanger Medal Award, October 8, 1970, General Dynamics Speeches, GEM-113-1.

Hugh O'Brian Youth Seminar, Kennedy Space Center, FL, July 17, 1970, General Dynamics Speeches, GEM-113-2.

Montgomery Area Chamber of Commerce, Montgomery, AL, January 26, 1970, General Dynamics Speeches, GEM-113-3.

Wincon 70 Convention, IEEE Aerospace & Electronics Group, Los Angeles, CA, February 12, 1970, General Dynamics Speeches, GEM-113-7.

Seventh Space Congress, Cocoa Beach, FL, April 23, 1970, General Dynamics Speeches, GEM-113-6.

Pioneers Banquet, Cocoa Beach, FL, April 24, 1970, General Dynamics Speeches, GEM-113-5.

Science Whiz Kids Space Seminar, Museum of Science and Industry, Los Angeles, CA, May 9, 1970, General Dynamics Speeches, GEM-113-4.

World Affairs Council of Pittsburg, Pittsburg, PA, May 26, 1970, General Dynamics Speeches, GEM-113-8.

Eighth Aerospace Systems Meeting, California Institute of Technology, April 1971, SDC Speeches, 1971-1974, Mueller's Personal Collection. (Also note that many of Mueller's speeches delivered while at SDC are in the SDC section of his papers at the Library of Congress, see boxes 131 to 135).

"Apollo: Looking Back," California Institute of Technology, Pasadena, CA, May 18, 1971, SDC Speeches, 1971-1974, Mueller's Personal Collection.

Commencement Address, Webb School 50th Anniversary, Claremont, CA, June 4, 1972, SDC Speeches, 1971-1974, Mueller's Personal Collection.

"R&D Top Priority," Human Factors Society, University of Southern California, Los Angeles, CA, October 18, 1972, SDC Speeches, 1971-1974, Mueller's Personal Collection.

"Report on the Status of US Space Effort," Madrid, Spain, January 15, 1973, SDC Speeches, 1971-1974, Mueller's Personal Collection.

Japanese Economic Research Council, Tokyo, Japan, January 23, 1973, SDC Speeches, 1971-1974, Mueller's Personal Collection.

Institute for the Advancement of Engineering, Los Angeles, CA, February 23, 1973, 1971-1974, SDC Speeches, 1979-1980, Mueller's Personal Collection.

"Total Systems Integration, Past, Present and Future," System Integration Symposium, Los Angeles, CA, October 17, 1973, SDC Speeches, 1971-1974, Mueller's Personal Collection.

IAF Communication, United Nations Science and Technology Subcommittee, Committee on the Peaceful Uses of Outer Space, New York, NY, April 16, 1974, SDC Speeches, 1971-1974, Mueller's Personal Collection.

Conference on Government Data Systems, Washington, DC, American Institute of Industrial Engineering, June 25, 1975, SDC Speeches, 1975-1978, Mueller's Personal Collection.

Testimony, Board of the Office of Technology Assessment, Washington, DC, June 14, 1976, SDC Speeches, 1975-1978, Mueller's Personal Collection.

"Future of Data Processing in Aerospace," University of Tennessee Space Institute, Tullahoma, TN, September 30, 1976, SDC Speeches, 1975-1978, Mueller's Personal Collection.

"Welcoming Address," 27th Annual Congress of the International Astronautical Federation, Anaheim CA, October 11, 1976, GEM-134-6.

"Data Processing for Truckers – Real Era Lies Ahead," Management Systems Conference, American Truckers Association, San Francisco, CA, April 19, 1977,

SDC Speeches, 1975-1978, Mueller's Personal Collection.

"Activities in Space in the Year 2027," Sea Space Symposium, Cozumel, Mexico, November 1, 1977, SDC Speeches, 1975-1978, Mueller's Personal Collection.

"New Dimensions in Management Information Systems," American Institute of Industrial Engineers, San Francisco, CA, November 7, 1977, SDC Speeches, 1975-1978, Mueller's Personal Collection.

"As I See It," Computer and Communications Industry Association, Beverly Hills, CA, November 14, 1977, SDC Speeches, 1975-1978, Mueller's Personal Collection.

"Real Cost of Social Control Legislation," Town Hall of California, Los Angeles, CA, January 24, 1978, SDC Speeches, 1975-1978, Mueller's Personal Collection.

"Rethinking Our National Priorities," The Commonwealth Club of California, San Francisco, CA, June 21, 1978, SDC Speeches, 1975-1978, Mueller's Personal Collection.

"Distributed Data Processing – The Real Promise Lies Ahead," American Institute of Industrial Engineers, San Francisco, CA, June 22, 1978, SDC Speeches, 1975-1978, Mueller's Personal Collection.

"Software Procurement – There Must be a Better Way," American Institute of Industrial Engineers, Washington, DC, June 27, 1978, SDC Speeches, 1975-1978, Mueller's Personal Collection.

"Beyond the Space Shuttle," American Astronautical Society, Houston, TX, October 1, 1978, SDC Speeches, 1975-1978, Mueller's Personal Collection.

"AIAA at the Cross Roads," AIAA Orange County Section, Santa Ana, CA, March 22, 1979, SDC Speeches, 1979-1980, Mueller's Personal Collection.

"Space – Everyone Benefits," Sixteenth Space Congress, Cocoa Beach, FL, April 27, 1979, SDC Speeches, 1979-1980, Mueller's Personal Collection.

Canadian Aeronautics and Space Institute, Ottawa, Canada, May 1, 1979, SDC Speeches, 1979-1980, Mueller's Personal Collection.

"AIAA at the Cross Roads," AIAA Tennessee Section, Oak Ridge, TN, May 10, 1979, SDC Speeches, 1979-1980, Mueller's Personal Collection.

"AIAA at the Cross Roads," AIAA San Diego Section, Sand Diego, CA, May 16, 1979, SDC Speeches, 1979-1980, Mueller's Personal Collection.

"Testing of Composite Structures, IAF Congress, Paris, France, June 8, 1979, SDC Speeches, 1979-1980, Mueller's Personal Collection.

Statement, Hearings of the Subcommittee on Space Science and Applications, United States House of Representatives, Washington, DC, June 26, 1979, SDC Speeches, 1979-1980, Mueller's Personal Collection.

AIAA Antelope Valley Section, Antelope Valley, CA, July 17, 1979, SDC Speeches, 1979-1980, Mueller's Personal Collection.

"Manifest Destiny," University of Alabama, Huntsville, Huntsville, AL, July 18, 1979, SDC Speeches, 1979-1980, Mueller's Personal Collection.

"Energy Self Sufficiency: The Need is Now," International Conference on Energy Use Management, Los Angeles, CA, October 26, 1979, SDC Speeches, 1979-1980, Mueller's Personal Collection.

AIAA Pacific Section, Seattle, Washington, November 1, 1979, SDC Speeches, 1979-1980, Mueller's Personal Collection.

National Contract Management Association, Los Angeles, CA, November 8, 1979, SDC Speeches, 1979-1980, Mueller's Personal Collection.

Addresses to the AIAA Sections in Los Angeles, St. Louis, Baltimore, Orlando/Cape Canaveral, November 26-29, 1979, SDC Speeches, 1979-1980, Mueller's Personal Collection.

"Patriotism in the 1980s, Santa Monica Chapter, American Businesswomen, Santa Monica, CA, February 29, 1980, SDC Speeches, 1979-1980, Mueller's Personal Collection.

"AIAA and its Members," Houston Section, Houston, TX, February 26, 1980, SDC Speeches, 1979-1980, Mueller's Personal Collection.

AIAA Washington, DC and New Orleans Sections, May 1, 1980, SDC Speeches, 1979-1980, Mueller's Personal Collection.

"Systematic Dismantling of our Nuclear Power Program," National Energy Resources Organization, Washington, DC, May 20, 1980, SDC Speeches, 1979-1980, Mueller's Personal Collection.

"Space Energy – Problems and Economics," IAF Congress, Tokyo, Japan, September 22, 1980, SDC Speeches, 1979-1980, Mueller's Personal Collection.

New Heritage of Wernher von Braun," National Foundation for Research, Athens Greece, October 14, 1980, SDC Speeches, 1979-1980, Mueller's Personal Collection.

"An Opportunity for Action," Commencement Address, Pepperdine University, Malibu, CA, December 13, 1980, SDC Speeches, 1979-1980, Mueller's Personal Collection.

"The Missing Term in the Equation," Los Angeles Council of Engineers and Scientists, February 17, 1981, GEM-134-11.

"Our Future in Space," Comision Nacional de Investigaciones Espaciales and the Air War School, Buenos Aires, Argentina, July 27, 1981, GEM-133-11.

"A World of Infinite Options," Address at the ARCS Dinner, Los Angeles, CA, November 20, 1981, GEM-133-6.

"The Potential of Knowledge-Based Systems," Information Institute Meeting on Education for Information Management, Santa Barbara, CA, May 7, 1982, GEM-134-5.

"Responding to a More Ominous Sputnik," American Astronautical Society, Washington, DC, June 24, 1982, GEM-131-11.

International Academy of Astronautics, Paris, France, September 28, 1982, GEM-134-7.

"Interstellar Space Travel," An Manned Space Exploration Conference, Brandeis University, October 21, 1982, GEM-134-14.

"A Giant Foothold for Mankind," Lunar and Planetary Conference, Houston, TX, March 16, 1983, GEM-134-12.

"The State of Space in the Year 2000," Committee on the Peaceful Uses of Outer Space (COPOUS) Legal Subcommittee, United Nations, New York, NY, March 25, 1983, GEM-134-3.

"The Era of Change," 20th Space Congress of the Canaveral Council of Technical Societies, Cape Canaveral, FL, April 26, 1983, GEM-133-10.

"Data Processing Implications of the Space Station," NASA/AIAA Symposium on the Space Station, Arlington, VA, July 20, 1983, GEM-131-14.

"Distributed Data Base Management," An Address at Computer Science at UCLA, Los Angeles, CA, September 26, 1983, GEM-134-12.

"The Next 25 Years: A View from 1984," *Aerospace America*, August 1984, GEM-269-9.

"Just a Few Thoughts Associated with the Development of the Shuttle," ASME, March 4, 1986, GEM-IAA Speeches-268-9.

"Man and the Moon," Innsbruck, Austria, 1986, GEM-269-9.

"From Apollo to Mars," An Lunar Planetary Institute, Houston, TX, March 15, 1989, GEM-IAA Speeches-268-9.

"Apollo Celebration, July 1989, GEM-IAA Speeches-268-9.

"What Did Happen on the Way Back From the Moon, or, From Apollo to Mars," April 1989, GEM-269-7.

"The Future of Space," July 1989, GEM-269-9.

"A Vision for the International Academy of Astronautics," 15th Annual IAA Annual Meeting, Toremolina, Spain, October 1989, GEM-268-9.

"Men to Mars," 8th IAA Man in Space Symposium, Tashkent, Uzbekistan, USSR, 1989, GEM-268-9.

"Management and Cost [of Spaceflight]", Undated, GEM-268-9.

"Introduction of IAA to Spanish Academy of Science, October 1989, GEM-268-9.

"Trials, Tribulations, Triumphs of the Apollo Program," May 1991 [also used in July 1994], GEM-268-9.

"On the 25th Anniversary of Apollo," July 1994, GEM-269-10.

"Apollo: 25th Anniversary," July 1994, GEM-269-10.

"One Vision of the 21st Century, July 1994, GEM-269-10.

"Apollo: Why & How, 25 Years Later," October 7, 1994, GEM-269-7.

"Bob Seamans at the Helm!" Seamans Memorial at MIT, Boston, June 10, 2009.

NEWSPAPER AND MAGAZINE ARTICLES

(By date)

New York Times Collection (1923-Current file); ProQuest Historical Newspapers The New York Times (1851-2006), Accessed multiple dates (*NYT*).

Midwest College Enrollment, 1940: "Report regarding Enrollment in Some of the leading Mid-Western Colleges and Universities," Document 746, Stated Meeting of the Board of Trustees of the Trustees of Purdue University, January 16-17, 1940, Purdue University Libraries, Archives and Special Collections, e-Archives.

Mueller Fellowship: Purdue University, Board of Trustees Minutes, June 8, 1940, Purdue University Libraries, Archives and Special Collections, e-Archives.

"New Defense Projects Costly, President Says," *Evening Star*, Washington, DC, November 14, 1957.

"World's Biggest Radio Telescope Tracks Rocket Aimed at Moon," St. Louis *Post-Dispatch*, October 13, 1958.

"Triumph From A 'Failure," *Life*, October 27, 1958.

"Ike Scores 'Stunt Race' in Space: Sees Federal Power Grab Threat to U. S.," New York *Herald Tribune*, August 7, 1962, 3.

Untitled, *Space Daily*, June 14, 1963, 766.

"Holmes Resigns, Refuses Curb on Manned Flight Authority," *Aviation Week & Space Technology*, June 17, 1963, 37.

"Mercury Gone, Holmes Going," *Missiles and Rockets*, June 17, 1963, 10 and "An American Tragedy," 54.

"Lunar Program in Crisis … ," Richard Witkin, The New York *Times*, July 13, 1963.

Book Review, K. A. Ponds, *Contemporary Physics*, Volume 6, number 4, April 1965, 306.

"Ike Discloses Bulk of Gaither Report, 'Missile Gap" Origin," *Washington Post*, September 21, 1965.

"Project Management in the Apollo Program: An Interdisciplinary Study," Eugene E. Drucker et al., Syracuse University, 1972.

"The Shaped-beam polyrod antenna," Schrank, H. and Herscovici, N., *Antennas and Propagation Magazine*, Volume 36, Number 2, IEEE, April 1994.

"Gen. Bernard A. Schriever: father of the ballistic missile program," *Air and Space Power Journal*, March 22, 2003.

"General Bernard A. Schriever: technological visionary," *Air Power History*, 22 March 2004.

"Local Engineering and Systems Engineering: Cultural Conflict at NASA's Marshall Space Flight Center," Yashshi Sato, *Technology and Culture*, volume 46, number 3, July 2005, 561-583.

SECONDARY SOURCES: INTERNET BOOKS
(By author)

Benson, Charles. *Moonport: A History of Apollo Launch Facilities and Operations*, Washington (DC): NASA SP-4204, 1996, http://www.hq.nasa.gov/pao/History/SP-4204.html, Access date, March 8, 2010.

Brooks, Courtney G., Grimwood, James M., and Swenson, Loyd. "The Slow Recovery," *Chariots of Apollo*, Washington (DC): NASA SP-4205, 1979, http://history.nasa.gov/SP-4205.html, Access date, March 8, 2010.

Bilstein, Roger E., *Orders of Magnitude, A History of NACA and NASA, 1915-1990*. Washington (DC): NASA SP-4406, 1989, http://history.nasa.gov/SP-4406/html, Access date, June 19, 2004.

Dunbar, Andrew J. and Waring, Stephen P. *Power to Explore: History of Marshall Space Flight Center, 1960-1990*, Washington (DC): NASA, 2000, chapter 5, "Conversation with Kurt Debus," http://history.msfc.nasa.gov/book/index.html, Access date, May 9, 2010.

Hacker, Barton C. and Grimwood, James M. *On the Shoulders of Titans*,

Washington (DC): NASA SP-3203, 1977, http://www.hq.nasa.gov/office/pao/
History/SP-4203.html, Access date, August 25, 2009.

Levine, Arnold S. *Managing NASA in the Apollo Era*, Washington (DC): NASA SP-4102, 1982, http://history.nasa.gov.SP-4102/ch4.htm, Access date, March 8, 2010.

Logsdon, John, editor. *Exploring the Unknown: Selected Documents in the History of the U.S. Civil Space Program, Volume I: Organizing for Exploration*, Washington (DC): NASA SP-4407, 1995, http://history.nasa.gov/sp-4407/vol1/chapter 2-3.pdf, Access date, March 30, 2010.

Newell, Homer E. *Beyond the Atmosphere,* Washington (DC): NASA, 1980, http://history.nasa.gov/SP-4211/cover.htm, Access date, February 14, 2010.

SECONDARY SOURCES: INTERNET ARTICLES
(By Subject)

Apollo: "The Apollo Spacecraft-A Chronology, vol. II, Foreword," NASA, http://www.hq.nasa.gov/office/pao/History/SP-409/frwrd2.htm, Access date, June 19, 2004.

Burnett: "Our People, Electrical and Computer," Purdue University, College of Engineering, J. Robert Burnett, https://www.engineering.purdue.edu/ECE/People/Alumni/OECE/1992/burnett.whtml, Access date, April 24, 2009.

Butler: "Ms. Sue Butler," Oral History of Kennedy Space Center, June 25, 2002, www.ksc.nasa.gov/kscoralhistory/documents/sbutler.pdf, Access date, May 21, 2010.

Crown: "Henry Crown, Industrialist Dies ...," Joan Cook, *New York Times*, August 16, 1990, www.nytimes.com/1990/08/16/obituaries/henry-crown-industrialist-dies-billionaire-92-rose-from-poverty-by-joan-cook.html, Access date, September 7, 2010.

Dunn: "Louis G. Dunn, JPL Director, 1946-1954," JPL, NASA, http://www.jpl.nasa.gov/jplhistory/learnmore/lm-dunn.php, Access date, April 21, 2009.

Eden: "History of Eden Theological Seminary," Eden Theological Seminary, http://eden.edu/About/History.aspx , Access date, March 18, 2009.

Explorer: "Explorer 6," NASA, http://nssdc.gsfc.nasa.gov/nmc/masterCatalog.do?sc=1959-004A, Access date, May 7, 2009.

Federal Acquisition Regulations (FAR) 37.104, "Personal Services Contracts," https://www.acquistion.gov/far/html/Subpart%2037_1.html, Access date, August 6, 2010.

Fletcher: "James C. Fletcher, NASA Administrator," NASA, http://history.nasa.gov/Biographies/fletcher.html, Access date, April 21, 2009.

Gemini-Titan 1: "GT-1," NASA, http://www-pao.ksc.nasa.gov/history/gemini/gemini-1/gemini1.htm, Access date, July 19, 2009.

Gemini-Titan 2: "GT-2," NASA, http://www-pao.ksc.nasa.gov/history/gemini/gemini-2/gemini2.htm, Access date, July 19, 2009.

General Electric: "Integration Role for GE," NASA, www.hq.nasa.gov/pao/History/SP-4204/ch9-1, Access date, May 1, 2009.

Getting: "Our History, Dr. Ivan Getting," Aerospace Corporation, http://aero.org/corporation/getting.html, Access date, April 28, 2009.

Hines: "Longtime NASA Reporter William M. Hines, 88 Dies," *Washington Post*, http://www.washingtonpost.com/wp-dyn/articles/A10570-2005Mar5.html, Accessed August 8, 2010.

Hornbeck: "John A. Hornbeck," National Academy of Engineering, *Memorial Tributes*, vol. 4, Washington, DC, 1991, http://book.nap.edu/openbook.php, Access date, July 19, 2009.

"IBM Launches the World's Fastest Chip," TG Daily, September 6, 2010, http://www.tgdaily.com/hardware-features/51405-ibm-launches-worlds-fastest-computer-chip, Access date, February 24, 2012.

James: "Adjustment to Marshall Organization," Organizational Announcement (Lee James), December 5, 1968, Marshall Space Flight Center, Huntsville, AL, http://history.msfc.nasa.gov/vonbraun/announcement_3.pdf, Access date, July 18, 2009.

Komarov, Vladimir: "Vladimir Komarov: The Unsung Space Hero," *Discovery News*, http://news.discovery.com/space/vladimir-komarov-unsung-space-hero-110321.html, Access date, March 30, 2011.

Kistler Aerospace: "Kistler Aerospace Corporation's K-1," GlobalSecurity.org, http://www.globalsecurity.org/space/systems/kister.htm, Access date, December 10, 2010.

Kistler Aerospace: "Lockheed Gets Kistler Rocket Tank Contract," June 18, 1997, Associated Press, "Coming Along at Kistler," September 11, 1998, by Karen Kaplan, and "Northrop Takes $30-Million Equity State in Start-Up Firm," March 19, 1999, by Karen Kaplan, *Los Angeles Times*, http://articles.latimes/keyword/kistler-aerospace-company, Access date, December 12, 2010.

Kistler Aerospace: "Private Rockets," *Discover*, April 1999, http://discovermagazine.com/1999/apr/rockets, December 10, 2010.

Kistler Aerospace: "Aerospace Company Targets 2002 Liftoff for New Satellite System," Marie McInerney, September 6, 2000, http://www.space.com/businesstechnology/business/kistler_satsystem_000906_wg.html, Access date, December 10, 2010.

Kistler Aerospace: "Kistler Aerospace Corporation Restructures its Finances … ," Press Release, July 15, 2003, http://www.rocketplanekistler.com/newsinfo/press releases/071503.html, Access date, December 10, 2010.

Kistler Aerospace: "Kistler Aerospace Files for Chapter 11 Bankruptcy Protection," Jim Banke, July 23, 2003, http://www.space.com/news/kistler_bankruptcy_030723.html, Access date December 10, 2010.

Kistler Aerospace: "Kistler Aerospace Corporation Announces Additions to Senior Management," Press Release, August 11, 2004, http://www.rocketplanekistler.com/newsinfo/pressreleases/081104.html, December 10, 2010.

Lehan: "Frank W. Lehan," National Academy of Engineering, http://www.nae.edu/nae/naepub.nsf/Members + By + UNID/638A7C0F65685F228625755200622D36?opendocument, Access date, April 20, 2009.

McKee: "The Administration: Lyndon Johnson Presents (McKee)," *Time*, May 7, 1965, http://www.time.com/time/magazine/article/0,9171,898722,00.htm, September 18, 2009, Access date, September 18, 2009.

Mercury capsule weight: www.museumofflight.org/spacecraft/mcdonnell-mercury-capsule-reproduction, Access date, April 18, 2011.

Minuteman: "Minuteman," Strategic Air Command, U.S., Air Force, http://strategic-air command.com/missiles/minuteman/Minuteman_Missile_History.htm, Access date, April 24, 2009.

Missouri School of Mines and Metallurgy: "Missouri School of Mines and Metallurgy," History of Missouri University of Science and Technology, Office of the Chancellor, http://chancellor.mst.edu/history/, March 18, 2009.

Mueller: "Dr. George E. Mueller ... Receives the IAA's 1999 Von Karman Award," IAA Press Release, Amsterdam, October 3, 1999, http://www.spacefre.com/news/viewpr.html?pid = 162, December 10, 2010.

National Reconnaissance Office of the United States, Center for the Study of National Reconnaissance: http://www.nro.gov/PressReleases/04_Pioneer_Fact_sheet.pdf, Access date, April 21, 2009.

Ohio State University: "History of the Ohio State College of Engineering," Ohio State University, http://engineeirng.osu.edu/overview/history.php, Access date, April 16, 2009.

Ohio State University: "The Ohio State University," Ohio History Central, A Product of the Ohio Historical Society, http://www.ohiohistorycentral.org/entry.php?rec = 785, Access date, April 17, 2009.

Ohio State University: "History of the Department of Electrical Engineering, Ohio State University," undated, Ohio State University, by Emerson E. Kimberly, http://www.ece.osu.edu/pdfs/Historyof EE.pdf, Access date, April 19, 2009.

Richard, Ludie: "Ludie G. Richard," College of Engineering and Applied Science, University of Colorado at Boulder, http://www.colorado.edu/engineering/deaa/cgi-bin/display.pl?id = 190, Accessed, May 21, 2010.

Packard Commission: Reports of the President's Blue Ribbon Commission on Defense (1986), Also Known As 'The Packard Commission,'" http://www.ndu.edu/library/pbrc/pbrc.html, Access date, April 22, 2009.

Petrone: "Rocco Petrone 1926-2006," NASA, August 30, 2006, http://www.nasa.gov/vision/space/features/rocco_petrone.html, Access date, July 18, 2009.

"Pioneer," JPL, NASA, http://msl.jpl.nasa.gov/QuickLooks/pioneer0QL.html, Access date, April 27, 2009.

"Pioneer," Mission and Spacecraft Library, JPL, http://msl.jpl.nasa/gov/Programs/pioneer.html, Access date, April 27, 2009.

Ramo, Simon: "Simon Ramo," Northrop Grumman Corporation, http://www.st.northropgrumman.com/siramo/biography1.html, Access, date, April 10, 2009.

Rudolph, Arthur: "A Matter of Conscious or Convenience? – War Crimes: Arthur Rudolph seeks Canadian vindication for his fall from grace in the U. S. One-top NASA researcher was tainted, decades after, by his Nazi past," Mary Williams Walsh. *Times* Staff Writer, September 3, 1990, The Los Angeles *Times*, http://articles.latimes.com/1990-09-03/news.vw-1589_1_arthur-rudolph, Access date, June 20, 2009.

Ruud, Ralph: http://www.aerospacelegacyfoundation.com/page22.html, Access date, July 14, 2009.

Saturn I lift weight: http://klabs.org/history/history_docs/jsc_t/sa06_launch_results.pdf, Access date, April 18, 2011.

Saturn V lift weight: www.nasa.gov/audiance/foreducators/rocketry/home/what-was-the-saturn-v-58, Access date, April 18, 2011.

"Saturn Terminology," NASA, http://history.nasa.gov/MHR-5/glossary.htm, Access date, July 1, 2009.

Schriever, Bernard A.: "General Bernard A. Schriever," U.S. Air Force, http://www.af.mil/history/person.asp?dec=pid=123057580, Access date, February 19, 2004.

Schriever, Bernard A.: "Gen. Schriever's visionary speech turns 50," *Air Force Link*, February 21, 2007, http://www.af.mil/news/story.asp?id=123040817, Access date, February 19, 2009.

"Skylab 2," NASA Kennedy Space Center, http://www-pao.ksc.nasa.gov/history/skylab/skylab-2.htm, Access date, January 8, 2010.

Space Launch Initiative: "The Space Launch Initiative," NASA, MSFC, April 2002, http://www.nasa.gov/centers/marshall/news/background/facts/slifactstext02.hml, Access date, December 13, 2010.

Terhune, Charles "Terry" H., U.S. Air Force, http://www.af.mil/bios/bio.asp?bioID=7361, Access date, May 6, 2009.

"Thor," Strategic-Air- Command, U.S. Air Force, http://www.strategic-air-command.-com/missiles/Thor/Thor_Missile_Home_Page.htm, April 21, 2009.

Thompson, Floyd L.: http://www.executivemediators.com/mediators-bio.php, Access, date, July 19, 2009.

Townes, Charles H.: "Charles H. Townes, Nobel Prize Committee, http://nobelprize.org/nobel_prizes/physics/laureates/1964/townes-bio.html, Access date, April 10, 2009.

Von Braun, Wernher: "Saturn the Giant," Wernher von Braun, an essay from *Apollo Expeditions to the Moon*, NASA, http://history.msfc.nasa.gov/special/pogo.html, Access date, June 20, 2004.

Voskhod 1: National Space Science Data Center, NSSDC ID: 1964-065A, NASA, http://nssdc.gsfc.nasa.gov/nmc/spacecraftDisplay.do?id=1964-065A, Access date, December 7, 2011.

Wooldridge, Dean E.: "Dean E. Wooldridge," California Institute of Technology, http://eands.caltech.edu/articles/LXIX3/wooldridge.html, Access date, April 10, 2009.

PUBLISHED BOOKS
(By Author)

Baum, Claude. *The System Builders, The Story of SDC,* Santa Monica (CA): System Development Corporation, 1981.

Benson, Charles D. and Faherty, William Barnaby. *Moonport: A History of Apollo Launch Facilities and Operations,* Washington (DC): NASA, 1978.

Bilstein, Roger E. *Stages to Saturn,* Gainesville (FL): University of Florida Press, 2003.

Bizony, Piers. *The Man Who Ran the Moon, James Webb, JFK and the Secret History of Project Apollo*, Cambridge (UK): Icon Books, 2006.

Buderi, Robert Buderi. *The Invention that Changed the World, How a Small Group of Radar Pioneers Won the Second World War and Launched a Technological Revolution*, New York: Touchstone Books, 1997.

Burns, James MacGregor. *Roosevelt, The Lion and the Fox,* San Diego (CA): Harvest Book, 1956.

Burroughs, William E. *Survival Imperative: Using Space to Protect the Earth*, New York: Forge Books, 2007.

Christensen, Lawrence O. and Ridley, Jack B. *UM-Rolla: A History of MSM/UMR*, Columbia (MO): University of Missouri Print Services, 1983.

Compton W. David and Benson, Charles D. *Living and Working in Space, A History of Skylab,* Washington (DC): NASA, 1983.

Compton, William David. *Where No Man Has Gone Before: A History of Apollo Lunar Exploration Missions*, Washington (DC): NASA, 1989.

Crainer, Stuart. *The Ultimate Business Library*, New York: John Wiley& Sons, 1997.

Dick, Steven J., editor. *Remembering the Space Age*, Washington (DC): NASA, 2008.

Dick, Steven J. and Launius, Roger D., editors. *Critical Issues in the History of Spaceflight*, Washington (DC): NASA, 2006.

Dyer, Davis. *TRW: Pioneering Technology and Innovation Since 1900*, Boston (MA): Harvard Business Press, 1998.

Elder, Glenn H. *Children of the Great Depression, 25th anniversary edition*, Boulder (CO)Westview Press, 1999.

Engerman, Stanley L. and Gallman, Robert E., editors. *The Cambridge Economic History of the United States, Volume III, The Twentieth Century*, New York: Cambridge University Press, 2000.

Gray, Mike. *Angle of Attack, Harrison Storms and the Race to the Moon*, New York: Penguin Books, 1992.

Glennan, T. Keith. *The Birth of NASA, The Diary of T. Keith Glennan*, Washington (DC) NASA, 1993.

Hansen, James R. *First Man: The Life of Neil A. Armstrong*, New York: Simon and Schuster, 2006.

Harland, David M. *How NASA Learned to Fly in Space An exciting account of the Gemini missions,* Apogee Books, 2004.

Hitt, David, Garriott, Owen, and Kerwin, Joe. *Homesteading Space, The Skylab Story*, Lincoln (NB): University of Nebraska Press, 2008.

Holmes, Jay. *Preliminary History of NASA, 1963-1968*, Washington (DC), NASA, 1970.

Hughes, Thomas P. *Rescuing Prometheus,* New York: Vintage Books, 2000.

Hughes, Thomas P. *American Genesis, A History of the American Genius for Invention*, New York: Penguin Books, 1989.

Johnson, Stephen B. *The United States Air Force and the Culture of Innovation, 1945-1965,* Washington (DC): U. S. Government Printing Office, 2002.

Johnson, Stephen B. *The Secret of Apollo, Systems Management in American and*

European Space Programs, Baltimore (MD): Johns Hopkins University Press, 2008.

Kevles, Daniel J. *The Physicists, The History of a Scientific Community in Modern America,* New York: Vantage Books, 1979.

Kraft, Chris. *Flight, My Life in Mission Control,* New York: Plume, 2002.

Lambright, W. Henry. *Powering Apollo, James E. Webb of NASA,* Baltimore (MD): Johns Hopkins Press, 1995.

Launius, Roger D. and McCurdy, Howard E. *Spaceflight and the Myth of Presidential Leadership,* Chicago: University of Illinois Press, 1997.

Launius, Roger D. and Jenkins, Dennis R. *To Reach the High Frontier, A History of U. S. Launch Vehicles,* Lexington (KY): University Press of Kentucky, 2002.

Levine, Arnold S. *Managing NASA in the Apollo Era,* Washington (DC): NASA, 1982.

Logsdon, John M., with Launius, Roger D., editors. *Exploring the Unknown, Selected Documents of the U.S. Civil Space Program, Volume VII,* "Human Spaceflight: Projects Mercury, Gemini, and Apollo," Washington (DC): NASA, 2008.

McCurdy, Howard E. *Inside NASA,* Baltimore (MD): Johns Hopkins University Press, 1993.

McCurdy, Howard E. *Faster, Better, Cheaper: Low-cost Innovation in the U. S. Space Program,* Baltimore (MD): Johns Hopkins University Press, 2001.

McDougall, Walter A. *The Heavens and the Earth,* New York: Basic Books, 1985.

Murray, Charles & Cox, Catherine Bly. *Apollo, the Race to the Moon,* New York: Simon and Schuster, 1989.

Neufeld, Jacob. *Bernard A. Schriever,* Washington (DC): Office of Air Force History, 2005.

Neufield, Michael J. *The Rocket and the Reich: Peenemunde and the Coming of the Ballistic Missile Era,* Cambridge (MA): Harvard University Press, 1995.

Nixon, Richard M. *The Memoirs of Richard Nixon,* New York: Grosset & Dunlap, 1978.

Orloff, Richard W. and Harland, David M., *Apollo, The Definitive Sourcebook,* New York: Springer, 2006.

Piszkiewicz, Dennis. *The Nazi Rocketeers: Dreams of Space and Crimes of War,* Mechanicsburg (PA): Stockpole Books, 2007.

Ramo, Simon. *The Management of Innovative Technological Organizations,* New York: John Wiley & Sons, 1980.

Rosenbloom, Richard S., and William J. Spencer, eds. *Engines of Innovation: U.S. Industrial Research at the End of an Era,* Boston (MA): Harvard Business School Press, 1996.

Sapolsky, Harvey M. *The Polaris System Development: Bureaucratic and Programmatic Success in Government,* Cambridge (MA): Harvard University Press, 1971.

Seamans, Robert C., Jr. *Aiming at Targets,* Washington (DC): NASA, 1996.

Seamans, Robert C., Jr. *Project Apollo: The Tough Decisions,* Washington (DC): NASA, 2007.

Sheehan, Niel. *A Fiery Peace in a Cold War, Bernard Schriever and the Ultimate Weapon,* New York: Random House, 2009

Slotkin, Richard. *Gunfighter Nation, The Myth of the Frontier in Twentieth-Century America*, New York: Harper Perennial, 1993.

VanNimmen, Jane, et al. *NASA Historical Data Book*, 1958-1968, volume 1, Washington (DC): NASA, 1976.

Ward, Bob and Glenn, John (FRW). *Dr. Space*, Annapolis (MD): Naval Institute Press, 2005.

Webb, Richard C. *Tele-Visionaries, The People Behind the Invention of Television*, Hoboken (NJ): IEEE Press, 2005.

Weitekemp, Martha A. *Right Stuff, Wrong Sex: America's First Women in Space Program*, Baltimore (MD): Johns Hopkins University Press, 2005.

Williams, James G. *Encyclopedia of Computer Science and Technology*, New York: Marcel Dekker, Inc., 1987.

Yahill, Leni, Friedman, Ina, Galai, Hayah. *The Holocaust: The fate of European Jewry, 1932-1945*, New York: Oxford University Press, 1991.

Yang, Guangbin. *Lifecycle Reliability Engineering*, New York: John Wiley and Sons, 2007.

"Space Research: Directions for the Future," A Report of a Study by the Space Science Board, National Academy of Sciences, Washington (DC): National Research Council, 1965.

"Wernher von Braun, 1912-1977," George E. Mueller, *Memorial Tributes, volume 1*, National Academy of Engineering, Washington (DC), 1979.

PUBLICATIONS BY GEORGE E. MUELLER (1963-1969)

"Apollo Capabilities," *Astronautics & Aeronautics*, June 1964, GEM-110.

"Managing Manned Space Flight," *DATA**, June 1965, GEM-110.

"Beyond Apollo," *Astronautics & Aeronautics*, August 1965, GEM-110.

"By 1970, What?" *Los Angeles Times*, June 19, 1966, GEM-110.

"Die Panung bemannter Raumflugmissonen," *Weltraumfahrt*, 1966, GEM-110.

"Report on Apollo," *Astronautics & Aeronautics*, August 1967, GEM-110.

"New Trick for Businessmen," *Forbes*, February 1, 1968, GEM-110.

"New Future For Manned Space Flight Developments," *Astronautics & Aeronautics*, March 1969, GEM-110.

"Post-Apollo Revisited," *Astronautics & Aeronautics*, July 1969, GEM-241-14.

"In the Next Decade: A Lunar Base, Space Laboratories and Space Shuttles," *New York Times*, July 21, 1969

"Investigation of the Moon," *Bulletin of Atomic Scientists*, September 1969, GEM-110.

"Report on Apollo 11," *Los Angeles IEEE Bulletin*, September 1969, GEM-110.

"Application of Computers to Manned Space Flight," *Computer Group News*, September 1969, IEEE, GEM-110.

"Commercial Applications of Space Research," *Flight International*, November 1969, GEM-110.

"Apollo – gateway to challenge," Electrical and Electronic Technician Engineer,"

Institution of Electrical and Electronics Technician Engineers (UK), November 1969, GEM-110.

"NASA is not collapsing," *The Sunday Telegraph (UK)*, November 16, 1969, GEM-110.

"An Integrated Space Program for the Next Generation," January 1970, *Astronautics & Aeronautics,* GEM-124.

"Space Benefits," May 1970, *Michigan Business Review,* GEM-124.

"Space Shuttle: Beginning a New Era in Space Cooperation," September 1972, *Astronautics & Aeronautics,* GEM-202-1.

"The trucking industry's future partner, COMPUTERS," *Today's Transport International*, April/May 1978, GEM-242-7.

  "Blueprinting a Workable MIS," *Administrative Management,* September 1978, GEM-241-14.

"The future of data processing in aerospace," *Aeronautical Journal of the Royal Aeronautical Society*, April 1979, GEM-242-1.

Abbreviations

AAP	Apollo Applications Program
AAS	American Astronautical Society
AASA	American Association of School Administrators
AEG	Apollo Executive Group
AES	Apollo Extension System
AFTE	American Federation of Technicians and Engineers
AIAA	American Institute of Aeronautics and Astronautics
AFB	Air Force Base
AFSC	Air Force Systems Command
AMU	Astronaut Maneuvering Unit
AP	Associated Press
APO	Apollo Program Office
ARCS	Achievement Awards for College Scientist (foundation)
AS	Apollo-Saturn Space Vehicle
ASPO	Apollo Spacecraft Project Office
AT&T	American Telephone and Telegraph Company
ATDA	Augmented Target Docking Adapter
ATM	Apollo Telescope Mount
BIS	British Interplanetary Society
BMD	Ballistic Missile Division (USAF)
Caltech	California Institute of Technology
CASI	Canadian Aeronautical and Space Institute
CEO	Chief Executive Officer
CIA	Central Intelligence Agency
CM	Configuration Management
CPIF	Cost Plus Incentive Fee
CSM	Command & Service Modules
DCA	Defense Communications Agency
DCR	Document Change Review (or Request)
DOD	Department of Defense
EOR	Earth Orbit Rendezvous

FAR	Federal Acquisition Regulations
FO	Flight Operations function
FY	Fiscal Year
GD	General Dynamics Corporation
GE	General Electric Corporation
GEM	George E. Mueller
GMT	Greenwich Mean Time
GPO	Gemini Program Office
GT	Gemini Titan space vehicle
IAA	International Academy of Astronautics
IAC	International Astronautical Congress
IAF	International Astronautical Federation
IATA	International Air Transport Association
IBM	International Business Machine Corporation
IEEE	Institute of Electrical and Electronic Engineers
JPL	Jet Propulsion Laboratory (Caltech)
JFK	John F. Kennedy
JSC	Lyndon B. Johnson Space Center, formerly MSC, also referred to as "Johnson"
KSC	John F. Kennedy Space Center, also referred to as "Kennedy"
LBJ	Lyndon B. Johnson
LEM	Lunar Excursion Module (renamed lunar module)
LC	Library of Congress
LM	Lunar Module
LOC	Launch Operations Center (later renamed KSC)
LOR	Lunar Orbit Rendezvous
LOX	Liquid Oxygen
LaRC	Langley Research Center, also referred to as "Langley"
MDA	Multiple Docking Adapter
MDS	Malfunction Detection System
MIT	Massachusetts Institute of Technology
MOL	Manned Orbital Laboratory
MPH	Miles per hour
MSC	Manned Spacecraft Center, also referred to as "Houston," and later named the Lyndon B. Johnson Space Center, JSC or "Johnson"
MSFC	Marshall Space Flight Center, also referred to as "Huntsville"
MSFP	Manned Space Flight Program (NASA)
NAA & NAR	North American Aviation, Incorporated (NAA) merged with Rockwell to become North American Rockwell (NAR)
NACA	National Advisory Committee for Aeronautics
NAE	National Academy of Engineering
NAS	National Academy of Sciences
NASA	National Aeronautics and Space Administration
NASC	National Aeronautics and Space Council
NASM	National Air and Space Museum

NAREB	National Association of Real Estate Boards
NRC	National Research Council
NSA	National Security Agency
NSIA	National Security Industry Association
NSSDC	National Space Science Data Center
NSC	National Space Club
NSF	National Science Foundation
NYT	*New York Times*
O_2	Chemical symbol for oxygen
OART	Office of Advanced Research and Technology
OMSF	Office of Manned Space Flight
OSSA	Office of Space Science and Applications
OTA	Office of Technology Assessment
PERT	Program Evaluation Review Technique
Ph.D.	Doctor of Philosophy
PUOS	Peaceful Uses of Outer Space
PSAC	President's Science Advisory Committee
PSI	Pounds per Square Inch
QC	Quality Control
RCA	Radio Corporation of America
R&D	Research & Development
R&QA	Reliability & Quality Assurance function
R-W	Ramo-Wooldridge Corporation
RN	Reference Number (NASA historical archives)
SA	Saturn-Apollo (refers to Saturn booster)
SCP	Samuel C. Phillips
SE	System Engineering
SETA	System Engineering and Technical Assistance
SETD	System Engineering and Technical Direction
S&ID	Space and Information Systems Division (NAR "space division")
SDC	System Development Corporation
SPO	System Program Office
SSB	Space Science Board, also called "the science board"
STAC	Science and Technology Advisory Committee for Manned Space Flight
STG	Space Task Group (a name used several times for different groups)
STL	Space Technology Laboratories, Incorporated
TIE	Technical Integration and Evaluation
TRW	Thompson Ramo Wooldridge, Incorporated
UCLA	University of California at Los Angeles
UN	United Nations
USAF	United States Air Force
USIA	United States Information Agency
UPI	United Press International

Index